BRITISH TRADE UNIONS
1707–1918

Volume 5

BRITISH TRADE UNIONS
1707–1918

Volume 5
1865–1880

Edited by
W. Hamish Fraser

Routledge
Taylor & Francis Group
LONDON AND NEW YORK

First published 2008 by Pickering & Chatto (Publishers) Limited

Published 2016 by Routledge
2 Park Square, Milton Park, Abingdon, Oxon OX14 4RN
711 Third Avenue, New York, NY 10017, USA

Routledge is an imprint of the Taylor & Francis Group, an informa business

BRITISH LIBRARY CATALOGUING IN PUBLICATION DATA

British trade unions, 1707–1918
Vols. 5–8
1. Labor unions – Great Britain – History – 18th century – Sources 2. Labor unions – Great Britain – History – 19th century – Sources 3. Labor movement – Great Britain – History – 18th century – Sources 4. Labor movement – Great Britain – History – 19th century – Sources
I. Fraser, W. Hamish
331.8'8'0941'09033

ISBN-13: 978-1-85196-831-2 (set)

Typeset by Pickering & Chatto (Publishers) Limited

CONTENTS

Introduction vii

George Howell, 'To the Trades' Unionists of the United Kingdom', *Operative Bricklayers' Society's Trade Circular* (1 August 1865) 1

Mr. Potter and the London Trades' Council [1865] 3

Trades' Societies and Lock-Outs. Report of the Conference of Trades' Delegates of the United Kingdom, Held in the Temperance Hall, Townhead Street, Sheffield, on July 17th, 1866, and Four Following Days (1866) 11

'The Tailors' Strike and the System of Picketing', *Operative Bricklayers' Society Monthly Report* (1 September 1867) 87

Edward Spencer Beesly, *The Amalgamated Society of Carpenters and Joiners* (1867) 89

Report of the Various Proceedings Taken by the London Trades' Council and the Conference of Amalgamated Trades, in Reference to the Royal Commission on Trades' Unions, and Other Subjects in Connection Therewith (1867) 111

Report of St. Martin's Hall United Kingdom Trades' Conference Committee on the Trades' Union Inquiry Commission till 7th August, 1867 [1867] 155

International Working Men's Association, *To the Trades' Unionists' of Great Britain and Ireland* [1868] 179

J. M. Ludlow, 'Sir W. Erle on Trade Unions', *Operative Bricklayers' Monthly Report* (1 March 1869) 183

Trades' Unions and the Cost of Labour. Speech Delivered by Thomas Brassey, Jun M.P. in the House of Commons 7th July 1869 (1870) 187

Robert Jobson, *Trades Unions: An Inquiry into Their Rules and Working, Based on the Evidence before the Royal Commission; Showing the Folly of All Attempts to Raise Wages by Violence and Strikes, and the Beauty and Excellence of the Divine Law Governing Workmen and Employers* [1870?] 251

W. P. Roberts, *Trade Union Bill, 1871. A Letter by Mr. W. P. Roberts, (of London and Manchester, Solicitor) to Mr. George Potter, Mr. William*

Allan, Mr. Alexander M'Donald and Others Interested in Trades' Societies (1871) 283
Report of the Conference of Trades Delegates, Held in the Town Hall, Leeds on December 2nd 1871. Called to Consider the Statement made by Wm. Newmarch, Esq., in His Address to the Social Science Congress Lately Held in Leeds [1872] 299
W. H. Wood, *The Advantages of Trades Unions* [n.d.] 313
Proceedings of the Miners' National Association Conference, Held in Leeds, on 18, 19, 20, 21 and 22 November 1873 [1873] 327
The Criminal Law Amendment Act. A Memorial to the Right Honourable Henry Austin Bruce, MP., with an Appendix of Cases under the Criminal Law Amendment Act [1873] 339
Henry Crompton, *Tracts for Trade Unionists No IV. The National Federation of Associated Employers of Labour. Report upon the Memorial Presented to the Home Secretary by the National Federation of Associated Employers of Labour, December 13th, 1873* [1874] 359
Frederick Clifford, *The Agricultural Lock-Out of 1874 with Notes upon Farming and Farm-Labour in the Eastern Counties* (1875) 381
National Association of Miners, *Extraordinary Meeting of the Council Held in the Mechanics' Institution, David St Manchester, 24 February 1875* (1875) 405
Glasgow Herald, *The Wages Agitation on the Clyde. 25,000 Men to be Locked Out* (1877) 419
Address from the Executive Council of the Amalgamated Society of Engineers to their Fellow Workmen Throughout the United Kingdom, the British Colonies, and Foreign Countries [1879] 421
George Howell, 'Trade-Unions: Their Nature, Character, and Work', *Fraser's Magazine* (January 1879) 429

Explanatory Notes 439
Permisssions 451

INTRODUCTION

By the mid-1860s there were signs of a new confidence among trade-union leaders. In the trades councils that had been formed in many of the major cities, they found a forum to discuss the wider issues affecting the working class. The leaders also found themselves being courted by political figures and key individuals who were keen to get even greater commitment by trade unions to the campaign for further political reform that was gathering momentum. The size and efficiency of some of the large craft unions was being presented as evidence that at least some of the working class, the so-called 'aristocracy of labour', were capable of self-government and, therefore, deserving of the vote. Middle-class reformers began to see the labour aristocracy as potential allies in their campaigns against the continuing domination of the landed class. Trade unionists began to involve themselves in campaigns in support of the North in the American Civil War and for Italian and Polish independence, where they made contact with middle-class radicals. On the other hand, there was a wariness on the part of many trade unionists about political involvement, fearing that it would prove divisive and a diversion from the industrial struggle. But, persuaded by the arguments put forward by George Howell and the Reform League, increasing numbers began to believe that a political voice at Westminster was necessary.

Confidence also arose from the fact that in the London Trades Council there was now a coterie of highly capable leaders of some of the new large amalgamated societies. In William Allan of the Amalgamated Society of Engineers, Robert Applegarth of the Amalgamated Carpenters and Joiners, Daniel Guile of the Friendly Society of Ironfounders, Edwin Coulson of the Operative Bricklayers Society, and George Odger, the secretary of the Council, Sidney and Beatrice Webb identified a increasingly powerful 'junta', who saw themselves as the natural leaders of the trade-union movement. Although they were never as united or as powerful as the Webbs suggest and their initial reactions to various disputes could be different, none the less there was clearly a great deal of discussion and collaboration between these men and they achieved a national standing. Attempts to dictate to the rest of the movement led to tensions. Personal animosities and offended *amour propre* played a large part in the battles between

George Potter and the Trades Council. But it was also a struggle between smaller societies and older union tactics, which George Potter stood for, and the more conciliatory policies being pursued by the larger unions.[1]

The same threats and pressures that brought about amalgamations also encouraged attempts by unions to get inter-union collaboration. The Glasgow Trades Council had taken the lead with a campaign to get amendments to the Master and Servant legislation. A well-attended national conference was held over four days in May and June 1864. The Sheffield Association of Organised Trades followed in 1866, after a series of bitter strikes, by calling a conference that established the United Kingdom Alliance of Organised Trades to provide mutual support during lock-outs. Unfortunately, it was swept away in the indignation aroused by a new outbreak of attacks on non-unionists by the Sheffield knife-grinders.

The outcry over the Sheffield outrages resulted in a Royal Commission on Trade Unions and Employers' Associations. There were fears that the outcome would be legislation restricting trade-union activities, which many of the press were demanding. The leaders of the largest unions were determined to seize the opportunity to get across to Parliament and to the public that trade unionism was changing. This was made more urgent by the fact that, hard on the heels of the events in Sheffield, the courts had found against the Boilermakers' Society in the case of Hornby *v.* Close, and declared that, because it had rules about who could be employed and on apprenticeship, it was 'in restraint of trade' in the eyes of the common law. This posed an immediate threat to funds of the amalgamated societies. Determined to get protection for their funds, they set up their own Conference of Amalgamated Trades to co-ordinate evidence to the Royal Commission. Once again, George Potter reflected widespread dislike of the attitudes of the big societies and called a very successful conference in March 1867, attended by unions and trade councils representing some 200,000 workers. The smaller unions that attended this conference were opposed to the idea of having trade unions register since it would expose their financial state.

The next conference, called by the Manchester and Salford Trades Council in 1868, is usually taken as the founding conference of the Trades Union Congress. It consisted of some thirty-four delegates, mainly from other trades councils and it backed the work being done by the Conference of Amalgamated Trades in front of the Royal Commission. Thanks to the carefully developed strategies of the London leaders working through their middle-class allies on the Commission, the positivists, Frederic Harrison and the Christian Socialist, Thomas Hughes, the evidence before the Royal Commission proved to be a showcase for the large societies. Harrison and Hughes managed to whittle away the more hostile comments in the final report and then refused to sign it, producing their own minority report, which gave a set of proposed legislation on which the unions could campaign.[2]

With the threats to their future from courts and possible legislation, by the end of 1866 trade unions in all parts of the country were participating in political reform demonstrations. The extent of their involvement convinced many politicians that a reform measure of some kind had to be pushed through. The fact that the Reform Act of 1867 went further than most had expected and gave the vote to a substantial proportion of the skilled urban working class meant that unions had a new influence. None the less, there was still much hostility to overcome.

The government was slow to act, but, in 1871, a Trade Union Act was passed giving protection to the funds of registered unions, and also apparently freeing them from the liability of charges of conspiracy in restraint of trade. But it did not, as had been hoped, remove special offences of workmen. The criminal offences of 'molestation' and 'intimidation' remained, and the Criminal Law Amendment Act of 1871 added to these the crimes of 'persistently following' and 'watching and besetting'. It further strengthened the case for political involvement, and a Parliamentary Committee of the Trades Union Congress was set up to act as a watchdog over Westminster.[3] The miners' unions in particular, struggling as always to maintain a nationwide organization, looked to the state for legislation to improve conditions. The activities proved effective and Disraeli's Conservative government amended the Criminal Law Amendment Act with the Conspiracy and Protection of Property Act of 1875, that protected unionists from charges of criminal conspiracy during disputes, as long as their actions were not in themselves criminal. The 1875 Employer and Workman Act, putting both on the same footing in breaches of contract cases, replaced the hated Master and Servant legislation, and the 1871 Trade Union Act was amended to cover all unions not just those that would have been deemed unlawful before 1871.

By the early 1870s, a number of the larger employers had come to appreciate the value of trade unions, recognizing their leaders as people with whom it was possible to negotiate compromise settlements. The talk of conciliation and arbitration that was widespread in the 1870s, really meant an acceptance of collective bargaining by employers. Boards of conciliation and arbitration were established in the hosiery industry of Nottingham and the iron industry of the North-East. The exceptional conditions of economic boom in 1871 and 1872 helped this process. It also gave opportunities for new groups of workers to unionize, in many cases for the first time. Most spectacular was the rapid spread of unionism among the agricultural workers of the Midlands and East Anglia, where Joseph Arch, a Primitive Methodist lay preacher, organized the National Agricultural Labourers' Union in 1872 and it briefly recruited 70,000 members.

As the economy turned downwards in the second half of the 1870s, with a deep agricultural depression setting in, the farm labourers' unions struggled to survive. A sliding scale, linking wage adjustments to fluctuations in prices, once seen as a way of avoiding disputes, was imposed on defeated Welsh miners as a way of keeping wages down. Most miners' county unions struggled and national unity broke down. Even the powerful Amalgamated Society of Engineers haemorrhaged members to an alarming extent and expended huge sums on unemployment and strike benefit. Employers began to collaborate to try to hold back union advances and to counter the influence that unions appeared to be having on government. Most of the union advances that had been won were pulled back by the employers. There was a readiness to sit out disputes. The Clydeside shipyard workers strike in 1877 ran on for six months before the strikers accepted defeat.

1. S. Coltham, 'George Potter, the Junta and the *Bee-Hive*', *International Review of Social History*, 9 (1964), pp. 391–432; 10 (1965), pp. 1–85.
2. Royden Harrison, *Before the Socialists: Studies in Labour and Politics 1861–1881* (London: Routledge and Kegan Paul, 1965), pp. 251–342.
3. H. W. McCready, 'Britain's Labour Lobby 1867–75', *Canadian Journal of Economics and Political Science*, 22 (1956), pp. 141–60.

To the Trades' Unionists of the United Kingdom.

FELLOW WORKMEN,

As a Trades' Unionist, like yourselves, I desire to speak freely and frankly to you on behalf of Manhood Suffrage.

During the last few weeks you have been called upon to witness a General Election. About 1,000,000 of our fellow countrymen have had the "*privilege*" of recording their votes in the election of the "*Law Makers*" of Great Britain.

Out of this number only about 100,000, or one-tenth of the whole, belong to the *Working Class* of our population. Besides which there are about 6,000,000 of the adult males without *any vote* or direct voice in the great council of the nation.

Yet these "Law Makers" enact laws binding upon us all; they impose taxes we all have to pay; aye, and spend these taxes most improvidently. Yet we have no voice to advise, no power to check, any extravagance in the expenditure, or mismanagement in the government, or injustice in the laws.

The privileged classes in this country tell us that we have "no rights without duties;" this we admit. But *they impose duties* and *withhold* the *rights*: of this we complain. We obey the laws in common, let us have a voice in their enactment. We pay taxes (far greater than the rich in *proportion* to the *means* we *have*, or our *property* to be *protected*), yet are denied any voice in the *adjustment* of our *taxation*, or in the spending of our revenue.

The consequence is, that our laws are partial; being made by a class, for a class; and the taxation most oppressive upon those least able to bear it.

As Unionists we have just reason to complain, inasmuch as there are laws not yet repealed which bear heavily upon the workmen. For example, those relating to Masters and Workmen. How often do we find some case which startles us by the manifest cruelty and injustice of the sentence. For if the case had been reversed, the employer defendant instead of the employed, it would have been dismissed. (*See evidence before Committee of the House of Commons this Session.*)

How often have we found some of our greatest and most combined efforts rendered nugatory by some antique piece of legislation in the shape of an unrepealed Act of Parliament, or by some action taken by the *Home Office*?

Is it not fresh in your recollection that the government lent soldiers to a private contractor at Chelsea to enable him to resist his workmen's demand? Have they not done so at Plymouth? Did they not do so at Aldershot? And in how many cases have they given assistance where the men could not bring positive proof, although they have felt morally certain? Scotland Yard provided "pickets" in blue coats for a large number of buildings in London during the Builders' Strike and Lock-out. Yet we were condemned if we sent a private citizen (without citizen's rights) simply to inform workmen in the most friendly and civil manner that a trade dispute existed.

Bearing all this in mind, let us now demand our enfranchisement; *not as a class*, but as subjects in the eye of the law, and as citizens of our common country, which we all love, all support, and all protect. You spend thousands of pounds annually to obtain some trifling advance in wages, and some small reduction in time; and are compelled to fight for its maintain ance over and over again, because no voice is heard in Labour's right in the "*Commons*," or people's House.

Men stand there and vilify our class, our motives, and our actions, because they know we cannot be present to rebut their charges.

Let us once be able to maintain by the force of intellect and truth our rights as workmen, in that house, and depend upon it we shall rise in the social scale, and show that we can discharge our duties as well as exercise our rights.

But they tell us that we are incapable of self-government, and unfit to be trusted!! This we deny. Our own organizations prove our power of government, our self-restraint, our financial ability and economy, and our law-making capabilities *likewise.* They show more, they prove that we understand justice, and are earnest in the protection of every member, whether he be assailed by an employer, or by a fellow-workman and brother Unionist. Your combinations are the most perfect in organization and the most powerful in influence of any in the kingdom. Let us use them for our rights as citizens, and you will soon see the rights of labour respected and the workman benefitted thereby.

We seek not to over-balance political power by the substitution of one class influence for another. We seek it for all men without distinction; for, on the principle of Manhood Suffrage, protected by the ballot, class legislation as well as class representation will be no more.

Support, then, *The Reform League* in your various localities, and our next parliament will merit your gratitude and respect, by passing a Reform Bill at once comprehensive and just to the People.

I am, fellow workmen,

Yours in unity, GEORGE HOWELL,

Secretary of the Reform League,

8 *Adelphi Terrace, Strand, W.C.*

P.S.—All information concerning rules, addresses, cards of membership, &c., and the formation of Branches in your respective localities, will be supplied by the Sec. as above, to whom subscriptions also may be sent.

MR. POTTER

AND THE

LONDON TRADES' COUNCIL.

THE Council held a special meeting on Wednesday, March 29th, at the Bell Inn, Old Bailey, in compliance with a request from the Operative Bricklayers—Mr. Mildred in the chair.

After the minutes had been confirmed, a letter was read from the Tinplate Workers, stating that they had voted £5 towards paying the expenses of the movement to obtain an alteration in the Masters' and Workmen's Act; also, letters from the iron trades now on strike.

The secretary said he had received an important paper from the Operative Bricklayers' Society; also, a verbal communication from the Amalgamated Engineers, on the conduct of Mr. Potter, in calling trades' delegate meetings on his own responsibility. He (the secretary) would read the paper on the subject.

TO THE LONDON TRADES' COUNCIL.

March 21st, 1865.

Sir and Gentlemen,—An announcement having appeared in the *Beehive* newspaper of March 4th and 11th, that a meeting of trades' delegates would be held at the Sussex Hotel, Bouverie-street, Fleet-street, on the 15th inst., to consider the lock-out in the iron trade, &c., and feeling the irresponsible character of such meeting, we forward to you the following resolution, viz.:—"That in the opinion of this Executive Council, the conduct of Mr. G. Potter in calling a delegate meeting of the trades, without first consulting the London Trades' Council, of which he is a member, is deserving of the severest condemnation, and calls upon that council to censure such assumption of undelegated power." It furthermore adds, that if the Trades' Council is prepared to allow its functions to be thus set aside with impunity by one of its own non-official members, it no longer deserves our confidence and support.

By order of the Executive Council,

W. CHASE, Chairman.

E. COULSON, Gen. Sec.

Mr. Howell would inform the council that he attended as special delegate of the Bricklayers' Society, numbering 4,000 members. He was instructed to ask the council whether it was intended to allow Mr. Potter, or any other person, to call delegate meetings on questions affecting the interests of trades' societies? If so, then he considered it time that the council ceased to exist, and handed its functions over to Mr. Potter to deal with them as he might feel disposed. He thought, when the council was established, that there was a guarantee against speculators interfering with the special character of trades' delegate meetings, and he would say that many things had been done by the council which secured for the societies the esteem and good-will of the most intelligent persons of the age; and he had hoped to see the same course continued. He would like to hear an expression of opinion on the subject.

Mr. Allan, secretary of the Amalgamated Engineers, said he was deputed by the engineers' executive council to appear before the meeting to express the indignation of that society at the wanton insult offered by Mr. Potter to the London Trades' Council. He had been on several councils in his time, but he never heard of one of the members calling a delegate meeting without consulting the other members of the council, and he thought that any one who would so act was not worthy the respect of those with whom he had been elected to do business. He would like to know from the secretary if he had received any communication from Mr. Potter on the subject.

The secretary said he had not.

Mr. Allan continued by saying that he had received letters from the men on strike, complaining of a delegate from the Trades' Council having been in their neighbourhood without calling on them for such information on the subject as would be useful to the trades of London, and prevent false reports being circulated.

Mr. Cope had stated his views on the subject at the delegate meeting, and felt that great mischief would result if Mr. Potter was allowed to continue the irresponsible course he was pursuing.

Mr. Coulson said that the bricklayers' executive considered that the conduct of

1

MR. POTTER AND THE LONDON TRADES' COUNCIL.

Mr. G. Potter, in convening a trades' delegate meeting on the iron trades' dispute, without consulting the London Trades' Council, or without the authority of any trade, was an unwarrantable liberty, and an insult to trade societies, which placed them in a false position; and it had appeared in the Press that the trades of London had specially sent a delegate down into Staffordshire to learn from the lock-outs the particulars of their dispute, which was without any foundation and entirely untrue. Therefore, it was highly esential that Mr. Potter's position should be clearly laid before the public, and he would move the following resolution:—

That this meeting regrets and condemns the course taken by Mr. G. Potter in calling a meeting of trades' delegates on the dispute in the iron trades without first consulting the London Trades' Council, of which he is a member; and consider that meetings so called by any unauthorised person or persons are dangerous, as resolutions may be passed in them pledging societies to principles that would be prejudicial to their influence and material prosperity, and place the trades' societies in a false position before the public.

This was seconded by Mr. Cope.

Mr. Applegarth, in supporting the resolution, regretted the absence of Mr. Potter, but as he had been informed of the nature of the business of the evening, and having waited till half-past 10 o'clock, it was no fault of the council that he was not preseut. He willingly accorded to every man the right of convening meetings or taking whatever steps he thought fit to assist the oppressed, so long as he acted in his individual capacity, and did not lead the public to attach more importance to such actions than they really merited. But he altogether deprecated the idea of any irresponsible and unauthorised individual arrogating to himself the right which belonged to a properly authorised and responsible body, such as the London Trades' Council, and more especially when such had been done by one of its members. One of the evils resulting from such proceeding was evidenced at the dissolution of the London committee appointed to receive subscriptions on behalf of the men resisting the "Discharge-note." After all legitimate expenses were paid, a gift of £8 in one instance, and £5 in another was awarded for "services rendered" during a few weeks, whilst, in hundreds of instances, far more onerous services have been rendered gratuitously. The committee were awarded 2s. 6d. per night for their services, when frequently we hear of men almost starving in defence of a principle on 2s. 6d. per week. He contended that

2

such proceedings, if not discountenanced and repudiated by the Trades' Council, were calculated to abuse the public mind still further against our societies, and afford just grounds for those grave accusations so frequently made against their leaders.

The resolution was unanimously carried.

Mr. Danter, President of the Amalgamated Engineers, considered the conduct of Mr. Potter as unworthy of the confidence of any respectable member of society; in fact, it appeared that Mr. Potter had become the aider and abettor of strikes. He thought of nothing else; he followed no other business; strikes were his bread and cheese; in short, he was a strike-jobber, and he made the *Beehive* newspaper his instrument for pushing his nose into every unfortunate dispute that sprung up. He (Mr. Danter) did not know the precise position in which the trades' unions stood with the *Beehive*, but they should take care that men of Mr. Potter's stamp should not commit them to any paper that made capital by trading on their misfortunes. He would move the following resolution:—

That the visit of Mr. George Potter to Staffordshire was undertaken without the knowledge or sanction of the London Trades' Council; and this meeting is of opinion that the only object he had in view was to promote the interests of the *Beehive* newspaper, and deem it our duty to make known the fact, and thus disabuse the minds of the men locked out, and likewise the public generally.

Mr. Applegarth seconded the motion, which was unanimously carried.

Mr. Applegarth proposed, and Mr. Cope seconded,

That this council expresses its surprise that the following resolution, which was unanimously adopted at the delegate meeting convened by the Trades' Council, March 23rd, should have been left out of the report of said meeting in the *Beehive* newspaper.

This was also carried unanimously.

The following is the resolution referred to:—

That the London Trades' Council be appointed a committee to receive subscriptions on behalf of the men engaged in the dispute in the iron trade.

Mr. Phipps, delegate of the North Staffordshire men, attended to give the latest information he had officially received from the executive. The delegates present considered that the information respecting the interview between the masters and men would be known more fully in the course of a day or two, therefore the meeting was adjourned to Friday night for the further consideration of the question, also the progress made regarding the Masters and Workman's Act.

G. Odger, Secretary.

1, Nassau-street, Marylebone.

MR. POTTER AND THE LONDON TRADES' COUNCIL.

THE STRIKE IN NORTH STAFFORDSHIRE.

MEETING OF LONDON TRADES' COUNCIL.

A numerously attended delegate meeting was held at the Bell Inn, Old Bailey, on Thursday, April 6, "to consider the conduct of the men on strike in reference to Ld. Lichfield's proposal to submit the question to arbitration on condition that the men return to their work and draw wages on account until the dispute was amicably settled." The "council," who are the regularly appointed representatives of some of the most influential and nationally organised trades in the kingdom, added to the circular convening the meeting the expression of their opinion, that "the present position of the men on strike was one of the most serious character possible, and demanded from every trades' unionist a frank expression of opinion."

Mr. Mildred (joiner) took the chair, and in opening the proceedings briefly referred to the causes which had led to the meeting, as stated in the circular, and said that he had before him a large number of certificates from the various trades who had sent delegates to that meeting, which showed that it was a real and *bonâ fide* meeting of representatives of organised trades. He then called upon the secretary to read the correspondence..

Mr. Ogder then read a letter from Mr. G. Newton, scretary to the Glasgow Council of the United Trades, which stated that so far as his inquiries had gone, he found that the conduct of the men who refused to submit to arbitration on the terms proposed by Lord Lichfield was held to be entirely wrong, and indeed that they had dealt a severe blow to the hopes of those of the Glasgow trades who wished to have all such disputes settled by arbitration. The trades of the United Kingdom had long wished to have the stamp of Royalty affixed to an act constituting courts of conciliation and arbitration. It was painful therefore to have the cup of hope rudely dashed from their lips by these strong-headed men in North Staffordshire. Could they now pronounce arbitration a cure for strikes when in a very dangerous case the patient had refused it? The writer regretted that the men of North Staffordshire, who professed their inexperience in such matters, had not acted on the wholesome advice tendered to them to refer the whole case to arbitration. At the same time a curious question suggested itself to his mind—the possibility of any man, or any number of men, having a desire to perpetuate the unfortunate state of affairs in the iron trade. Was it true that George Potter, on the day after the first meeting in the Sussex Hotel, went down among the men and told them they would be liberally supported, and that one trade was ready to advance £500 for the purpose? He believed that in fact the men in North Staffordshire had been misled and deceived; and he hoped that from the meeting of the London Trades' Council there would go forth such an expression of opinion on the conduct of the men in refusing arbitration on fair terms as would amount, if not to a denunciation, at least to an utter and complete repudiation thereof, as a breach of faith and a disgrace to the working classes of our common country. Mr. Dronfield, secretary to the Association of Organised Trades of Sheffield and the neighbourhood had written to say that resolutions had been unanimously passed at a meeting of the executive of the association held on the 5th inst., 1st. That this Executive Committe approves of the action of the London Trades' Council in protesting against any member calling delegate meetings on his own authority, as not only dangerous but calculated to place the trades generally in a false position. Secondly, That in the opinion of this Executive, the resolution passed at the London Trades' Council of delegate meeting, recommending the North Staffordshire men to submit their case to "unconditional arbitration," was wise and proper under the circumstances; and further, that the men have acted most unwisely in not accepting the arbitration, or resuming work under the terms named by Lord Lichfield—this Executive having from the first recommended "arbitration." Mr. Dronfield added that "the more he saw the more he was convinced that the men were not alive to their own interests, or must have been very badly advised, otherwise they would have resumed work, and drawn wages on account until the question was settled by arbitration, inasmuch as they had failed to establish their claim to the advance sought." A letter from the chairman of the Amalgamated Society of Carpenters and Joiners enclosed a resolution cordially approving of the action taken by the London Trades' Council with reference to the unwarrantable proceedings of Mr. Potter with respect to the dispute in the iron trade. Mr. Odger having read these letters, briefly recapitulated the past proceedings of the council in this matter, the effect being that Mr. Phipps and other representatives of the North Staffordshire men had agreed to leave the question entirely in the hands of the Trades' Council, on the simple condition that the question should be referred to arbitration, entirely unclogged with any conditions, because it was believed that the proposal of conditions on one side would open the door to the proposal of conditions on the other, and thus the whole question of arbitration would fall through. The contrary course had been taken by the men on strike, and the result they all too well knew. The men had declined to arbitrate on what he called fair and reasonable terms, and he held that the resolution they had passed, pledging themselves to give these men support if arbitration was refused by the masters, ceased to have any effect.—(Hear, hear.) He believed that if the men had accepted the terms proposed by Lord Lichfield they would have been now in work, that the evils of protracted strike would have been avoided, and, above all, that the stigma likely to be cast upon the trades generally, by a refusal of fair and reasonable arbitration, would not, as he feared it would, be cast upon the trades of this country. For one —while not aftaid of strikes when absolutely necessary, and justified by grave and substantial reasons—he had long desired that all such disputes should, if possible, be referred to impartial arbitration; and that, he believed, was the general feeling of the organised trades of the country which had their centres in London. (Hear.)

Mr. Howell (bricklayer) expressed his deep regret that arbitration had been refused by the men in North Staffordshire. He was convinced that every trade had more to gain by arbitration than fighting. (Hear, hear.) Of course he did not mean to say that there were not occasions when they must stand up for great principles or essential questions, but in point of fact English society as a whole, was a series of compromises, of mutual concessions, and the submission of any particular claim to arbitration did not in fact preclude further and ulterior action, if the claim itself was founded in reason and justice.—(Hear.)

Mr. Stronach (printer) expressed his entire concurrence in the opinion that arbitration was by far the best mode of settling all industrial disputes.—(Hear.)

Mr. Applegarth (joiner) then moved, "That this meeting views with deep regret the refusal of the men in North Staffordshire to agree to the proposal of Lord Lichfield to resume work, drawing wages on account pending the whole question being submitted to arbitration; but, believing the men to have been ill-advised, and deceived by false promises of support, we most earnestly call upon them to accept the proposal of Lord Lichfield, feeling as we do that to accept such an offer is a duty equally incumbent on them as trades' unionists as to those depending on them for support."

Mr. Bockett, delegate from the London and Westminster Bookbinders, seconded the resolution. He cordially agreed in the principle of arbitration, and could state that if that principle had been adopted it would have prevented the loss of at least £5,000 in his own trade, which consisted of only 400 or 500 members.

Mr. Harris (carpenter) and Mr. Nightingale (carpenter) differed from the mover and seconder of the resolution. They thought the men were quite right in refusing to resume work unless at the old prices. How could they tell what amount they would be allowed to draw on account?

Mr. Cremer (Amalgamated Society of Carpenters and Joiners) cordially supported the resolution, and believed that no more fair or reasonable proposal than that of Lord Lichfield could possibly have been made. As a sincere trades' unionist, he deeply regretted that the refusal of arbitration should have come from the men. In any case, he believed that the men could not have been damaged by the reference.

Mr. Benney, Mr. Clarke, Mr. Hancock (bricklayer), and other delegates having spoken in support of the resolution.

Mr. Harris moved, as an amendment, "That this meeting strongly approves of the conduct of the men of North Staffordshire in not accepting Lord Lichfield's terms to draw wages on account pending arbitration, because such a course would have tended to show that the men were going in at a reduction."

Mr. Nightingale seconded the amendment.

Mr. Allan (secretary to the Amalgamated Engineers) said he certainly could not support the amendment.—(Hear.) Nothing astounded him more than to hear any trades' unionist repudiate arbitration. When in his own trade they had disputes—and all present knew upon how extensive a scale these had occurred in the body he represented—their great object had been to secure fair and impartial arbitration. In the builders' lock-out, the most strenuous efforts had been made to secure a settlement by arbitration, and he was surprised to find delegates connected with these trades now repudiating arbitration. He thought that some of those present did not clearly understand the meaning of drawing wages on account. Whatever the men might

MR. POTTER AND THE LONDON TRADES' COUNCIL.

draw did not in the least prejudice their claims, whatever they were. If those claims were bad, undoubtedly the decision of the arbitrators would be against them, and justly so. He believed that the scale on which the puddlers were to be paid was based on a fallacious principle in itself; but, nevertheless, it was a fair and open bargain between the men and the masters, and as a working man and as a representative of the working men, he said that any mutual arrangement of that kind ought to be honourably carried out. The question was branching out widely; the South Staffordshire men and the iron-workers farther North had withdrawn, leaving the North Staffordshire men standing alone; but in Scotland the masters threatened a reduction in the iron trade, and it was deeply to be regretted that the course pursued had tended to place the men in a false position, as well as to exhibit trade societies in the light of opponents of arbitration, which he was certain that meeting would agree with him in saying they were not.—(Cheers.)

Several other delegates addressed the meeting in support of the original resolution, many of them explaining that their views had been altered by what they had heard since entering the room as to the real nature of Lord Lichfield's proposal.

Mr. Applegarth replied in a speech which was loudly cheered, and Mr. Harris having withdrawn the amendment, the resolution was unanimously agreed to amidst cheers.

Mr. Heslop, of the Amalgamated Engineers, then moved, "That this meeting tenders its best thanks to Lord Lichfield for the able and impartial manner in which he conducted the trust confided in him in relation to the strike and lock-out in the iron trade, and hopes that the conduct of the men in refusing his suggestion will not deter him from using his good services in so noble a cause.—(Cheers.)

Mr. Hughes seconded the motion, which was carried unanimously, amidst great applause.

Mr. Howell moved, and Mr. G. Winspring (Amalgamated Engineers) seconded the following resolution:—"That the two resolutions passed this evening be forwarded by the secretary to Lord Lichfield and to the executive of the men on strike, with their earnest wishes that Lord Lichfield's efforts will be renewed, and become successful."

The resolution was unanimously agreed to.

Mr. Phipps, delegate from North Staffordshire, who, with Mr. Billington, had entered the room shortly before the resolution was put, said he really could add nothing to the statement he had made at the Sussex Hotel on the previous evening.

In answer to a question put by Mr. Allan, Mr. Odger, secretary to the council, explained that the so-called delay on their part in calling a meeting together was caused by a deep conviction that the interference of them and of outside parties, in the initiatory stages of a dispute between masters and men, was not by any means judicious. If the men could not of themselves adjust the dispute, an appeal either for advice or assistance might fairly be made, and he had no doubt would have been answered at once. It might be that the letters sent to Mr. Potter were such as to induce him at once to go down to the districts in which the dispute took place and to summon a meeting as he had done, but after all it would have been very easy for him to have communicated with the secretary of that council, and if that had been done, the whole of the organised trades of the country—not of London alone—would have been brought to bear on the question, and thus the resolutions passed would have had money at their back, which he feared would not be the case with resolutions passed elsewhere.—(Hear). Mr. Phipps had, no doubt, done his duty well to his constituents, but if he stopped in London six months he would not so well advance the interests of his trade as by going down to his own district and inducing his fellow-workmen to agree to arbitration.—(Hear). As the secretary of that council, he warned the trades in the country against putting forth resolutions passed by meetings called by irresponsible persons, and supported by those who did not present any credentials of their claim to represent trade societies.—(Cheers).

This statement led to an animated debate, in which Mr. Odger repeated the charges against Mr. Potter made at the meeting in the Sussex Hotel, Bouverie-street, April 5th, namely, that upon Mr. Connolly's own acknowledgment Mr. G. Potter and T. Connolly (mason), had been hired by Mr. Brassey, contractor, and received £10 for their services to go to Birkenhead and advocate his election as M.P. during the Masons' Strike against the Hour System, when at the same time Mr. Brassey was supplying the stone for the erection of the buildings the men had struck from; and many other places Mr. Potter had gone for a consideration. And during the first two years that Mr. Potter had been manager of the *Bee Hive* the whole of the transactions that passed between him and Mr.

MR. POTTER AND THE LONDON TRADES' COUNCIL.

Troup, the Printer and Editor, no receipts or vouchers had been produced, whilst Mr. Potter had represented the "Bee Hive Company" as having sunk £1,500, and being £1,100 in debt, without invoices or vouchers to prove that these debts were correct, or in what way they had been incurred. Further, Mr. Potter had been accused publicly by Mr. Facey, on the authority of Mr. Tilling (printer), of receiving 20 per cent. on the money that was paid for printing during the contest for the Nine Hours and Lock-out in the Building Trades in 1859, amounting to a very large sum. When accused of this reprehensible act he (Mr. Potter) promised to take steps to clear his character; but although his friends had offered to pay all his expenses, he had failed from that day to this to keep his promise, and, therefore, the charge against him had never been disproved. And at the starting of the *Bee Hive* Mr. Dell advanced to Mr. Potter, on behalf of the Trades Societies, a loan of £120, upon the condition that he (Mr. Potter) should consult with the London Trades' Council and the leading representative men; but he neither consulted the Trades' Council or Bricklayers' Society; neither was any public meeting called of the trades for their advice or sanction to bring out the *Bee Hive*, and appoint directors; but, on the contrary, Mr. Potter collected men together who made themselves directors, and Mr. Potter's salary was fixed at £2 2s. per week before any of the shareholders had met in general meeting. Mr. Potter's desire was not to do business legitimately, but to arrogate to himself all power and all control. And the majority of those who made up Mr. Potter's meetings were carpenters and painters, who have no influence in the building trade, men who had no delegated authority, but whipped up for the occasion by a special notice from Mr. Potter. He would here state that it was Mr. Potter's practice, whenever he wanted a majority, to bolster up his position, to issue circulars to his known supporters (of whom a complete register was kept) urging them to attend and vote for him. The speaker held in his hand one of the circulars with Mr. Potter's signature, and challenged Potter to deny the fact that it was by such means that he swamped the votes of elected delegates. He reiterated his offer to submit these charges to the investigation of any fair and impartial committee of the trade delegates. He proposed, further, to put these charges in writing in time for Mr. Potter to prepare himself fully to give an answer. Mr. Potter was present at the Sussex Hotel when he made this offer. No answer was made, and therefore, he said, judgment went by default.

Mr. Cremer wished to explain, with respect to the constitution of the trades' council, that it had 15 members, all of them elected annually by organised trades, numbering directly 45,000 members, and indirectly, at least 100,000; what they did bound the bodies they represented, whereas the meetings at the Sussex Hotel were meetings which were composed of persons not merely not delegates, but even of persons not members of any trade society whatever. They were packed meetings, convened by private circular for personal purposes; but the council of the trades and the delegates who had come with their credentials, he contended, really represented the trades of London.

Mr. Allan moved "That this meeting highly approves of the proceedings of the London Trades' Council in the course it has adopted throughout in relation to the unfortunate dispute in the iron trade in North and South Staffordshire, and, further, that we in no way recognise the proceedings of the meetings convened by Mr. Potter, and held in the Sussex Hotel in the name of the trades of London."

The resolution was unanimously carried, and the proceedings closed at midnight, with a vote of thanks to the chairman.

MEETING OF THE TRADES' DELEGATES WITH LORD LICHFIELD.

A meeting of delegates from the London trades was held on Tuesday evening at the Bell Inn, Old Bailey, for the purpose of hearing from Lord Lichfield "an explanation as to the proposal for arbitration he had made to the puddlers of North Staffordshire, with a view of bringing the strike to a close, and as to how his proposal was received, both by the masters and the men." About 150 delegates were present, and a special deputation was also present from the North Staffordshire men; Mr. Mildred, carpenter, in the chair.

Mr. Odger stated that Lord Lichfield had called on him respecting the resolution passed at their previous delegate meeting, and said he felt happy to think the delegates had appreciated his conduct in the matter, and that they had looked more fully into the whole question than some others had done. (Hear, hear.) He told his lordship he thought there was some misunderstanding regarding the course he had taken in connection with the important question of arbitration, and he (Lord Lichfield) replied that if, by meeting the London trades, he could be

MR. POTTER AND THE LONDON TRADES' COUNCIL.

of service in explaining the matter more fully than he had done, he would be happy to attend a delegate meeting for that purpose. At the council meeting on the previous Tuesday, he (Mr. Odger) explained what had transpired in the interview he had had with his lordship, and the council expressed regret that he had not invited him to attend at once, and there was much conversation about sending a deputation to him on the next day. He then said that if he should see his lordship again, and if he was equally willing to meet the council, he should certainly call a meeting, to which the whole of the members present said they hoped he would. On the following day (Wednesday) his lordship again called upon him, and expreesed the pleasure he should have in meeting the delegates of the London trades, and explaining the whole matter. Mr. Odger then read some correspondence he had had with Lord Lichfield.

Mr. Potter said that in order that the meeting should hear both sides of the question, he had written to the North Staffordshire men to send delegates, and he was happy to say that two of the men Lord Lichfield met at Hanley were present as a deputation. (Hear, hear.) This meeting had not been sanctioned by the council.

Mr. Coulson said that the council that met on the previous Tuesday gave the secretary power to convene a meeting to meet his lordship, if necessary. (Hear, hear.)

Mr. Allan (Amalgamated Engineers), remarked that full permission was given. Mr. Odger called upon him, and he advised him to the effect that the sooner the meeting was called the better, so that no misunderstanding should exist amongst the men. This was a perfectly legal meeting, and he was glad there were delegates present from North Staffordshire, as there could be no objection to them. (Hear, hear.)

Mr. Guile said that, as a lover of unity, he would move that this meeting be considered legally constituted.

Mr. Odger said the Council instructed him last Tuesday night to call this meeting, and therefore he should have been wanting in duty if he had not done so.

After some further discussion on this point, which was not conducted in the most orderly manner possible,

Mr. Danter, the chairman of the council, begged most distinctly to state that Mr. Odger had full power from the council to call a meeting of the London trades' delegates, whenever he could get Lord Lichfield to come to it. (Cheers.) Therefore, he would propose that, as his lordship was present, they should proceed to hear him at once. (Loud applause.)

Lord Lichfield, who was very warmly received, then went on to detail at some length the reasons which had induced him to interfere in the matter. Both the masters and men had agreed to accept the principle of arbitration, but the masters were only willing to arbitrate provided the men would resume work at the reduction, and the men refused to arbitrate unless they were allowed to resume work, pending the arbitration, at the old prices. A middle course suggested itself to his mind, and that was, to refer the wages question to arbitration, the men, pending the proceedings of that arbitration, to draw wages on account. He proposed, this, in the first instance, to the masters, by whom, after several objections, it was accepted, with the addition that he (Lord Lichfield) should appoint the arbitrators, whose names should be finally sanctioned by both parties. He then submitted his proposal to the men, by whom, he regretted to say, it was rejected, and who, in his presence, affirmed their previous resolution not to consent to arbitration unless allowed to resume work at the old prices. He endeavoured to induce the men to reconsider their determination, but in vain; and all his efforts had been fruitless. It had been said he had failed from his not having made the men fully acquainted with his proposal, but this he begged to deny. They thoroughly understood him, and he well understood them. (Cheers.) His lordship then proceeded at length to express a strong opinion as to all trade disputes being settled by arbitration, which would do much to prevent the ruinous system of strikes and lock-outs, and thought some legislative enactment should take place in furtherance of such object, towards the accomplishment of which he was willing to lend his aid. His lordship sat down amidst great applause.

Mr. Edwards, one of the deputation from North Staffordshire, after thanking his lordship for the interests he had taken in this question, and the London trades for the subscriptions towards the men on strike, said the men had made a great many concessions since the struggle commenced, and were willing to do so again, but all they had done had been to no purpose. The men had never refused to let their case go by arbitration, but they thought they ought not to resume work upon the reduction. The men had been orderly from the very commencement of

MR. POTTER AND THE LONDON TRADES' COUNCIL.

the strike, but the masters had been using every means to entrap the men by drink and otherwise to get them into difficulty.

Mr. Waddell, another of the deputation from North Staffordshire, did not blame the masters, but it was the managers and others they had about them that did all the mischief. He personally had nothing to say against his lordship's proposal, and the men were proud that they had his interposition. If they went into work at the reduction, it might be three months before the question was decided by the arbitrators, and he for one was not going to work for that time without knowing how much he was to obtain for his labour. (Hear, hear.) If the men had been asking for on advance of wages and agreed to arbitrate, they would not expect to receive those advanced wages pending the arbitration proceeding, and, therefore, they ought not to be called upon to go in at the reduced wages offered by the masters whilst arbitration was going on. They, however, had offered to divide the proposed reduction with the masters, and, instead of being reduced 1s. per ton, to resume work at the reduction of 6d. per ton, but the masters would not consent, and the men had now fallen back on their old resolution and not to go to work except at the old prices. (Cheers.)

Mr. Potter rose to move a resolution to test the meeting on the subject matter of the dispute. The lock-out on the part of the masters was a complete and signal failure, and, after being in existence for four weeks, it was abandoned by the masters. And now the North Staffordshire men were again receiving support from the Brierly Hill district, *and they would not, he thought, require extraneous aid.* He then moved, "That this meeting regrets that the offers of the North Staffordshire men to resume work either at the terms they left off, or at the rate of 10s. per ton pending arbitration, were not accepted by the employers; but after having heard the statement of Lord Lichfield and the special deputation from the men on strike, this meeting is further of opinion that the terms and mode of settling the present dispute in North Staffordshire should be left to the masters and men in that district, *and that this meeting pledges itself to support the men by all the pecuniary means in their power until a settlement is effected.*" (Cheers.)

Mr. Davis seconded the resolution.

Mr. Howell felt persuaded that if arbitration was to take place it must not be shackled with conditions. (Hear, hear.) The question was what middle way could be suggested whereby neither party would concede the point at issue. The men had more to lose by refusing the arbitration than the masters could possibly gain. ("No," and hear.) As an amendment he would move, "That this meeting of the delegates of the trades of London, having heard the statements of Lord Lichfield and the North Staffordshire delegates with reference to the present state of affairs, strongly advise the men on strike to submit their case to arbitration on the terms proposed by his lordship." (Cheers.)

Mr. Guile, in seconding the amendment, made a powerful speech in favour of arbitration.

Mr. Allan, secretary of the Amalgamated Engineers, said the latter part of the resolution was useless, because Mr. Potter had told them the Brierley Hill Executive had now received the men into the union again, and would grant them support. He had not come from the society he represented to pledge them to support the men in North Staffordshire, and he had no doubt others had come with no such instruction. (Hear, hear.) He saw in the *Beehive* newspaper of a week since a statement that Mr. Potter had given evidence before a Parliamentary Committee on arbitration; but from the blue-book containing the minutes which he had before him, he found Mr. Potter did not give evidence at all at that committee. There were certain persons who got the credit of doing their best to bring about courts of arbitration; but so soon as they had the opportunity of rendering good service, they did their utmost to pass resolutions to prevent anything being done to bring forward such a consummation. (Cheers and hisses.)

As it was now getting very late, Lord Lichfield said a few parting words, and on the motion of Mr. Applegarth, a hearty vote of thanks was accorded to him for his attendance. The noble lord briefly responded, and quitted the room.

After a very noisy and personal discussion about things in general and nothing in particular, Mr. Odger stated that he thought it would be better not to pass any resolution at all. They had met simply to hear Lord Lichfield, and having done that, the business of the meeting was at an end. (Cheers and dissent.)

Another noisy discussion ensued, and ultimately Mr. Stoner moved "That this meeting do now adjourn." To this there was considerable objection, and after a time the chairman vacated the chair, and the meeting broke up in the greatest confusion, without either the amendment or resolution having been put.

J. Kenny, Printer, 40, Parker Street, Little Queen Street, W.C.

Trades' Societies and Lock-Outs.

REPORT

OF THE

CONFERENCE

OF

TRADES' DELEGATES

OF THE

UNITED KINGDOM,

HELD IN THE

TEMPERANCE HALL, TOWNHEAD STREET, SHEFFIELD,

ON JULY 17TH, 1866, AND FOUR FOLLOWING DAYS.

Specially Reported for the Committee, by MR. J. FARRELL, *of the "Sheffield Daily Telegraph."*

WITH

A REPORT OF THE PUBLIC MEETING,

List of Delegates who attended the Conference, the Trades they represented, &c.

SHEFFIELD:
LOXLEY BROTHERS, PRINTERS, ATLAS OFFICES, FARGATE.

1866.

REPORT, &c.

FIRST DAY.

THE Delegates assembled in the lower room of the Temperance Hall, at twelve o'clock, for the appointment of officers and the transaction of routine business of the Conference, including order of procedure, limitation of the time for Delegates speaking on propositions, Committee for preparation of the same, and other business of an ordinary character, involving considerable discussion on general details.

The credentials of the various Delegates were examined by the officers, and it appeared that 138 Delegates were present, representing in the aggregate nearly 200,000 members.

[*For List of Delegates, see end of Report*].

In consequence of the number of Delegates present being much larger than was anticipated, the room engaged by the Committee was not sufficiently commodious; it was therefore resolved that for the future the Conference should meet in the large room.

RESOLUTIONS PASSED :—

"That Mr. C. Bagshaw, President of the Association of Organised Trades of Sheffield, be appointed Chairman of the Conference."

"That Mr. M. Lawrence, of the London Protective Tailors' Association, act as Vice-Chairman."

"That Mr. W. Dronfield, Secretary of the Organised Trades of Sheffield, be appointed Secretary of the Conference, and that Mr. W. H. Wood, of the Manchester Typographical Society, act as Assistant-Secretary."

"That Mr. W. Broadhead, of the Sheffield Saw Grinders' Society, be appointed Treasurer."

4

Mr. T. Holland, of the Filesmiths' Union, offered his services as doorkeeper, and they were accepted.

With a view to printing the deliberations of the Conference for general circulation, it was decided that a special report be taken of the proceedings, the selection of the Reporter to be left in the hands of the Secretary.

"That the Representatives of the Public Press be admitted to the Conference."

"That the Meeting continue open until the preliminary business is arranged, and then adjourn for the day."

A Committee of seven was formed to draw up standing orders for the government of the Conference, preparation of resolutions, &c., consisting of Mr. C. Bagshaw, Sheffield; Mr. M. Lawrence, London; Mr. W. Dronfield, Sheffield; Mr. W. H. Wood, Manchester; Mr. W. Broadhead. Sheffield; Mr. J. Dixon, West Yorkshire; and Mr. G. Newton, Glasgow.

"That the Delegates meet each day at ten a.m., and continue until five p.m.; an adjournment for dinner taking place from one to two p.m."

"That each speaker be allowed fifteen minutes on any one question, and only to speak once; the mover of the resolution to have five minutes to reply."

"That the mover of any resolution submit the same to the Chairman, in writing, signed by the mover."

"That the Minutes of each day's proceedings be printed for the Delegates, the first day's to contain the list of Delegates, number of members, and the towns represented."

SECOND DAY.

The names of the delegates having been called over, the meeting proceeded to the business of the day.

The minutes of the previous day's deliberations were then read, when it was

RESOLVED—"That the minutes as read be confirmed and adopted, as a faithful record of yesterday's proceedings."

Mr. Hinchliffe, of the Sheffield Joiners' Society, desired to be admitted to the Conference, alleging that, although not yet accredited, a meeting, to be held this evening, would authorise his attendance.

RESOLVED—"That Mr. Hinchliffe be admitted, but not allowed to vote until his credentials are forthcoming."

Credentials were received from the following Trades, viz.: the Scissor Grinders—Messrs. R. Holmshaw and J. Gale; Nottingham United Operative Brickmakers—Mr. Charles Oakland; and also from the General Council of Operative House Decorators and Painters, London—Mr. G. Shipton; and a letter was read from the Iron Workers' Association, Gateshead, announcing the appointment of Mr. J. Kane.

Letters were read from the Trades' Council of Edinburgh, Cardiff Trades' Committee, the International Working Men's Association, &c.

RESOLVED—"That the correspondence be received."

MR. SWEETING asked how it was intended to regulate the voting. If every delegate had a vote irrespective of the number of members represented by him, Sheffield would have a preponderance, there being no less than 02 delegates from Sheffield.

THE CHAIRMAN: Have you a proposition on the matter? It is quite right that you should ask the question.

MR. FOX hoped there would be no discussion on the point. He was sure that every delegate present had a desire to arrive at the same good end.

MR. LEIGH moved "That every delegate be allowed one vote on each subject, irrespective of the number of members represented by him." The mere fact of a delegate representing 5000 members did not prove that he had five times the brains of the man who only represented 1000. Though he and his colleague only really represented 1400, they had the interests of 7000 operatives at heart, and it was not right that they should have more than one vote each, on any subject that might come before them.

Mr. T. BANKS seconded the motion.

Mr. BAILEY, who believed that their motives were all alike, supported the proposition.

Mr. ALLEN: I am willing to extend to every delegate present an equal amount of power, yet I feel called upon to lay before this meeting a matter which is causing me some little anxiety. I quite agree that a delegate representing 100 members should have as much power at this Conference as the delegate who represents 1000, and I believe we shall all aim at one point, and that is the general support of the oppressed. According to the prospectus that was issued, the meeting was to be composed of persons representing trades' societies, members of societies that have for their object the protection of labour. Whether such is the case or not, I am not prepared to say; but I desire some information on the subject, and the information I shall elicit may remove the anxiety to which I have alluded. I find the society of Engineers represented, and the letter introducing him expresses sympathy and good wishes. I accept them; but it occurred to me, whether the delegate could take upon himself to say that the different branches of the extensive society to which he belongs would render support to the objects of the Conference. That brought to my mind the society or association from which I find a letter has been received—the International Association of Working Men. Am I to understand that this International Association consists of the members of some particular society formed for the protection of trade? If they are, then we must respect them as such; but I should like the point to be cleared up for the information of the delegates generally. What are we to understand when they call themselves members of the International Association? Are they members of a trades' society? And if members of a trades' society, are they organised as a trades' society belonging to the British dominions? Can those who represent them here, tell us this? If we form ourselves into an association of labour, we certainly must have a fund for the protection of labour when it is oppressed.

A DELEGATE rose to a point of order.

The CHAIRMAN said that the question before the meeting was whether every representative should have one vote.

Mr. ALLEN: If they are representatives of trades' societies, I do not object to them; but can we depend upon them contributing to the funds of this society when it is established, if it requires support?

Mr. ODGER: There is no delegate here who can commit his society now to any amount of money; therefore our friend's question is purely an invidious one. It is not in my power to commit either of the trades I represent as connected with the

7

London Trades' Council to any amount of money whatever. I am sent here to do my best to further the object you have in view; but not to commit these trades to any amount of money towards your object, nor do I know that that question is before the chair at this moment, but whether a delegate shall have one vote or not, according to the number of members he represents. If the last speaker had confined himself to the legitimate question before us, he would have concluded his remarks long since, and we should have been dealing with the next order. He has provoked a discusssion. I presume he will take the conduct of every delegate to amount to this—that his society is well disposed towards you, and that on his return, when you have completed your business, he will use his best efforts to get his society to endorse and to support you pecuniarily. The International Association is supported by working men generally. It is a *bona fide* society, extending itself throughout Switzerland, France, Belgium, Italy, and a great number of places in Germany. If the International Association must now declare whether they will give more money or not, it is an invidious, unfair, and partial proceeding, and I hope the friend who has suggested it, will withdraw it, and allow the business to proceed in a proper manner. Whatever plan is drawn up, the delegates cannot pledge themselves to bind their societies to it.

The proposition as to voting was then put and carried.

LOCK-OUTS.

MR. GEORGE NEWTON, Glasgow, moved, that—

> "Whereas, within the last few years, great and disastrous Lock-outs have been frequently resorted to by the employers of labour, as a means to compel the workmen to implicitly obey their wishes, and whereas these Lock-outs are alarmingly on the increase, and are therefore likely to produce unparalleled injury to the trade and commerce of the country, and consequent suffering and misery on the children of labour, it is desirable to enquire into the causes which give rise to these Lock-outs, with the view of providing a remedy for so serious an evil"

On the meeting of the special committee, last night, to endeavour to sketch a few resolutions to submit to you to-day, we were met at the threshold of our labours with one vast difficulty, and that arose from the fact that we were not in possession of the views of the delegates as to the nature of Lock-outs, their causes, or any other consideration in reference to them; and it seemed desirable to frame a resolution so worded as to raise a general discussion on the matter, with a view to elicit the history of Lock-outs in past times, and, as far as possible, to obtain your opinions thereon, so that, in submitting any subsequent resolution, the committee may be guided by that historical information, and by the views of the parties present. The resolution says, mainly, three things. The first is, that "within the last few years, great and disastrous Lock-outs have been frequently resorted to." He might point, in support of that, to the great Lock-out in London, the Lock-out, last year, in the Iron trade, in Staffordshire, and the immense disturbances in Sheffield, the Lock-out on the Clyde, and the Lock-out in the Staffordshire Potteries. All these show that Lock-outs have been frequently resorted to, and are very disastrous. I am probably not aware of half the Lock-outs that have taken place; but as there are so many present from so many quarters, we shall no doubt receive a vast amount of information in relation to the matter. That they "are alarmingly on the increase" is true, for there is scarcely a movement by workmen but the opposing weapon of Lock-out is resorted to; and it becomes us to consider whether we can arrive at the correct cause, and make a proper diagnosis of the disease, and then probably the cure will follow. Let us examine ourselves, and see if there be any wicked way in us that contributes to this unsatisfactory state of things, and if we discover that we are not blameless, then we ought, first of all, to set our house in order, and to avoid, as far as we are concerned, any such contingency arising in the future. Then let us examine the opposite side of the camp, and see how they stand; and if we find that they have not done all that they ought to have done with a view to prevent these serious evils, let us undisguisedly, and in plain language, point out where we consider they have erred, and, by increasing public opinion, in a healthy way, against—tyranny come people call it, but perhaps a milder word would be better—against the unwise policy used, it will do much to repress it in future. A great many Strikes, and perhaps Lock-outs, too, have arisen from a stubborn refusal, on the part of both sides, to look the question honestly and fairly in the face. The habit has been to look only on their own side of the question; the proneness of human nature to be selfish has shewn itself in a very remarkable manner, and we have found out that we are all selfish more or less, and very apt to be blinded by that selfishness. In approaching this question in future, it will be well to divest ourselves of the operation of this feeling; for no word can be too strong to designate the position of men who are entire slaves to feelings of selfishness. If we approach this question with improved feelings—I do not say that the men are always wrong—but it will be found that we have not acted in all cases with that wisdom which was best calculated to secure the objects we have had in view. The misery and suffering which accrue from vast Strikes and Lock-outs, I think, warrant any body of men in taking serious steps with a view to prevent their recurrence. There can be many justifications advanced on behalf of Strikes; and a Strike is a simple matter, easily understood and easily defined. A Strike is a resolution of men only to work on such conditions, and does not affect those who are not pledged to that resolution. A Lock-out, on the other hand, is a resolution that amounts to this—that if Tom Jones does not do this or that, Harry Robinson shall have no bread to eat. I cannot on any pretext whatever justify a Lock-out, because it compels men to suffer who have no hand in the affair. (Cheers.) I think it would be very impolitic in me to set the bad example of occupying a great deal of time. There are, no doubt, many here, who are perfectly competent to speak on the point; and if they direct their attention to searching out the causes, and giving us historical information of the causes, and the effects thereof, this day will not be spent in vain if it is entirely devoted to the discussion of this great and important question. I hope you will divest yourselves of every feeling but that of sound justice. Don't indulge in sentimental nonsense. It is a matter of business, and not of sentiment. Discuss it in a business like spirit, and I have no doubt great good will result from the deliberations of this day. (Cheers.)

Mr. T. SWEETING, London, seconded the motion.

Mr. PETER HALL, Sheffield, asked if it would be considered a Lock-out if the employers gave the legal month's notice. In the trade to which I belong, the Spring Knife Cutlers, one firm is almost equal to a town, and when we are locked out there, we are virtually locked out of all the others.

Mr. THOMAS SMITH, Sheffield: There are various ways of locking out. I say it is a Lock-out to give the men a month's notice and to drive the trade away. I think, myself, that the men are themselves the cause of Lock outs in many cases. My honest conviction is this, that if there was a proper understanding between master and man, many matters might be settled that are not settled. I think there ought to be mutual arrangements. The masters have the capital, and the men the labour, and their interests are inseparable. There ought not to be any collision whatever between capital and labour; and I have no doubt that if the File Trade dispute had come before the Organised Trades in the first instance, it would not have been a Strike or a Lock-out. These are my honest convictions. I should like to see some means adopted for getting a better understanding between masters and men. I have had the honour of waiting on manufacturers in my own, the Saw trade, and I must say, we have been received and treated as gentlemen; and if this mode of procedure was adopted on all occasions of difference, I have no doubt that Strikes and Lock-outs would be prevented, and that three-fourths of the Strikes and Lock-outs which have taken place never would have occurred. (Applause).

Mr. G. AUSTIN, Sheffield: Mr. Smith has not properly defined the question of Lock-outs. My opinion is this: that if, in consequence of a dispute arising between the masters and men in a given department, the masters deprive the other branches of the trade in the factory of employment, that is a Lock-out. (Hear, hear). Mr. Smith has made some allusion to the dispute in the File Trade. This is not the place to call to account anything that has been amiss in the conduct of that dispute. Probably had different steps been taken at first, the Lock-out might have been prevented, and that remark will apply to a good many cases in which Strikes and Lock-outs have been resorted to.

10

Mr. A. McDOUGALL, Sheffield: The simplest way of answering the question, "What is a Lock-out?" is to give an illustration. A short time ago, the Iron Workers of the Clyde asked that the hours of labour should be reduced from 60 to 57 hours per week. The employers stated that compliance with the request would involve them in a loss of £300,000 a year in machinery alone, and they asked how that money was to be made up to them. Several of the delegates endeavoured to answer the question, but failed to do so satisfactorily. The employers were willing to comply on these terms: that the wages be reduced in the same proportion as the hours; but that the workmen refused, and the employers then said, 'We will lock you out for three months." They closed their doors against all the men in their employ, refusing the services of those who were ready to sell their labour on their own terms; and that I think, meets the question of "What is a Lock-out." Employers have the power legally to close their doors, and in the case of what they regard to be an unreasonable demand, they immediately put into practice what by law they had power to do.

Mr. R. JOYNSON, Potteries: I feel a deep interest in this question, and I hope its discussion will tend to do good throughout the whole of the kingdom. If we inquire into the cause of Lock-outs and Turn-outs, I think we shall find they are traceable to a great variety of circumstances. The principle reason, I am sorry to say, so far as I have seen, is that the workmen generally have been too ready to resort to Turn outs. (Hear, hear) I feel sure there has not been that readiness to argue and reason the question in dispute, and to refer it to arbitration that there ought to have been; and until we have a better feeling amongst workmen generally, and until it is felt that recourse to Strikes ought not to be had on any account, except as a last resource, and unless the odds are a hundred to one in our favour, they will continue to occur. The amicable settlement of disputes has, in my opinion, been too much neglected by workmen generally. They have not considered the peculiar circumstances under which they have been placed, not considered the odds against them, and they have been crushed in consequence.

Mr. H. ROBERTS, Manchester: I think it would have been better if the last speaker had confined his attention to the question before the meeting—Lock-outs—and not given us a dissertation on Strikes. I would compliment the proposer of the motion for the manner in which he introduced the question, and I hope that every speaker will adhere to facts In the business which I represent, Lock-outs are invariably resorted to by the masters, whenever the men ask for an advance of wages. If the legal notice be given to the men, I cannot conceive that it can strictly be termed a Lock-out; but I cannot conceive that the doing of what they are compelled by law to do, at all really affects the question.

Mr. W. BROADHEAD, Sheffield: This great assembly has

met for the purpose of enquiring into the cause of Lock-outs. I think the cause is obviously to compel implicit obedience on the part of the workmen, and to destroy Trades' Unions by the expenses which a Lock-out entailed upon them. Lock-outs are alarmingly on the increase, and I can see no better mode of suppressing them than by the trades of the country being prepared to fight the Lock-outs unitedly, instead of dealing with them in sectional detail. Lock-outs have found their way into Sheffield, and it was here that the subject of this Conference was first mooted, because it was seen very clearly that if some steps were not taken in the matter, it would be impossible for any individual trade to stand. It requires the aid and assistance of all the trades to fight this great battle. (Hear, hear.) It is true that manufacturers in Sheffield, and in other parts of the country, say that it is not their intention to destroy the Unions. They said so in the recent struggle in the File Trade. But there was a struggle in the File Trade in 1836, and there was then no Lock-out. (Hear, hear.) It was not seen necessary then to lock out all the branches because one branch struck for an advance of wages, and why should the necessity have arisen in this case? In 1836, the branches with which there was no dispute continued at work, and if work could not be found them they were employed in other occupations, so that they might live. (Hear, hear.) It was not so on the recent occasion. (Hear, hear.) The Lock-out system has been introduced, in the first place, throwing the men on the trade's fund, and, by so doing, swamping their efforts and their means of getting assistance. This is the great object in all Lock-outs; and this being apparent, I think it is our duty to take steps for the purpose of bringing together one vast body for the purpose of resisting these endeavours, when they take place. (Hear, hear.) Lock-outs frequently arise out of Strikes, it is true; and it should not be the object of this assembly to promote Strikes, or to find assistance beyond what is seen to be necessary; but our great point should be to resist Lock-outs whenever they are had recourse to.

Mr. W. BURNS, London: It has been said that there is a disposition among the working-men to rush too hastily into Strikes. There are some men present who have been concerned in some of the greatest Strikes and Lock-outs that have taken place; and I appeal to them to say whether the utmost efforts have not invariably, in the first place, been resorted to for the purpose of bringing about an amicable settlement. (Hear, hear.) We don't want to strike if we can help it; but what we want is, a fair amount of remuneration for our labour, so that we may live as hard working honest men ought to live. The best way to prevent Strikes and Lock-outs is to have a fund of £50,000 or £100,000. If we had such a fund, our differences would be readily settled by arbitration. In the trade to which I belong the masters tried to make us like slaves, and work for them upon their own terms. We waited upon them, but they would listen to nothing but unconditional surrender, and as we could not agree to that, we were locked out for sixteen weeks. But we beat them, and made a proposition to them as to a three months' notice, to prevent the necessity of resorting to such an extremity again.

Mr. J. Davies, Hyde: The people of the locality in which I reside will coincide with me, when I say that the generality of our masters resort, in many cases, to conduct that is low and mean. For instance, a man makes himself rather conspicuous in trades' society business. He has the good of his trade and his fellows at heart, and the masters take advantage of him. In the cotton manufacturing districts we have a great many petty evils to contend with. Supposing there is a dispute amongst the spinners, and that some of the spinners have members of their families working in other departments of the factory, if the spinners turn out, the other members of their families are locked out; and that practice, in some cases, leads to great hardships. If we could amalgamate all the trades into one general society to protect each other, it would be a great advantage.

Mr W. Harry, London: In my opinion, Lock-outs frequently arise from the workmen demanding an increase in their wages, or a decrease in their hours of labour. It has been admitted by political economists of the present day, and also by those of the past, that man has a perfect right to appraise his own labour. The shopkeeper has a right to place his value upon the article of which he disposes, and the workman ought to have the same right with respect to his labour. When this right is about to be exercised, the employers in many instances deny the right, although the workmen know very well that, from the state of trade, the demand can easily be conceded. In order to resist the demand, the masters close their shops, and that is a Lock-out. What is it that enables masters to do this? If we look from one end of the land to the other, we find that non-society men are in the predominant. In some trades, I admit, the bulk of the men are connected with the Society; but the masters see that they have only a particular section to deal with. How would the case stand, if the bulk of the trades were associated together in the manner which, I hope and trust, will be agreed upon by this Conference previous to its closing? The masters would find, then, that there was not one section only to deal with, and they would see that the organization was so perfect and sensitive, that they could not prick one part but it would pass through the entire system like an electric shock. (Hear, hear.) That is what we must endeavour to realize.

Mr. T. West, Nottingham: The trade I belong to was locked out in 1860. We had a new, or partially new class of work, and the employers would not give us what we thought was the value of the work. Three large firms locked the men out, though out of the number not one-sixth was engaged on the new work. An endeavour was made to come to an arrangement; but the masters at the time declined. It is quite clear to me that the objects of Lock-outs are to destroy the working man. We ought to have a National Council to inquire into these matters, and, if possible, to settle them by arbitration. Lock-outs are selfish in the extreme, and always aimed against the working man.

Mr. P. SHORROCKS, Manchester: The system of Lock-outs is not a newly invented one, but dates as far back as Strikes. A Lock-out is a Strike on the part of the masters against the men, and is in opposition to an effort of the men to improve their position. It was so in the great Preston Lock-out. There is now a large Lock-out in the Iron Trade of the north. In this morning's paper I find the following on the subject:—"At Walker, as in the surrounding districts, the puddling furnaces and the rolling mills are inactive, and some 400 or 500 men are idling about the streets. Upwards of 100 of the employés of the Walker Iron Company occupy houses belonging to the firm, and the company have served notices to quit upon them." This is a diabolical proceeding, to throw them out of house and home because they will not take a reduction of ten per cent. Working-men must unite. Without this, it will be impossible for us to get the full market value of our labour. There is a want of union amongst ourselves. (Hear, hear.) We want union, not only in individual trades, but a union of all trades together to resist Lock outs. I commend the men of Sheffield for taking up this question in the spirit they have done, and if it is carried out, they will get the thanks of the working-men of the whole country for it. (Hear, hear.) I think the Conference ought to pass a resolution of a conciliatory character, such as this:—"That this Conference recommend to all parties concerned in a Lock out or Strike, to inquire into the question thoroughly before proceeding to so disastrous and lamentable a proceeding as a Strike or Lock-out, and if necessary, to refer the subject to a third party for settlement."

Mr. ELI DYSON, Manchester: Lock-outs mean capital against poverty; or capital against labour. It is a common thing to talk about freedom; but I say the working man is not free only so far as capital will allow him to be. We certainly have not the power that we ought to have. When the capitalists are in prosperous circumstances, they never let us know, and they think we have no right to participate in their good fortune, though the expenses of living are daily increasing. Until we are able to fight capital with capital, we shall do no good. There must be a combination of all the unions, and then we shall be able to meet and overcome Lock-outs.

Mr. E. MEMMOTT, Sheffield: A Lock-out means no work; no work means no wages; no wages means no bread. I consider Strikes and Lock-outs a very clumsy, awkward, and stubborn way of settling disputes. Prevention is better than cure; and I see nothing so well calculated to prevent Strikes and Lock-outs as a combination of all the trades for mutual assistance. A great Lock-out is impending in the North, because the masters want to take ten per cent. off the wages of the men. Without entering into the question of whether that reduction is justifiable or not, I may say that Mr. Stevens, in the American House of Representatives, said "that the reason why England could send her iron over there so cheaply as she did was because the rate of wages in England was one-third what it is in the United States." It is said that food and clothing are very dear in the States; but if the workmen get three times as much wages as they do in England, they can afford to pay high prices. Besides, there is no disguising the fact that prices are becoming very high in England, but wages are not advancing in a corresponding degree. (Hear, hear.) Let us become one united body throughout England, and support each other when the time of need arrives.

Mr. T. BANKS, Preston: I believe the term Lock out originated in 1822, in consequence of a step taken by the Preston manufacturers. In 1836, another Lock-out took place. I will give you the cause of it. Bolton, Chorley, and Preston are rival places in the cotton trade for fine counts—fine and middling. At Bolton, the spinners were receiving ten shillings, twelve shillings, fifteen shillings, and in some cases twenty shillings a week more than the spinners of Preston, and an effort was made to obtain more equal terms for the Preston operatives. A strike was threatened, and the masters offered an advance of ten per cent., and at the firm where I worked, 12½ per cent. was offered. The leaders urged us to stand for the Bolton prices, and to accept nothing less. A Lock-out took place. We were out seventeen weeks, and at the end of that time we were obliged to succumb for want of funds. On that occasion I consider we acted wrongly. They offered us part; and I think we ought to have taken it for the time, and then agitated for the difference. (Hear, hear.) In 1847 a reduction, I think of ten per cent., took place in consequence of bad trade. We tried to prevent it; offered to work short time, but nothing would do but a reduction of ten per cent. We submitted, on this understanding, that when trade revived they should give us the ten per cent. back again. We naturally expected they would keep their word. Things improved; the panic passed away; and '49 was a brisker year than any of its predecessors. We reminded the manufacturers of their promise, and asked for the ten per cent; but it was no use. We went struggling on, working and agitating till something like 1853, and then we were determined to have the ten per cent. We were all prepared to give in our notices, and it had the desired effect. We got the ten per cent., and enjoyed it for about five weeks, when there came a great talk about the Russian war, and trade began to slacken. The masters wanted to take the ten per cent. off again, and as we would not let them, there was again a Lock-out. In the year 1853, 25,000 of us were locked out—spinners, minders, weavers, carding hands, throstle hands, and so on. That lasted 36 weeks, and on that occasion we were right, because we were still under Bolton and other districts. The whole country responded to our call; London, alone, sending no less than £4000. During that 36 weeks not less than £100,000 came to Preston for our relief, and that shows what trades can do when they are united. Not only were we able to continue the Strike, but when it ended we were a thousand pounds better off than we were at the commencement (Cheers.)

MR. GEORGE ROSS, Glasgow: Various causes have been given of Lock-outs, and various causes may be given; but the very fact that they have existed, and are at this moment existing around us, is a sufficient reason on the part of the working men of Great Britain to make a strenuous effort to remedy the evil. They are an evil to society at large, and to us as individual members of society. I cannot, however, agree that working-men are the great cause of Lock outs. I do not pretend to believe that working-men are infallible; but I admit that they are equally guilty of errors as other people. Nine-tenths of the Lock-outs have been caused by the tyranny and selfishness of employers. It is quite true that the Lock-out on the Clyde was brought about because the great body of the workmen refused to accept the employers terms; but, as I understand, the question is one of hours and not of wages. The employers have not acted honestly with the men in the matter, and I hold that the evil lies with the employers in nine cases out of ten. The great question of Lock-outs must be settled, not by the mere system which has hitherto been adopted, of being dependent on eleemosynary aid; it must be by combination, and by such a combination of the working men of Great Britain as shall effectually do away with both Strikes and Lock-outs. If such a combination were honestly carried out, the moral power we may wield, and the influence we may be able to bring to bear upon the public at large, will tend greatly to lessen this great evil. A great fund has been spoken of; but if a proper organisation took place, the money could be advanced when required. While we as a National Association would not interfere with the local working of any institution connected with us, or with their Strikes or trade disputes, we would be prepared to find means through these trades by a regular levy upon the members in proportion to the number registered, so as to meet a difficulty. A levy might be made in proportion to the wages received; and if that principle were adopted, trades receiving support might receive it in proportion to the wages they received, and on which they paid. If they were taxed at 2½ per cent. on their wages, the income from the members represented by the delegates present, would be about £5000 weekly; and if that could have been collected during the Lock-out in the File Trade, the disaster would have been of shorter duration, and would have had a more satisfactory termination. (Hear, hear.) That, I think, is a very fair, and a very equitable principle, and its adoption would, at least, stem the torrent of starvation, if it did not stem the torrent of selfishness and tyranny on the part of the employers; and it would stem the torrent of pauperism, and the torrent of terrorism held over our heads in these Lock outs. (Applause.)

16

Mr. W. DRONFIELD, Sheffield: In consequence of the position I hold in connection with the Trades' Societies of Sheffield, I think it right that I should give you my views on this important subject. I fear that a false impression has been, or may be, created by remarks that have been made by previous speakers on the late disastrous Lock-out in the File Trade. I do not in the least blame the men for the part they took in that dispute. I conceive that there was not the slightest necessity for the masters to lock out the whole of the branches of the File Trade because a certain portion of the men took the course they did. I do not now say whether the course taken in the first instance was right or wrong; but I do say this, that the cause of the Strike was the refusal of the masters to meet the grinders, who had given notice for an advance. I have said before, and I do not hesitate to say again, that where men are fit to work for an employer, they are fit to confer with an employer. (Hear, hear.) The grinders sent in a respectful request to be allowed to discuss the revised list with their employers, and the employers refused to meet them, and steps were taken to lock out the whole of the branches, numbering no less than 3,500 persons. Had the masters agreed to meet the grinders in the first or second instance, they might have been brought to see that the demand was unreasonable, and that it was not to their interest to persist in it. But the masters resorted to the too prevalent system of locking the men out. We are all very well aware that Lock-outs have increased greatly of late years, and I hope this Conference may hit upon some scheme by which these evils may be counteracted. I believe the best remedy is, to create a better feeling between employer and employed. I agreed with Mr. Newton when he said the men were not always in the right. We are all liable to err; but I do not think the men are more unreasonable than the masters. If it can be shown that their demand is unreasonable, they are not so unreasonable as to persist in the demand. The File Trade had not much money to fall back upon, but they found (thanks to the trades of the town and of the country generally) a great deal of money to support them during the sixteen weeks that the Lock-out lasted. The concentration of trades will have a tendency to render these Lock-outs unnecessary. I hope we shall not provoke counter action on the part of the masters; but if it comes to that we must be prepared to meet it. (Hear, hear.) I believe the working-men are sufficiently strong to be able to meet whatever is brought against them, and that their only great fault is apathy. The question of mutual support is one of great importance, and it should be the object of this organisation to devise plans to resist Lock-outs, and to promote a better understanding amongst the trades generally. Let us be prepared to assist those who are less fortunate than ourselves. We do not know how soon it may be our turn. All trades are liable to a serious Strike or Lock-out; and I hope the gentlemen now assembled will impress this strongly upon their constituents, and prevail upon them to give in their adhesion. We

17

shall, I hope, have the question of arbitration brought before us, and I will only now say upon that point that, so far as I know at present. employers, as a rule, always object to arbitration. I do not know the reason, unless they are afraid to meet the men, and submit the matters in dispute to an impartial person. We always urge arbitration in Sheffield, and, as far as our own association is concerned, we know it has prevented many Strikes since its establishment, by acting as mediators between masters and men. I should like this Association to be on that model; and I would advise you not to attempt to do too much in the first instance, but do the thing slowly and surely. (Applause.)

Mr. S. MEEKIN, Basford: Though I have never been personally concerned in a Lock-out I can see very clearly that the general cause of Lock-outs are the masters themselves. When a man demands a proper amount of remuneration for his labour, they are determined, in many cases, not to give it; and I am exceedingly glad to see that there is now to be a powerful Association to meet such cases.

Mr. W. WILLBOND, Nottingham: From my own experience, and from my observation in other trades, I am quite sure that Lock-outs have been caused chiefly by employers refusing to receive deputations, and to show that conciliatory disposition towards workmen that workmen have shown towards them. I have known means resorted to by the men to conciliate matters, but I have not seen that disposition shown by the masters before they have resorted to a Lock-out of the men. The great object of the masters is to suppress the power and influence of Trades' Unions, and, if possible, to suppress them. The best preventative would be to form a National Union of Workingmen. If such a union were formed, I think the masters would pause before they resorted to the abominable system of Lock-outs.

Mr. W. LEIGH, Hyde: Strikes are in many instances the first cause of Lock-outs. Before we resort to Strikes, we ought to use every endeavour to obtain an amicable settlement of the differences. I don't think there is an employer in Sheffield who would lock his men out, if he did not think he was going to punish them; and one of the best means of depriving the weapon of its sting would be, so to organize as to be able to allow to the men locked out as much as they would earn were they at work. Give the men twenty-five shillings a week while they are walking about the streets, and I don't think the employers would lock them out. In one case, eighteen shillings a week was agreed upon for the men, and so much per head for their families, and, in four days' time, the Strike was brought to a close. I am in favour of a general organization of the trades of the country, and, if possible, that the association should be international.

Mr. W. H. WOOD, Manchester: I should have refrained from offering any remarks, but I have not yet noticed that any gentleman has mentioned what I regard as the real source from whence emanate Lock-outs and Strikes. If we trace the question to its true source, I believe we shall find that masters are competing against each other for the production of their work at the cheapest possible rate, and it is this system of competition which I conceive to be the real foundation from which all our grievances flow. My idea is, that in order to check this system of competition, we should seek to promote, as much as possible, the equalization of wages over the length and breadth of the country, and not confine our labours to our own country alone, but extend our ramifications everywhere—America, Europe, and the most distant of Her Majesty's dominions. I have had opportunities, as Secretary of the Manchester Branch of the Provincial Typographical Association, of consulting with employers, and I have ever found that they have been ready to listen to us in the most courteous manner, whenever we have had grievances to lay before them, but I have noticed that we have always been met with this great difficulty—we have to contend with other employers who pay a less rate of wages than we are paying at the present time, and we should do our work at a disadvantage if we grant your request and compete with them in the market. To meet this state of things, it is to our interest to form an extensive organization, to bind ourselves together in one bond of unity, and to endeavour to bring about the equalization of wages, and to place them on a footing more just and satisfactory than that they occupy at the present time. We must not close our eyes to the fact that working men occupy a very different position to what they have done in times past. Employers of labour, as a general rule, are but working-men grown wealthy, and they understand all the machinery by which we seek to obtain advantage over them as capitalists. There is no secret in the working out of our system by which we seek to compete with capital, and, therefore, we must lay our case clearly before them, and, in a business-like manner, ask whether they are not prepared to redress such grievances as we believe we are labouring under, and come to what I believe to be the only sure preventative of Lock-outs and Strikes—Courts of Arbitration and Conciliation. In the adoption of rules for the guidance of this Association, we must discard all selfish motives, and adopt rules upon a broad and expansive basis, so as to meet with the acceptation of the general body of Organised Trades. (Applause.)

Mr. B. Oran, London: Attempts are sometimes made by the masters to reduce our wages, and, when we resist, they turn us out. In my trade, they attack us in the depth of winter, when they think we can least bear the pressure, and when they think they can the more readily obtain an advantage. It is no use fighting for one branch. We must fight for all; and all must be put upon a footing on which they can live. Get a fund in hand, for when men are well organized, and have funds of their own, the masters will not dare to interfere as they have done. A levy of four shillings a man would yield £40,000. I don't ask this to be done, but I suggest it for your consideration.

19

MR. G. ODGER, London: The resolution submitted to us is a very appropriate and justifiable one. A more judicious resolution could not be devised for the consideration of the question. It pledges you to no specific action in endorsing the principle for which we are all here assembled, and we should stand apart from our duty, if we were to waste much longer time in discussing the resolution as it is. The form of carrying it out will be another thing. Lock-outs, their cause and their remedy, are the three points of the resolution. A gentleman, who addressed the meeting this morning, wished us to understand that Lock-outs should be discussed as apart altogether from Strikes. I wish they could; but Strikes are part and parcel of Lock-outs. Lock-outs are the weapons with which the master too often very successfully fights us. What have working men themselves done to avert Lock-outs? As far as I could gather, Mr. Newton said he believed they had done their utmost to avert them; but in some cases, no doubt, they had made mistakes; and I presume there is not a man in this room who is not prepared to admit that his society, either in its private character, or when advised by another, has committed some mistake or another in its time; and it has been a most redeeming feature in this Congress to hear so many speakers address the chair in the calm and dignified manner they have, and tell us of our faults, as well as speak in our praise. While, I say with Mr. Newton, that we have done the best we could in most instances, I am afraid the Strike in the Staffordshire Iron Trade was an exception. The London Trades' Council declared against them, and did all they could to induce them to accept the arbitration offered to them by the Earl of Lichfield. The men, I believe, were animated by pure motives, and with everything justifiable on their part; but they took a different view than we did, and a mistake was the result. They have paid dearly for that mistake. Mistakes find their own remedy, and the sufferings that come from them are sometimes exceedingly painful. There is a special point in this case, and that is, that the men are to receive so much reduction in their wages when iron delines so much per ton. If we have an agreement with a number of employers that under certain conditions our wages shall rise and fall, it is highly essential that they shall be fully complied with; but if we find them prejudicial to us, we can give notice that after a certain time, we shall not consider ourselves bound by that condition, and go on working peaceably till that time. It is sometimes said "our stocks are swollen, and work slack," and that is given as a reason why wages shall be reduced. Of all reasons in the world that is the most shallow, because it is in direct opposition to everything that is moral and just, though it is consonant with the laws laid down by political economists. I have yet to learn that some of the principles of political economy cannot be shaken by practical fact. If we are to decline our wages as work gets slack and slacker, and still more slack, what will our wages be when depression comes upon the trade? It must result in what we cannot live upon, and that is a piece of degradation, not only to the individual, but to the nation. It is an anomaly that because work declines, wages should decline. There are more men on the market wanting assistance, and if anything should happen, it is that more money should be paid for the little work that is done. It is a fact, that within four years, the Revitting Trade has gone to one-half the price at which it commenced. Leicester had a monopoly, because of its cheapness. Why should the wages fall, while the masters have a monopoly in the market? For the simple reason, that they were continually introducing men who were not shoemakers at all, but became useful for the rougher part of the work, putting the more experienced workmen to the more particular parts of the work; and the men having no combination on their side to resist the first encroachment, wages gradually declined. The men of Leicester formed themselves into an association, and the masters immediately re-acted upon the men, and the result has been one of the most cruel and unjustifiable oppressions on the part of the masters that has been known in the country, and the poor fellows are comparatively defeated. We are told that we must accept Lock-outs as a legitimate result of Strikes. I am quite prepared to accept that. The masters say, if you use your weapon of Strikes, we will push our Lock-outs. I do not suppose it will come to the Kilkenny cat business. For the last twenty years at least, trade has undergone a remarkable change. Moral suasion is the thing aimed at by the bulk of trades' unionists at the present time. Glasgow and Sheffield have acted with us as to the Masters' and Workman's Act, and Arbitration, and we have manifested to the world, I think, that we are desirous of using moral force, instead of anything of a harsh character, and we have a right to expect to be received in a friendly spirit. The men offered moral suasion in the dispute in the File Trade, and the masters declined it. Having done that, they were responsible for a great deal of the sorrow and suffering that followed, and I hope that in future, disputes will be settled in a less disastrous manner, when the opportunity is offered. (Applause.)

Mr. G. TROUP, London: I represent the London Working Men's Association, and we decline to give assistance in cases of Lock-out or Strike, where the men decline arbitration. We assisted the Sheffield File Trade. We were very desirous that the North Staffordshire men should leave their case to arbitration; but they were rather afraid to do it.

Mr. W. LINDSEY, Preston: If the trades had been amalgamated on the occasion of the last Lock-out at Preston, we should have been able to keep the masters from getting the men they did into the town, and by that means would have saved much suffering.

Mr. JAMES CARLISLE, Brierley Hill: Now that so much has been said as to the cause and history of Strikes and Lock-outs, I think it would be as well if we turned our attention to the devising of means for preventing them in the future. In my opinion, the best thing we can do will be to engender a good and friendly feeling between employer and employed, to remove all feelings of bitterness, and to let the things of the past be forgotten.

MR. G. FOX, Sheffield: I wish to protest against the remarks that have fallen from some of the speakers. It has been distinctly asserted that it was the masters, and they alone, who had brought Lock-outs into operation. That opinion seems to have been entertained by the Committee who framed the resolution, which says—"Whereas, within the last few years, great and disastrous Lock-outs have been frequently resorted to by the employers of labour as a means to compel the workmen to obey their wishes." From the information I have derived from this meeting, and from the information I gathered from the newspaper reports, I believe, as a general rule, it is not the masters who have originated Lock-outs; but they have been caused by the men in different ways, ways not in accordance with their best interests, or with the best interests of their employers. There is a very prevalent opinion amongst working men that when there is a great demand for their labour they should immediately Strike, and endeavour to obtain an advance of wages, and to increase them to the highest possible point. Taking only one view of the question, that seems quite right. When there is a great demand for labour, and when manufacturers are selling the labour at a great profit, it seems only right that working-men should, in some way or other, participate in the fruits. But there is the other side of the question that ought not to be lost sight of. If working-men are to receive the highest wages when trade is the briskest, they must naturally expect that at another time, when the pressure is not so great, they must receive less.

MR. C. WILLIAMS, Liverpool: I believe that Lock-outs as a rule, and I speak advisedly, arise from the action of my brother operatives. That is the result of my experience and observation. There are isolated cases in which the employer takes unwarrantable action, and closes his place against those who are willing to gain a living for themselves and their families. One grand national organization will enable us to meet these cases, and there are no means so well calculated to prevent them as the working-men being banded together in one consolidated union, having one interest in view. They will then be able to resist any encroachments that may be made wantonly by any employer. In Liverpool, we are far better than we were in that respect. We have organized a United Trades' Protection Association, and we tax the members by a monthly levy. The employers know it, and feel very delicate indeed about producing Strikes and Lock-outs.

MR. G. SHIPTON, London: All the moral suasion in the world will never have the weight and influence with the employers as such an association as that spoken of by Mr. Williams exercises. The trades of the country are now ripe for amalgamation, and as the want of organization has led to Strikes and Lock-outs that have resulted in great suffering, I hope this movement will have a successful issue. I should like some resolution of a substantial character to be brought forward—a resolution for the formation of a powerful organization—and any such resolution I shall be very happy to lay before the body I now represent.

Mr. J. CONSTANTINE, Halifax: It has been suggested that we may provoke counter action on the part of employers, a sort of federation of employers to resist movements for the advance of wages and the reduction of the hours of labour. If such a federation be formed, the weakest must go to the wall, and if we adopt the same weapons as the masters, there can be no doubt as to what the result will be. I think that the only safe method of dealing with Lock-outs is a complete federation of the whole of the trades, and I hope that will be adopted.

Mr. J. NORMANSELL, Barnsley: I have been locked out about twenty-four months in six years. There can be no doubt that the greatest portion of the Lock-outs originate in consequence of the operatives resorting to Strikes. I can speak more particularly in reference to the miners of South Yorkshire, for whom I am secretary. We have had some serious Lock-outs in that district. In 1860, we had one of no less than 4000 or 5000 men, and that was in consequence of the miners at three collieries striking for an advance of wages, and for the abolition of old customs that were considered to be oppressive to the workmen. The most disastrous Lock-outs have followed disputes in which the men were compelled to strike. In 1864, about 3000 of us were locked out, because the miners employed by one firm were on strike for an advance of wages. I don't think there would have been the Lock-out in the File Trade, if the grinders had not given notice for an advance. I do not say that Strikes are wrong, for had not the men of South Yorkshire struck, they never would have attained the position which they now occupy. In South Yorkshire, we have from 4000 to 5000 members, and they are in a better position than the miners of any other part of the United Kingdom. (Hear, hear.) Before we agree to support Lock outs we must have something to say about the Strikes which cause them, for we know very well that the men in some of the districts would rush into a Strike in order to bring about a Lock out which would be followed by support from this Association.

Mr. J. ALLEN, Liverpool: My definition of the word Lock-out is this—that it is an unhallowed act of injustice on the part of the employers. and that it carries to the hearths and homes of the poor that poverty and misery which should alone visit the criminal. We are told that Lock-outs are brought about by the men resorting to Strikes; but, before the men are blamed, we ought to know whether the Strikes were justifiable or not. If unjustifiable, they do not justify Lock-outs. If a Strike is unjustifiable, let those who commit the error be visited with the punishment, and not those who had no voice in the first step, and who are by no means accountable for, and receive no benefit from, the Strike. The Lock-out on the Clyde cannot be justified by the employers. During the thirty-six weeks of that Lock-out, disorganized as the trades were, £100,000 were paid in for their support. If that could be done then, when a proper organization is formed, it could be done in a fortnight. If men would associate themselves together, and work in their own shops, and on their own tools, and with their own capital, it would go far to let the masters see that the men were taking up the right end of the cudgel, and were able to lace them with the other. I would give to the employer everything that was just, and I would demand the same for the working man. Are we to be told that if a Strike takes place for an advance of wages, it is unjustifiable, and that it leads to a Lock-out? In that case we become the mere slaves of capital. I would suggest that we should contribute monthly or quarterly so much per member, and oftener if necessary.

Mr. G. SHIPTON, London: I hope that the proposition will be cordially adopted, and that the next motion, which will provide means for successfully resisting Lock-outs, will meet with the same reception.

Mr. J. HUMPHREYS, Wolverhampton: The society to which I belong have adopted the principle of Courts of Conciliation or Arbitration, as a means for the settlement of trade disputes, and the prevention of Strikes or Lock-outs.

Mr. J. RUSHFORTH, Sheffield: One effectual remedy for Lock-outs will be found in each individual trade thoroughly organising itself. The trades can meet in a conciliatory spirit, and they can use their efforts for the prevention of Strikes and the promotion of good feeling between masters and men. Any National Association will stand very little chance of success, unless each individual trade thoroughly organises itself, and depends in a great measure upon its own resources and its own efforts.

Mr. NEWTON replied: I am glad that this discussion has taken place. Much valuable information has been elicited, and without discussion we never can have that perfection of knowledge indispensable to the progress of society. I deny that the resolution conveys any aspersion upon the employers; but we cannot say that it is the workmen who Lock out, because they don't. If any statement is true, it is that the employers resort to Lock-outs as a means of compelling obedience to their wishes. The whole blame, therefore, of Lock-outs must lie at the doors of those who keep the keys.

The resolution was then adopted unanimously.

ORGANISATION.

Mr. J. HUMPHREYS moved—

"That in the opinion of this Conference an Organisation composed of the Trades' Societies of the United Kingdom be formed for the purpose of enabling the Workmen to effectively resist Lock-outs, and that it be remitted to the Committee to prepare resolutions in accordance with the expressed opinions of the Conference."

The arguments that have already been advanced show that such an Organisation ought to be formed. There is nothing in the resolution that pledges us to any detail, and it will, therefore, not be necessary for me to occupy further time in moving the proposition.

MR. THOMAS BIRTWISTLE, Accrington, seconded the proposition.

MR. W. HARRY moved as an amendment, "That this Conference having the most sincere desire to avoid Strikes and Lock-outs, and the evils necessarily attendant thereon, do earnestly recommend to society and non-society men the necessity of forming themselves into one general Organisation, to be governed by an Executive Council, elected by the majority of those belonging to the General Labour Protective Association."

MR. W. GILMORE, Potteries, seconded the amendment.

MR. G. ROSS, Glasgow, proposed "that non-society men be excluded until the constitution was appointed."

MR. W. H. HEY, Halifax, seconded Mr. Ross's amendment. If non-society men had not the ardour to support their own trades' unions, why should that Association support them?

MR. W. BROADHEAD contended that it was utterly impossible to accept non-society men as members of that Association. Let them join their own trade societies, and then there would be a guarantee that they might be controlled like others. Without that, it was an utter impossibility.

MR. E. MEMMOTT, Sheffield, and MR. T. MAUDSLEY, Manchester, having spoken,

MR. A. STANSFIELD, Sheffield, said he thought the recommendation as to non-society men a very good one. Their object was to get non-society men to be society men; and if they agreed to combine for the protection of labour, they of necessity became society men.

MR. W. H. MARKHAM, London, set his face dead against the admission of non-society men. If Mr. Harry's amendment were submitted to the society which he (Mr. Markham) represented, he did not think there would be one hand in the 10,000 held up in its favour,

MR. E. GREEN, Sheffield, thought that a scheme for promoting emigration would be the best for the prevention of Strikes and Lock-outs. The moment an employer saw that his men were about to leave the country, he would open his gates.

Mr. W. LEIGH, Mr. P. HALL, Mr. E. GRAYSON, Mr. J. ALLEN, Mr. J. TEASDALE, Mr. W. H. WOOD, and Mr. BURNS having expressed their opinions,

MR. ODGER suggested that as the delegates appeared to be opposed to Mr. Harry's amendment it should be withdrawn.

MR. HARRY adopted the suggestion, and Mr. Humphrey's proposition was carried.

After making some arrangements for the Public Meeting, the Conference adjourned for the day.

PUBLIC MEETING.

A public meeting was held in the Temperance Hall, on Wednesday evening, to hear addresses from the delegates to the Conference. There was a large attendance. MR. BAGSHAW occupied the chair, and called upon the speakers to confine themselves as much as possible to the immediate subject before them, as, owing to the number of speakers, only fifteen minutes could be allowed for each address.

STRIKES AND LOCK-OUTS.

MR. R. JOYNSON, Potteries, was first called upon, and he besought the courtesy of the meeting, as this was his maiden speech. They ought, he said, to give some reason why a Conference had met at Sheffield this week, and he apprehended that the principal reason was, that they wished to do their part in solving what had been a great difficulty for years and years past—namely, reconciling capital and labour. Because they called these conferences, they were not to be looked upon as parties who were wishing to set class against class. (Cheers.) If their feelings and objects were rightly interpreted, it would be found that they were wishful to reconcile classes—to make better fathers and better citizens. He thought that Strikes were in themselves a very great and serious evil. He recollected—and it was almost as far back as he could remember—a Strike in the Potteries in 1836 and 1837, and one incident in that Strike in his own experience was this: One morning, his mother was giving her children their breakfast, but there was no butter on the bread, and he asked how it was. She stood and looked at them a moment, and then said, "It is because the gates are shut." Now, a system that would bring about such feelings as were in his family at that time, and to a much greater extent in other families, a system that would bring about a condition of things like that, was a system that they ought to inquire into, and, if possible, find some remedy for. (Cheers.) This was their object in meeting in conference this week, and he thought they ought not to be left alone in this matter—it ought not to be left to the working-men, as a class, to try to solve this very great and important problem. (Hear, hear.) The masters were in duty bound as much as, nay more, than them, to help in solving the problem. In connection with the Strike to which he had referred, he should not be doing his duty if he did not return thanks again to the friends at Sheffield, for the earnest support they gave to the

B

Potteries at that time. (Cheers.) This was a rather unpleasant point for him as an individual to touch on, because he believed that the Potteries did not return that kindness in the way that they ought to have done—he meant some part of the Potteries—for many did their duty, and returned, he thought, the money to double the extent that, individually, they were bound to do. (Cheers.) The attempt that the men were making at that time was, from various causes, a failure; and they knew that when a defeat took place it was like the rout of an army. He hoped Mr. Odger would excuse him if he spoke a little upon his subject—the means to remedy the Strikes they complained of. They ought to be careful how they entered on Strikes. Every effort ought to be used to avoid them, by reason, by arbitration, by every means it was possible to conceive of; for a Strike was a calamity they should dread. He hoped that because they, as working-men, were trying to better their condition, and raise a fund for that purpose, they would have more discretion than to be puffed up when the fund was created. He hoped the fund would not have the effect of making them rush into Strikes without great caution, but that they would be particular in guarding against them; for they must bear in mind, that if they had thousands and thousands of pounds, it was as necessary to avoid running into Strikes as if they had nothing in hand. (Hear, hear.) Therefore, he hoped that this fund would not cause parties to be reckless, but that they would use the same discretion and means to avoid Strikes as if they had no funds to fly to. Although he said as much as this, he apprehended that, bad as Strikes were, there were some evils that were worse. He should not like to see the time when the citizens of England would submit to bad treatment from their employers. He should like to see the employed, while willing to use every means to avoid a conflict, having an independent spirit, and, when no other course was left, they must resist an injustice—that was, a plain and direct injustice—to the uttermost of their power, come from whatever quarter it might. (Cheers.) He would just relate some little experiences, to illustrate a mode of preventing Strikes; but, first, he would remark, that if the masters were too stubborn to enter into this question with the men, that feeling of disregard would drive the men into certain things that, in the end, would be more injurious to the masters than trying with the men to settle it amicably. He meant to say, that if Strikes and Lock-outs continued as they had hitherto done, the inevitable result would be, that the men would have to look to manufacturing goods themselves, and to see if that would stop them. (Cheers.) This would inevitably come to pass in time, but it would be a work of time. In 1859 there was a dispute in the potting trade. The usual time of making agreements in the Potteries was November, and a certain portion of the hands thought that an advance was required, and they named it to their employers, who objected. The thing went on till December, when it was proposed to refer the matter to arbitration. The master potters chose five arbitrators on their part, and the men chose five on theirs, and on the 12th of December, 1859, they met at Tunstall. Before they went to any business, an umpire was mutually agreed upon, and the result of the inquiry was, that the men gained nine points, he believed, out of the ten they asked for, and the thing was amicably settled, and a Strike averted. Well, three years later, there came on another case, in which one single firm was concerned. The masters and the men appointed three arbitrators on each side, and an umpire was fixed upon, and the result of this inquiry was against the men. He believed the men themselves were satisfied in their minds, when the case came to be properly investigated, that they had not a sufficient reason to ask for the advance that was proposed. (Cheers.) These two cases would show that, when an earnest effort was made, these things could be settled; for whether the decision went in favour of the one party or the other, each side was substantially satisfied. (Cheers.)

MR. WILLIAMS, Liverpool, was the next speaker, and, after complimenting Mr. Joynson upon the success of his first speech, he proceeded to consider the question of Lock-outs. Every evil, he said, must have a cause, and a Strike or a Lock-out was an evil in society. It was all very well to saddle certain wrongs on the wrong horse, and all very well to try to escape blame. He admitted that there were many evils connected with his brother unionists, but, while admitting that, he maintained that there was greater tyranny among the capitalists. (Hear, hear.) He would not run away from the evils that were among themselves: they were indifferent about the welfare of themselves, and indifferent a'out the prosperity of their trade. He said, "Away with the sympathy that is only in words. Give me the Quaker's sympathy; give me co-operation; give me the earnest work of a good hard-working son of toil." He regretted as much as any man when a contest began between employers and employed, and was quite aware that it was detrimental to both parties; but, while he admitted this, he said that, in a great many instances, it was the only alternative the working-man had to free himself from serfdom and slavery. (Cheers.) The remedy was a great national organization. Let this be brought about, and Lock-outs would be a thing of the past, and their children would read of them in the page of history with amazement that their progenitors should be so long in ignorance of the power they could wield. (Cheers.) It was easy for the capitalists to make out a case for themselves. They were patronized—he hoped the press would excuse him—he did not know much about the local press of Sheffield, but where he came from, the press flattered, they pandered to, the employers. They would not report the men faithfully. If the men gave a good reason for a certain object that they had in view, the press invariably cooked it. (Laughter and cheers.) They tried to turn public opinion against the men. Well, in the trade he was particularly connected with, the men offered to their employers six months' notice for an advance of 6d. a day. The masters wanted to cut up a chicken as they liked, but the men were well united together and organized; they understood themselves; they wanted to improve themselves socially, morally, and politically; and they won. (Cheers.) Then the masters said, "You can go back to your society, and say we are compelled to give it you because your union is so strong." (Cheers.) Ten years ago, it would have been—"If you don't go to work—if you offer to attempt to take us in detail—we will lock you all out, and starve you." He advised them to co-operate. England was a noble country, and had great institutions; but there were great improvements wanted, which would give the men the power of commanding respect, to show to the world—and more especially to the capitalist employer—that they were not the machines of labour; that they had other duties to perform than eating, drinking, working, and sleeping; that they were composed of the same material, and had the same intellect, as the masters, and that they wanted time to cultivate that intellect. (Cheers.) The farmer might as well leave his field unsown, and then call upon God to give him a harvest, as for working-men to say they wanted improvements without working for them. He called upon them as honest hard-working sons of toil, to band themselves together for the great object of emancipating themselves from the serfdom and slavery they had suffered under for a series of years, and to rise in their might, put on their strength, and do battle with all the evils that afflicted society. (Loud cheers.)

Mr. JOHN NORMANSELL (South Yorkshire Miners) said, the interest which he represented knew well the suffering entailed by Lock-outs, he himself having been locked out about twenty-four months in six years. In 1864, the Lock-out in South Yorkshire lasted 19 weeks; there were 2000 in the Union at the time, and as the Strike was only at one firm, and 2000 could very well keep four hundred men out for an advance, 3000 men were locked out, the relief given averaging 5s. 6d. per week. But the men stood it, and the employers were compelled to withdraw the Lock-out, and the men resumed employment just as they came out with the very men on Strike still remaining out. It had been well said, that if food could be found, Lock-outs would cease, for employers generally hoped to succeed by depriving those who were contending of the necessaries of life. Nothing had benefitted the trade in South Yorkshire so much as the Lock-out of 1864. Then they were only 2000 strong, and 3000 were locked out; therefore, there were 1000 locked out who had nothing to do with the Association. But now there were no less than 4000 financial members on the district books, and about 1000 more on the lodge books. This had taught them a lesson which they would never forget; they did not know their duty to other trades before. Within the last six months they had had a Lock out on, and the men who had families received about 15s. per week on the average. They had been locked out six months, and the society had been able to support them with a very slight levy—3d. per week—and had been able to contribute to their Sheffield friends £400; to the glass cutters, £25 or £30; and grants to other trade societies similarly situated, and they had not sunk more than £100 in the whole time, though they had 400 men on their funds. (Cheers.) The men of South Yorkshire had suffered, so much from Lock-outs, and it had done them so much good, that they were quite willing to go heart and soul with their friends here to organize a National Association, but he thought an International Association would be a step too far. Prevention was better than cure, and it would be foolishness for them to sit there and deliberate on such a great question, and never go into the cause of it. They knew that Strikes had causes, but they ought to know what the cause was; and if there was a Lock-out in any trade, the Board of this Association ought to know why the Lock-out had come about, or there would be some dispute as to the contributions to be paid. (Hear, hear.) Mr. Normansell then went on to show that the South Yorkshire Miners had got, by their organization, the system of weighing established, instead of measuring, and so justice was done to the miner on the pit banks. In Lancashire they had not this system, and the men worked twelve hours, while the South Yorkshire men worked eight hours, and got more money. The Lancashire men had been trying to get their grievances righted by law, and their masters said they were contented because they did not Strike. It was his firm opinion that it was Strikes that had shown the world and the capitalist that they must do something for the working classes, otherwise there would be nothing but tumult throughout the length and breadth of the United Kingdom. He did not condemn Strikes, but he would have a just cause for a Strike, then Strike hard enough. (Cheers.)

COURTS OF CONCILIATION OR ARBITRATION.

Mr. GEO. ODGER, Secretary of the London Trades' Council addressed the meeting on the subject of Courts of Conciliation and Arbitration. He did not, he said, know of any subject more affecting the working-man than this. A great deal had been said about Strikes and Lock-outs. With the principles of Strikes he had no sympathy, but he looked upon them as a sad necessity—as a two-edged instrument which was dangerous to use, and which ought to be avoided. Strikes were to the social world what wars were to the political world. (Hear, hear.) They became crimes, unless they were prompted by absolute necessity. Such being his sentiments, the meeting would not be surprised to learn that he had devoted the greater portion of the past few years of his life in endeavouring to induce his fellow-workmen not to resort to force, in the slightest degree, to obtain the ends they desired. They had, he thought, so far as obtaining a good wholesome opinion was concerned, been

30

successful. It was true the great object was not yet attained. It was true they could not look around, and see or hear of any of the great effects of those principles of conciliation. The work had to be done, and a great deal of time had to be spent, before it becomes palpable and feasible. Then it behoved all those who were prepared to fight the great battle in the path of progress to wait patiently. The speaker then read the following letter, which had been addressed to him by Lord St. Leonards:—

"Boyle Farm, 16th July, 1866.

"SIR,—Since my last interview with you I have been in communication with many persons in relation to my bill, and I find that the masters no longer desire the bill. They act, I think, under a mistake, inasmuch as the bill is not compulsory, but is simply permissive; although, of course when accepted, orders could be enforced. I stated to the House of Lords when I introduced the bill, that I should not offer it for a second reading unless both the masters and operatives desired the measure. As the masters do not desire it, I have put it down for Thursday next, when I shall state what has passed and withdraw the bill for this session. I have ascertained that both masters and operatives are anxious for a Court which could settle future wages. My bill provides for no such thing. I think that it would increase the present evils. An order for fixing future wages might be acceptable at the time it was made, but when the time arrived when the demand and the supply of labour showed to the operatives that their wages ought to be increased, I fear that the order would not be obeyed, and measures to enforce it would add to the present difficulties But it appears to me that as both parties desire such a provision, power might be given to the New Court to establish by consent of both parties a rate of wages in any given trade *for a limited time*, for example one year, and that might lead to a better understanding than now prevails The operatives as well as the masters cannot lose sight of the alarm which exists in the public mind at the vast spread throughout the land of Strikes and Lock-outs.

"Your faithful servant,

"ST. LEONARDS."

He then explained the interest which his lordship took in the welfare of the working-man. He had taken up the question with the full intention of getting it settled; and he introduced a bill into the House of Lords. He then discovered that the masters were opposed to arbitration. (Hear, hear) The workmen might use all the moral suasion they had, but it was hard to get the masters to admit a principle, in fact, far more difficult than getting the men to admit one. He hoped that their employers would come to the conclusion that arbitration was better for both parties. His own opinion was, that it would be far more beneficial to them than to the workmen. The questions which arose affected labour and capital, and were questions of a most difficult character, and of the highest importance, and any system for the advantageous adjustment of those questions should be gladly accepted by all parties. The speaker then referred to the objects of the bill, and expressed his regret that Lord St. Leonards should be compelled to withdraw it. Great praise was due to his lordship for the manner in which he had taken up the question. It showed to the working-man that there were those who took an interest in his welfare, and who were ready to help him in his struggles to gain for himself those rights to which he was entitled. There were, he had no doubt, men who thought that Unionists, as a rule, liked Strikes. Such expressions were made use of with sinister motives, and were readily published in the newspapers. Such reflections on the working-classes were undeserved, and those who uttered them were capable of much worse. (Applause.) He would advise those who read articles in the newspapers, or in pamphlets, denouncing Trades' Unionists as a class, to do them the justice to enquire into the accuracy of those articles before accepting them as truth. (Hear, hear.) He concluded by again impressing upon those present the necessity for Courts of Conciliation and Arbitration. Such courts would have the effect of keeping up the wages of those who were creditably paid, and be the means of lifting up those who were ill-paid to a degree more befitting that of a human being, and also be the means of preventing their employers keeping them in that position which appeared to them suitable to their avaricious bargaining. (Applause.)

MR. J. HUMPHREYS, Wolverhampton, spoke on the same subject. After the speech of Mr. Odger, it would, he said, be almost unwise for him to go more into the question. The system which he was advocating had been in operation, and had worked well as far as it had gone. Birmingham and Coventry had adopted the system of arbitration prior to their adopting it, but it had failed. He then alluded to the masons' "discharge" note, and to the great indignation it caused; then to its withdrawal by the masters, and their agreement to conciliate. The workmen were invited to meet them, and they did meet them, and, at the meeting, a board was appointed as arbitrators, and an umpire was appointed to the satisfaction of both parties. Since that time there had not been a single dispute. (Applause.)

MR. GEORGE ROSS, Glasgow, after expressing the pleasure he felt at being present, said thorough unity would secure the interests of the workmen of the United Kingdom. (Applause.) He had, he said, at one time been opposed most strongly to the bill, the subject of which they were discussing, but he felt now that he ought to support it, and he hoped there would not be a single man who would oppose it. (Hear, hear.) Unless it were to be permissive, and at the option of the employers and the employed, let them not have anything to do with it. (Hear, hear.) He felt sure that Courts of Conciliation and Arbitration would be of the greatest possible advantage to them in settling all those vital questions which affected the different trades from time to time. If the movement only brought the masters and workmen to see it, it would do a great deal of good. He would not, he said, refer the dispute of a builder to a shoemaker, or the dispute of a shoemaker to a builder; but would refer each dispute to arbitrators in the trade in which the dispute arose. In conclusion, he earnestly entreated those present to exert themselves individually to obtain those Courts of Conciliation and Arbitration, and he was sure of success. (Applause.)

32

THE LAW OF MASTERS AND WORKMEN.

Mr. GEORGE NEWTON, of Glasgow, the Secretary of the Masters' and Workmen's Executive, next addressed the meeting, In his opening remarks he said there were scarcely any matters which more deeply concerned the social relations and private character of the people than did those which had been referred to. He entered at some length into the subject, premising that as far as employers were concerned, their share in the contract only involved civil obligations, whilst, on the other hand, on the workmen's part, it was considered in a far different light, breach of contract leading him to be treated as a felon of the deepest dye. Many attempts had been made to reform the criminal population of the country; they had been pitied, had had places built for them (whether as a testimony to their superior genius or not he could not say)—(laughter)—but certain it was they had been fed and clothed, and that perhaps in a manner excelling that of the average of working-men; and this with a view of reforming those who cut throats and robbed the people of the country, whilst at the same time those who built up the nation's wealth, and contributed so much to the industry and commercial greatness of the country, had until lately had no time or attention spent upon them; their manacles were preserved for them, and when they were sent to prison, it was always with hard labour. After further alluding to the injustice which he considered the working classes suffered, the speaker proceeded to explain that the three statutes enacted in the reign of the Georges provided that it should only be required that employers or their agents should state upon oath that the servant had been guilty of breach of service, and a warrant could be at once issued for the apprehension of the man. One of the highest judges of Scotland had declared that it was illegal to summon a man in such cases; and they could only be proceeded against on warrant. Then, on a case being proved, the sentence must be hard labour, and actually on a man being sentenced to thirty day's imprisonment only, the sentence was declared illegal, because hard labour was not included. There had been some talk of improving the condition of working-men; but he certainly did not think a prison, where they would be herded along with murderers, assassins, and thieves, was a fit school wherein to teach them frugality, temperance, or moderation. (Hear, hear). The speaker went on to describe the measures which during the past three years had been carried on with a view of bringing about a change, but he remarked that notwithstanding the "pressure" which had been brought to bear, the committee appointed by the House of Commons for gaining information on the subject had not yet issued their report, and though that report would shortly be issued, it would be requisite that all the influence which could be brought to bear should be called to aid in prosecuting the movement, so that the House should be induced a remand, but his application was not granted, and he was to pass some suitable measure. He urged the men of Sheffield to take the matter into their serious consideration, and by their pens, their pence, and their voices, to get the law altered. It was high time that the workmen should be raised above the position of serfs, and that they should be freed from the tyranny and oppression to which they were now subjected. Such a state of things might be in keeping with ages of ignorance, but as education spread and knowledge was diffused, such things ought to be banished. In the the course of his remarks the speaker adverted to the injustice of masters sitting upon the bench to adjudicate in cases relating to workmen, and he called upon his fellow men to be alive to their own interests. With regard to the part which Lord Elcho had taken in aiding the movement for bringing about a better state of affairs, he spoke in terms of the highest commendation.

Mr. WM. DRONFIELD said, he congratulated Sheffield on being the first place in which such a large number of trades' delegates had ever assembled—representing about 200,000 members. He repudiated the idea that they were acting in a spirit of antagonism to the employers, but urged that it was their wish to devise some good means of preventing Strikes and Lock-outs. He then proceeded to speak on he subject introduced by the last speaker, and further explained the present state of the law relating to masters and workmen. In adverting to the steps which were being taken for the purpose of, if possible, bringing about a removal of the grievances complained of, he called upon all working-men to do their utmost to aid the movement for their benefit which had been so ably inaugurated and carried on by their Glasgow friends in conjunction with London, Sheffield, and other places. He had pleasure in stating that Sheffield had taken an active part in the agitation—upwards of forty petitions to the House of Commons from the different trades of the town and neighbourhood having been numerously signed, most of which had already been presented to the House by Mr. Roebuck, and others. In addition to signing these petitions, the trades of Sheffield had subscribed liberally towards the expenses of the agitation. Mr. Dronfield further stated that at the last Borough election this question was brought prominently before the candidates, and that at least one of their members—Mr. Roebuck—had promised them his assistance in getting an alteration of the obnoxious law of which they complained. As an instance of the glaring acts of injustice which were perpetrated even in this neighbourhood under the existing state of things, he said that there was a delegate on that platform who had on a certain occasion been dragged from his home, and taken in search of a magistrate. He had to be taken about six miles before a magistrate could be found, and then when they had arrived at the house, he could not get a fair hearing. His solicitor, who had been hastily summoned, without time having been afforded for his friends getting up any defence, applied for obliged to submit to the severe indignities which the law so unjustly allowed. The speaker held that those cases ought to be treated as civil contracts both with regard to master and man, and urged the duty of working-men being alive to their own interests in the matter. (Applause.)

CO-OPERATION.

MR. W. HARRY, London, was next called upon to speak upon the subject of Co-operation. After remarking on the importance of Co-operation, he enlarged on the benefits which the working classes might derive from availing themselves of those advantages which they would obtain from associating together, so as to make the most of their resources. He contended that the working-men of the present day did not do their duty, and urged them to individual effort. He considered that if they did not do their best to ameliorate their condition they had no right to complain, and he reminded them that they must none of them by any means assent to any oppression. He also contrasted the condition of the people of the time of Alfred the Great, when they only worked eight hours per day, and those of the present day, when some of the working-men of London actually worked fifteen or sixteen hours per day.

MR. H. ROBERTS, Manchester, showed that the principle which was brought to bear in joint stock companies, railway companies, and other associations in which capitalists combined, was just as applicable to the wants of working-men; and urged them to avail themselves of the privilege of associating together so as to get the full benefit of their labour. He considered it an imperative duty to advocate the principles of Co-operation, and in the course of a sensible speech, adduced arguments in its favour. Men could do far more in many ways unitedly than individually, and the old maxim of the bundle of sticks, "United we stand, divided we fall," was as true as ever. He believed education was doing much for working men; but he considered that they ought to avail themselves of the advantages of the press, and have a daily newspaper of their own to act as their organ—just as the present newspapers were the organs of the employers. He hoped the day was not far distant when the advocates of trades' societies and measures for the advancement of the working-classes would have an organ of their own, for he considered that the capitalists only wielded the power of the fourth estate. With all deference to individual opinion, he thought that at present working-men were not fairly represented. But it was necessary that they should be, and for this reason he urged that they should have a daily press of their own. (Cheers.) The speaker further adverted to the other objects of the meeting, and called upon all present to weigh well the important matters laid before them.

The meeting terminated about half-past ten o'clock, with a cordial vote of thanks to the chairman, for the able manner in which he had presided.

THIRD DAY.

The minutes of Wednesday's proceedings were confirmed.

Credentials were received from the following delegates:—Operative Millers' Friendly Society, Sheffield,—Mr. J. Allen; Branch of the General Union of Carpenters, Sheffield,—Mr. T. Hinchliffe.

Mr. Clare, filesmith, of Warrington, and Mr. J. Holmes, of Leeds, were admitted to the meeting on application, the meeting resolving that all well-known members of Trades' Unions be admitted (with the permission of the Chairman) to listen to the deliberations, but not to take part in the proceedings.

A letter from Mr. T. J. Dunning, of the London Bookbinders' Society, was read, and ordered to be received. In speaking of Lock-outs, Mr. Dunning said:—

"There is great necessity for the subject of Lock-outs to be widely ventilated, for they are indefensible on any principle of economics, of justice, or even of the usage of civilized existence They take us back to the law of the strongest, the principle of savage life, which exchange the principle of civilized existence was established to repress, and, if possible, to set aside. A Lock-out on an extended scale is a conspiracy against common rights, as utterly indefensible as an act of robbery, or, at the very least, of revengeful deprivation on other people, because you cannot get your work done at your own price. It is enacting, in grave reality, and with deliberate intention, on the principle of the following burlesque lines, and with equal burlesque of justice and moral honesty:—

'As when two dogs, erst fighting in the street,
One of those dogs another dog doth meet,
With angry tooth he bites him to the bone,
And this dog suffers for what t'other dog had done!'"

Mr. CUTTS, of Sheffield, announced that the Amalgamated Society of Engineers had just forwarded the sum of £1000 to the Sheffield File Trade, to assist in the payment of the obligations arising from the recent Lock-out.

Resolved,—"That the thanks of this Conference be tendered to the Amalgamated Society of Engineers and all other bodies who have responded to such application."

The Committee presented the following

Preamble:—The Conference having at great length considered the question of Lock-outs, with the view of ascertaining the cause thereof, and providing such remedies as seem called for to obviate such in future, as far as possible, and, in cases where they do occur, to make such provisions for the wants of the suffering workmen as will neutralise the hitherto painful effects thereof—

Find—That these social perturbations arise from a spirit of opposition engendered in employers by the efforts of workmen to obtain a greater share, than hitherto has been granted them, of the wealth and comfort accruing from the industrial pursuits of the mass of the people of this country.

That, whilst in many instances the demands of the workmen have undoubtedly superinduced these Lock-outs, it is painfully apparent that, in a vast number of cases, they are resorted to by the employers, as a means of constraining workmen to act in such a manner in reference to Trades' Unions, rate of remuneration, hours of labour, and kindred questions, as they, the employers, choose to dictate.

That Lock outs are an outrage on the most clearly defined and generally accepted principle of social and political economy, and are calculated seriously and permanently to injure the trade and commerce of the country.

That the time has come when all classes of society should direct their earnest attention to the question, with the view of preventing the recurrence of these disastrous events, by the adoption of such wise and conciliatory measures as will obviate their necessity in future, and that it behoves all workmen to make special provisions for the support of such of their number as may still be exposed to the evil effects thereof, and that this can only be effectively accomplished by a National Amalgamation of all Trades.

Mr. G. NEWTON: The Preamble is our first business, and as it is merely an expression of the opinions delivered by the delegates yesterday, and framed with the view of giving the result of Lock-outs, their history, and cause, there will be very little discussion upon it, and I move its adoption. It is very certain that if workmen were always content to work for what employers chose to give them, there never would be any Strikes or Lock-outs. There will, therefore, be no difficulty about finding No. 1. The second finding is this: "That whilst in many instances the demands of the workmen have undoubtedly superinduced these Lock outs, it is painfully apparent that, in a vast number of cases, they are resorted to by the employers as a means of constraining workmen to act in such a manner in reference to trades' unions, rate of remuneration, hours of labour, and kindred questions, as they, the employers, chose to dictate." The honesty of this finding, I think, none can dispute. We admit that we are ourselves to blame not unfrequently, and that the employers are also to blame, which, I believe, is also in consonance with the views expressed by the delegates yesterday, and with the facts of the case. I don't see that there can be any dispute about the next finding; and the last sums up the whole, declaring that it is the duty of all classes to direct their attention to this important matter, with a view to putting an end to Lock-outs, and recommend that conciliatory measures be adopted with the view to prevent them, and provides, in conclusion, that if it be impossible to prevent them, it is the duty of workmen to make provision for those of their fellows who suffer by them.

Mr. W. H. WOOD seconded the proposition. After leaving the general meeting last night, the committee went immediately to work, in order that we might be prepared to lay a very clearly defined series of resolutions before you for consideration, and I may add that we did not arrive at our conclusions until we had lengthily discussed their merits. It was a quarter-past two o'clock before these resolutions were prepared for the printers, so that you can easily understand that your committee have not been lying on a bed of roses since your deliberations yesterday. The judgment and discrimination manifested by my colleagues in the preparation of these resolutions, I must say, astonished me, and I cannot refrain from congratulating the meeting upon the unity of feeling manifested.

Mr. H. MITCHELL, Sheffield, thought the question as so whether the association should be "national" or "international," ought to be discussed.

The CHAIRMAN replied that the subject would be brought forward in the next resolution, and could then be discussed.

Mr. HARRINGTON, Sheffield, supported the proposition.

Mr. McDOUGALL: The question before us is a natural and an important one. The Committee appear to have overlooked that it is a question that affects the United Kingdom. We have representatives from the northern parts of Scotland, and though it is all very well that we should remember ourselves, we must not forget our neighbours.

The CHAIRMAN: The first resolution will meet your wishes.

The VICE-CHAIRMAN: It meets exactly what you intend to say.

Mr. McDOUGALL: I will move that, instead of "by an amalgamation of all trades having their seat of government in the United Kingdom," the following words be employed:—"By an amalgamation of all trades in the United Kingdom."

Mr. WILLIAMS: Perhaps you would like it to be the United Kingdom and Ireland? In my opinion a national institution embraces every nationality.

Mr. ROSS: I feel a difficulty in rising to speak to the amendment, but I feel honestly called to do so.

The CHAIRMAN: Do you intend to second it.

Mr. ROSS: Yes, I have pleasure in doing so. If the institution is simply to be a national one, it will apply to England alone, and exclude the part I represent. (No, no.) I am from the North, and that is my opinion. If it is so altered as to include the United Kingdom, or any term is employed implying that, I shall be prepared to support the proposition.

MR. RUSHFORTH thought the alteration ought to be made.

MR. BROADHEAD: I don't see that we need carp about this. The resolution applies to the United Kingdom, and there has been no such idea as to keep out Scotland or any other part of the kingdom.

MR. NEWTON: I am amused with the idea. If I had said association of English trades, there might have been something in it. This is a nation made up of England, Wales, Ireland, and Scotland. It is all stuff that "national" does not include every corner of the empire.

MR. SHORROCKS: I think some opinion ought to have been expressed in the Preamble with regard to masters who are owners of cottage property, turning the men they have locked out on to the wide world without shelter. It is a very serious matter, and a degredation to England, and if the committee could draw up something that would call forth an expression of indignation against such conduct, it would meet with universal approval.

MR. HARRY supported the amendment. Though they all lived under one government, the laws of Scotland were different to those of England.

The CHAIRMAN announced that the Preamble would be altered as suggested in the amendment.

MR. NEWTON: Being a Scotchman myself, it is not likely I would give the go-by to my own country, if I knew it.

MR. CONSTANTINE moved that the word "national" be altered to "international." He thought that alteration would be better than that just proposed.

MR. SHAW seconded the amendment.

The VICE-CHAIRMAN: The committee took that matter into consideration. We are only now seeking to form an association of all the trades of the United Kingdom. Every attempt in this country hitherto has failed, and let us try to effect that before we seek to spread ourselves to the continent. If we can succeed in this country, then the motion to unite with other countries will be in good order; but until we have formed a union of our own, how can we ask other countries to join us? We may keep up a friendly intercourse with societies in other countries; but to call ourselves international, when we don't know that we shall be national, is going too far. (Hear, hear.)

MR. ODGER: The original proposition might be adopted, and then our friend, if he should feel disposed, or I myself, may propose a resolution to the effect that this association, after it is formed, sympathises with the objects which the International Association has in view, and will do its utmost, or something of that sort, to assist the International in developing its principles throughout Europe, and all over the world, if it be so needed. That will not be committing yourselves fully to the International Association, but merely an expression of opinion. I would suggest the withdrawal of the amendment, and that you allow the proposition to pass, with one little alteration. The second paragraph says, "that these social perturbations arise from a spirit of opposition engendered in employers by the efforts of workmen," and so on, so that we seem to take upon ourselves the onus of creating these Lock-outs of which we so much complain. I would have it read, "that these social perturbations arise, in many instances, from the spirit of opposition," &c. The next paragraph puts it correctly, "that whilst in many instances"; but this paragraph says they arise altogether. I think the committee will agree to make that little alteration.

MR. CUTTS: I think it would be unwise to go so far as to embrace an international confederation. It is our best policy to have a national alliance at present, and if, in future years, we can see how to form it into an international association, so much the better. We have a great deal of work to do yet, before we have a thoroughly effective national association. I do not see how it would exclude the Amalgamated Engineers, for although they have branches in all the colonies and in other countries, still they are part and parcel of one association, and, consequently, if they joined it, would be embraced in this confederation.

MR. E. GREEN: It would be very much to our benefit to have an international organization; but the proposition is premature. I hope we may ultimately see our way to form an international association.

MR. W. DRONFIELD: The trade to which I belong would be as much benefitted by the introduction of the word "international" as any other, for printers work all over the civilized globe. I might, therefore, be in favour of its introduction, but I think it would not be politic at present to introduce it. We should get too much entangled in the affairs of the working-men of other nations. It would be very proper to keep up a correspondence with trade societies throughout the world, so far as we know them, for the purpose of rendering information to them, and receiving information from them; but to attempt, at the outset of the association to extend our ramifications throughout the world, would be attempting a scheme that it would not be possible for us to accomplish. I think we had better go slowly and surely, and not attempt more than we can accomplish. The less complex we are, the more support shall we get from the trades of the United Kingdom. Though I am in favour of the laudable work of promoting a feeling of brotherhood throughout the world, I am for the present opposed to the introduction of the word "international."

MR. BURNS coincided with Mr. Dronfield.

MR. MEMMOTT: Will the branches of a trade society, having its chief seat of government in this country, be allowed to enter this association, though they are in the colonies, or in other parts of the world?

The CHAIRMAN: Certainly.

MR J. ALLEN: It might be all very well to make our society international, if we could do so conveniently and advantageously. We know what difficulties have had to be over come in this country during the last thirty or thirty-five years to get trades well organized, and we have not succeeded yet. We are now about to form the whole into one, and it will take time to do that. Let us get established on a firm basis, and then we may introduce the subject to our friends in every quarter of the globe. (Applause.)

MR. ODGER'S suggestion was met by the introduction of the word "occasionally" after the word "perturbations" in the second clause of the Preamble, and

The Preamble was agreed to.

TITLE OF THE ALLIANCE.

MR. G. ROSS moved—

"That it be a recommendation of this Conference that the principle of a national amalgamation be formed of the Trades' Societies of the United Kingdom, to be designated 'The National Alliance of Organised Trades.'"

MR. G. HOUSLEY seconded the proposition.

MR. McDOUGALL moved as an amendment "That it be a recommendation of this Conference, that an amalgamation of the Trades' Societies of the United Kingdom be formed, to be designated 'THE UNITED KINGDOM ALLIANCE OF ORGANISED TRADES.'"

MR. R. YATES seconded the amendment.

MR. C. WILLIAMS moved, "That this Conference assembled feel it their duty to recommend that an amalgamation of the Trades' Societies be at once formed, to be designated, 'The United Kingdom Trades' Protective Association.'"

MR. J. TEASDALE seconded.

The CHAIRMAN announced that Mr. McDougall's suggestion had been adopted by the mover and seconder of the original proposition.

MR. T. HINCHLIFFE suggested that the proposition should read thus: "That it be a recommendation of this Conference that a grand amalgamation be formed of the Trades' Societies, to be designated the 'Alliance of Organised Trades;' that the various Trades' Societies be invited to join such organisation and render such pecuniary and moral support as may be necessary."

MR. ODGER: I hope, when the resolution of the committee is passed, that we shall all feel the responsibility that it involves. It will be one of the greatest farces that ever was played in this country, if, after a Conference such as this is——one of the largest of the trades that ever assembled, perhaps, and I must say, as far as I can judge of it, one of the most intelligent, and, at the same time, business like, I ever attended in my life, no practical good followed. I don't object for one moment to any sentence I have heard expressed in this assembly. We have men congregated here from different parts of the country, and we have heard the opinions and remarks of these different men, and it is a proof of the progress that men are making, that they can sit quietly and listen to these various remarks, and, in the end, have but little difference, and adopt a resolution expressive of the conclusions which have been arrived at. I have seen in my time several conferences having the same object in view that this has, but they were not attended with success. There was a national association formed some years ago, and no doubt there was some ability manifested by those who had charge of that association, but it fell away in time to nothing, and though there is a remnant of it, it is a perfect myth, so far as its recognition by societies at large is concerned. That society was instituted by men of mark and intelligence, men of integrity, and men well able to do the work. It was presided over by the late Mr. Thomas Duncombe, a man of sterling integrity and industry, and had for one of its advisers the late Mr. Wakley, coroner for Middlesex, and member for Finsbury. There were others also connected with it who were well known for the efforts they had made to remove abuses. There was an effort made to form a society of this sort at Derby. It was initiated by Mr. Potter, and there was an assembly of many delegates from the principal towns and large trades. There was a committee, an executive, and everything of that sort, but the association died from the want of energy and want of funds; in fact, I heard it distinctly stated by a member of the committee, that after its formation they never once met. This is a sorrowful thing to have to contemplate; but it is a fact, and facts must be looked in the face, or we may make another mistake. (Hear, hear.) Somehow or other it seems that when the rules and regulations were drawn up, and the preliminary work finished, the members fell back into their local habits and customs, and lost sight of the great principle involved in their society. But I think from what I have seen, that such is not your intention. The carrying out of the programme will involve an immense amount of deliberate labour; there will have to be brought into exercise energy and persistent perseverance, and unless that is done, we shall add one more failure to the many that have gone before. That would be a disgrace to us. (Hear, hear.) As a proof of what I am stating, let me call your attention to the support which the aggregate bodies receive at the present time. The Trades' Council of Glasgow never had a shadow of the support which its energies and perseverance merited. From all the members who are in union with the London Trades' Council at the present time, we don't get £30 a year to carry it on with. I am stating facts to you, so that you shall know there is work to do besides merely carrying resolutions. When trades are in trouble in London, they come to us to assist them, and there is nothing that is more agreeable to me than to talk matters over with them, and to try to get them such assistance as is necessary; but while I can point to societies ever ready to subscribe with vigour to persons on strike, as a rule, they are not prepared to carry out any grand scheme by which Strikes can be averted. In a confer-

42

ence with Mr. Bright, and another with Mr. Gladstone, the Chancellor of the Exchequer, I was enabled to show our trades' report, containing our aggregate expenses, and, notwithstanding their smallness, we were called, in the *Standard*, I think, a set of "bloated trades' delegates." The support given to large bodies, such as I represent, is ridiculous in the extreme, and the reason for the lack of support is this—that when men meet in their own societies, they interest themselves immediately in their own concerns, and neglect the calls for their immediate sympathy and support. I cannot say that feeling arises from any bad motive on the part of the men. The very fact that our societies give money to the extent they do, is the best possible proof that there is no motive of an improper character in this state of things. If I go to my society at the next meeting night, perhaps I may be disposed to throw aside, until the next meeting, the consideration of this national association. Next week, there is probably some other thing in the way, and the subject is deferred again, and the executive become disheartened, and the result is overthrow. If you do not give prompt attention to whatever is sent to your society by the executive, you may depend upon it the carrying of this resolution will be perfectly useless, and worse than useless, because it will entail heavy preliminary expenses, and in the end, will throw some amount of discredit upon those who have initiated the affair, and are prepared to carry it out, if you would only do your part of the work.

Mr. ROSS replied: I think it is a misconception on the part of Mr. Williams to introduce the word protective or protection, because I hold that we are already protective associations. If this institution were formed under such a heading, it would presume that we are not already protection societies. I maintain that we are so, and that we are here in that capacity, and that there is no necessity for such a term being applied. One word in reference to a remark by Mr. Odger about the character of the Glasgow Trades' Council. I was a member of that body, holding, perhaps, an important position, and now a movement is going on in Scotland to associate Dundee, Aberdeen, Greenock, and Glasgow, and to establish a true representation of the trades of the different localities. Our own trade, though falling away from the Trades' Council of Glasgow, has again come forward, and has given the liberty to its branches to become connected with it; and I hope that the movement will extend itself throughout this country. I don't think this great association will materially alter the position of Trades' Councils, any more than it will alter the position of our private Trades' Societies.

Mr. ODGER: I said the Glasgow Trades' Council never received that support which it merited.

Mr. ROSS: Almost every trade in Glasgow supported it.

The original proposition was carried.

43

SUPPORT FOR LOCK-OUTS.

MR. ODGER moved—

"That the various Trades' Societies of the United Kingdom be invited to join such organisation, for the purpose of resisting Lock-outs in any trade so connected, and in rendering pecuniary and moral support to such branches as are necessitated to seek the same."

You will have applications, no doubt, from some of the most prosperous trades, and applications from some of the most impoverished; and it will be well to consider frequently the advisability of withholding encouragement in cases that are likely to result in Lock-outs, when there are so many trades in the country whose remuneration is of a frightful and degrading character. If we understand one another upon that point, we shall be able to show to the world that we are equal to the work we undertake. Take such a trade as the shoemakers. There are certain branches of that trade, in certain parts of the country, very well paid. Take, for example, the West end of London, where I work. There the wages are good; but two miles into the city you find the same trade getting labourers' wages. It has been my desire to lift the worst paid branches of the trade up, instead of always making efforts to push the best paid trades further on. (Hear, hear.)

MR. W. FINLINSON seconded the proposition. MR. W. BURNS and MR. E. GREEN supported.

MR. WILLIAMS said the Protection Society of Liverpool, always gave the preference of its support to those trades which received the least amount of money for their work, and there was not a trade society in connection with the association but what had made great progress.

MR. T. SMITH quite approved of assistance being first rendered to ill-paid trades. Where the manufacturers saw a trade down, they tried to give it another shove deeper down. If the funds of that Association were judiciously spent, many poorly paid trades would be benefitted.

MR. JOYNSON said, that whatever that Association promised, or led others to believe they would do, they ought faithfully and sincerely to carry out. I am therefore compelled, he proceeded, at this moment to state the position in which I am placed. The society to which I belong passed a resolution to the effect that a delegate be sent to this Conference, and another to the effect that the delegate shall attend the Conference to hear the proposals for future action, give a faithful report of the business transacted, pay a proportionate share of the real necessary expenses, but he shall not commit the society to any extent, or in any form, but merely bring the proposals for the future government of the society. Of course, other societies are prepared to go much further than that; but it would be wrong in me not to explain the position I am in.

The CHAIRMAN: You are just in the same position that we are all in.

MR. W. BROADHEAD: I stand in that position; and I presume there is not a delegate present who is not in the same position. All our deliberations will be subject to the approval of the various trades we represent. The advice which has been given by Mr. Odger is both sound and good. It too often happens that apathy follows efforts like this. Associations of this character frequently attempt too much at the outset, and, in some cases, aid and assistance are asked in an unreasonable fashion. Trades should be careful before rushing into any contest, and should not be misled by the idea of the vast assistance that this Association can lend. Persons should not expect too much; but we should all put our shoulders to the wheel, and be determined to do more than we promised at the outset.

Mr. W. LINDSEY: I beg to call the attention of the Conference to the fact that Lock-outs only are named in this resolution, and I wish to ask whether or not assistance will be rendered by this organization, if formed, to trades seeking an advance of wages, by strike if necessary? I look upon this as an important question. Lock-outs in the trade to which I belong are comparatively unknown, and, for the sake of the society I represent, and the satisfaction of the Conference generally, I should like the question I have put answered.

Mr. HARRY: I had drawn up a resolution on the point, and as I think it makes up an omission, I will move it. "That the various Trade Societies of the United Kingdom be invited to join such organization for the purpose of resisting Lock-outs, and all attempts made on the part of employers to reduce wages, and to render such moral and pecuniary support to such societies as may need the same."

Mr. G. AUSTIN: seconded the amendment. He alluded to double-handed trades, and a better equalization of the wages, in some cases the second-hand receiving barely sufficient to support life. The labourer ought to be paid a sufficient sum of money to enable him to bring his family up respectably, and to provide a suitable education for his children.

Mr. W. LEIGH supported the original proposition. He thought the latter part of the proposition included all that was embodied in the amendment.

The VICE-CHAIRMAN: You will observe that we came here not to form a trades' combination for the advancement of wages either one way or another, but to resist these great evils of Lock-outs and to discourage Strikes. We do not wish to create a fund to assist societies who have neglected their own interests; but we have met to see if, by a large association, we can take this weapon of Lock-outs out of the hands of our employers. If an employer locks out his men to force them, by reducing them to starvation, to submit to his terms, we must assist to keep them respectably. By that means the sting will be taken out of the Lock-outs, and all that a Lock-out can come to will be an improvement in the health of the men locked out. (Hear, hear.) If there was anything indicated by the discussion yesterday, it was that we were not to encourage Strikes, not to devote ourselves to the support of societies that have neglected themselves. If the Sheffield File Trade had had respectable support when locked out, the masters would have had to take them in; but when the masters saw that the men had not enough to keep soul and body together, they had the glorification of knowing that the men must ultimately give way.

The CHAIRMAN said, the amendment being withdrawn, the only resolution before the meeting was the second on the paper.

MR. J. DIXON said he noticed that Mr. Odger and other friends had an idea that they should try to get some increase in the wages of those who were badly paid. For his own part, he thought the resolution was comprehensive enough, and ought to be passed. If they tried to influence the employers who were lowering wages, they would be actually involving in fresh conflicts a body of men who had gone through Strikes, at more suffering than he could tell, to get into a good position, and it was hardly fair to call on them to suffer a second time in behalf of men who perhaps were obliged to take lower wages because they had not joined the Association to better their position—men who, in some way or other, were wanting in looking out for themselves. Besides, such a policy would compel the Alliance to embrace Strikes in all directions, while the only object they had in view was Lock-outs. He wanted to begin at the beginning. He hoped the foundation at this Conference would be a plan to protect them from Lock-outs; but he should like this to be only the foundation, and for them to improve it, and lay one stone fairly on another. If they began at the top of the tree, it was possible that Strikes, in different ways, would break them up at the commencement But they had not had half the discussion upon the question that there was in it.

MR. J. ALLEN did not think Mr. Odger intended to give the meeting the impression that they ought to support Strikes, as well as Lock-outs, or that the meeting should take up the question of badly paid trades as a single matter, but merely to refer these trades to their own society, and get them in the Association; and then, that where appeals from oppressed or low-paid classes were made, they should have priority over appeals from the societies of men in better circumstances—that was to say, that when a case of great distress came before the authorities of the Alliance for consideration, and also one from men in better circumstances, the case of those who were worst off should first of all be taken into consideration. If this was Mr. Odger's view, he agreed with it fully; and he believed that such were the objects of most trade societies. There were some that embraced three or four branches of a trade, and all the branches were not paid alike, and when distress prevailed, and appeals were sent in, the object was to bring the lower-paid up to the higher-paid—to elevate the men's position, and bring them on a footing as nearly as possible. If they held that the organisation was to support Strikes in general, as well as to resist Lock-outs, they would have to ask, in every instance—"Are these Strikes justifiable?" And to whom were they to apply for information? Why, to the heads of the respective societies. Let each society confine the business of its members to themselves, and elevate the position of the lowest paid branches belonging to it, and when this was done, this organisation would have no trouble with it, except the efforts of the society to advance wages, or what not, as the society thought right and just, should produce a Lock out. In that case, let this Association take the Lock-out, and leave the society to manage the Strike. Strikes might often be prevented by moral influences, perseverance, and judicious management; but it was sometimes better for the working-classes to accept a Strike than forego it through fear of the evil consequences to follow. In many instances they were bound to accept a Strike; for if they did not accept it in the first instance, they would have embargoes laid on them, until, in the end, they would be reduced below the pitch that a Strike at the first would have brought them to. Moral aid, combined with pecuniary aid, was very grand; and if there was a man in the meeting who had a right to advocate the principle of moral intervention between societies he thought it was himself, for he found that, within the last five years, one of the greatest societies, and a supposed protection society, had infringed upon the society to which he belonged to the tune of £1500. He hoped and trusted that when this Association was properly organized, they would exert a moral influence on all the trades with which they had to deal, confining them to the limits of their own trades, and not allowing them to overstep that line and interfere with other trades, as had been the case. Societies, whether large or small, should consider whether it would not be a benefit to them to join this Association, for they were but men, and the day might come when the greatness of the greatest would be brought down; and small societies should consider it their duty to support the Association, and teach that duty to all within the pale of their unions. He prayed that the Association might go on offering pecuniary and moral aid until, in the end, they all became one, and stood on one platform on an equal footing as men, for they were no more than that.

Mr. J. Carlisle said the ironworkers did not earn £3 a week, and he should not be doing his duty if he let that statement go forth uncontradicted. In the branch he represented they never worked more than five days a week, and one day a fortnight besides they did no work. Having said so much, he thought they were getting far away from the subject. A resolution passed the day before stated that it was desirable to inquire into the causes which gave rise to Lock-outs. He believed the opinion they arrived at was unanimous, that Strikes were the cause. Was he to understand that they turned their backs on the cause, and began to deal with the effect, or should they not in the first instance deal with the cause? He said it was their duty to deal with Strikes, because they were the next great evil to Lock-outs, and the fruitful source from which all Lock-outs sprang. Mr. Lawrence, told them that when members knew they had the support of the Conference they would be rushing into Strikes. Then what were the Standing Judicial Council and Executive Committee for, if not to inquire into the justice or injustice of a Strike or Lock out? That once ascertained, he contended that the duty of that Council was to render effective aid to any body of men on Strike or Lock-out; and if the Conference brought its influence to bear on Strikes, they would not be so bad as they would be if the Conference turned its back entirely on them.

Mr. H. MITCHELL, Sheffield, said he agreed to some extent with what had been advanced as to raising the worst paid branches; but at the same time it was just possible—indeed, likely—that trying to raise the low paid would cause the well paid branches to be Locked out, and they should have the support of the Conference. He rose to propose an amendment. The word "Locked-out" was what he objected to, because he believed that it would prove a barrier to some just cause. They should try to avoid that; and at the same time he should not like to add the word "Strikes," which might induce them. He proposed the following simple resolution:—"That the various Trade Societies of the United Kingdom be invited to join such Organisation for the purpose of enabling them to receive such assistance as they may be necessitated to seek."

Mr. HOLDEN, Sheffield, seconded the amendment.

Mr. J. KINGDOM, Bristol, proceeded to speak upon the changing condition of the world, the more extensive intercourse of thought, and the more powerful action of mind upon mind. The intelligent and well disposed of all parties seemed to be brought nearer together, and were exerting an influence now, which, though it might be said to be feeble in its infancy, would, nevertheless, in his opinion, be felt throughout the world. Public opinion, that helm which directed the affairs of nations, peoples, and empires, had received a new direction, and was tending upwards and onwards, shaking off—as that meeting was evidence—the errors and the prejudices of the past. Both mankind and the laws and principles by which they were governed seemed about to be redeemed from slavery through the moral and intellectual—

The CHAIRMAN interrupted the speaker, and called upon him to keep to the question.

Mr. KINGDOM said he was arriving at it, and continued to speak in the same strain for some time. He read extracts from a paper written by him at the request of the Cordwainers' Society of the city of Bristol, when Locked-out and on Strike seven months ago. The extracts were similar in nature to the speech, and as the meeting manifested extreme impatience, the Chairman at length said he could not allow the time to be so taken up, and Mr. Kingdom sat down.

Mr. P. CASEY, Barnsley, thought that if they could agree to something clear and short, it would be better than a resolution of this description. They had already, in the first resolution, agreed to recommend that a National Amalgamation be formed. They should in their next, in his opinion, state the object of the Association rather more clearly than it was stated at present. For instance, in the second resolution there were three points, the first of which was, that the various trade societies of the United Kingdom should be invited to join such Association. That, in his opinion, had no business in a resolution stating the object. An invitation could be issued by a committee in some other form, better than in a resolution. The Association was for the purpose of resisting Lock-outs in any trade so connected, and that was the whole point of their assembling there. The third part of the resolution related to rendering pecuniary support to such trade societies as were necessitated to seek the same. He was afraid the Conference could not carry this out so successfully as a national amalgamation ought to do. From the discussion already passed, it seemed to be the opinion of some that they should leave Lock-outs entirely out, but he believed that if they did so, the Association would not last to assemble for another Conference. The gentleman who spoke just before dinner clearly showed that they had met there for the purpose of devising means for resisting Lock-outs, and for nothing else; and he thought that this rendering pecuniary and moral support, &c., was putting it too vaguely, and would lead them into other questions in which they could not give general satisfaction, and this would be the very means of breaking up the Association. Thinking so, he proposed this in place of the second part of the resolution:—"That the object of this Association be to render support to all its members who may be locked out from their employment, for the purpose of breaking up their local trades' associations, or compelling them to submit to a serious sacrifice of principle."

Mr. T. SWEETING perfectly agreed with the whole of the resolution. It appeared to him there was everything contained in it that they were met for, and he thought the committee had acted very wisely in so wording it. He certainly objected to some of the remarks made by Mr. Odger with respect to first taking into consideration the low-paid societies. He knew very well, that in the shoe trade in the East end of London, scarcely one out of ten belonged to any association at all, and it might be that the low wages they got was one reason why they did not; but another was that in the West end, where they got better paid, they had not been down to the East end to get these men to join. If they got better paid, it was their duty, in his opinion, to help their brethren to get better paid, and endeavour to organize them, because all large trade associations, notwithstanding that they might get a larger amount of wages, paid a larger contribution to their own society, and unless the shoemakers paid a larger amount than now—sixpence a week—it was not to be expected they could do more. It was their duty to pay larger contributions to their own society to fully organize their own trade. He considered that low contributions meant a very poor association, and ever would; and he hoped his brethren in Sheffield, who were numerously represented there, would not be offended if he told them that their contributions to some of their associations were not sufficient, and he hoped that, to thoroughly organize themselves, they would endeavour to pay more.

MR. DRONFIELD thought the question was really, whether they should embrace both Lock-outs and Strikes in the Association; but so many matters were talked of to-day that should have been talked of yesterday, that it was difficult to say what was before the meeting. He understood the last speaker to say that the trades of Sheffield paid low contributions. In reply to this he might say, for the information of the delegates, that there was not a town in the kingdom where such high contributions were paid, and he was at a loss to know where Mr. Sweeting had got his information. As to embracing both Lock-outs and Strikes, he thought it was foreign to their purpose. Their resolutions all through had gone on the basis of supporting men who had been locked out, and if they got on the question of Strikes as well, they would be defeating the object they had in view when they called the Conference. Messrs. Mitchell and Holden misunderstood the object of the Conference and their position. If this amendment was carried, they would not get the support from the country that they wanted, to make this a national organization of the trades. It was no use going into the question of Strikes, and Lock-outs had been discussed over and over again. If they embraced both, the Association would not be so successful as it would if they confined themselves to Lock-outs. He quite endorsed what had been said by some speakers, that if they at all endeavoured to interfere with Strikes in the way of giving pecuniary support—for it was no use to talk about moral support—if they led trades to believe that they would get the support of the trades of the country on all occasions of Strike, Strikes would be precipitated by the action of this organization, and it would be a complete failure in a short time. Let them confine themselves to the iniquitous and obnoxious system of Lock-outs, and they would act wisely in so doing. The committee had drawn up the resolutions with the intention of confining themselves simply to Lock-outs, believing that that was the object for which the Conference was assembled. He might say, in reply to Mr. Casey, that these were not rules or orders, but resolutions; and the first part of the resolution was quite in place. There was a further resolution on the sheet, that a committee should be appointed to draw up rules for the government of the Association. This was simply an expression of opinion, and he held that it was correctly worded as far as it went. He urged the meeting to confine itself merely to Lock-outs. This had been his belief throughout, and he had not seen anything to induce him to alter that opinion.

MR. McDOUGALL did not wish that a statement made

by Mr. Allen, that all Lock-outs sprang from Strikes, should go forth to the public as a fact. There was at present a Lock-out in Scotland among the Ironstone Miners. They were threatened with a reduction of sixpence a day from their wages—but he saw from a paper that they were beginning to work again—and they were Locked out. There was also a Lock-out in Greenock, among the Joiners, brought about without entering on a Strike—without an intimation of a disagreeable feeling on the part of the men employed.

The CHAIRMAN here called the speaker to order, and said they should never get business done if these matters were gone into.

Mr. BAILEY thought it was the duty of other branches to get the wages of the low paid branches raised. But if they paid in a certain sum to the Association, they were entitled to the full benefit whether they got good wages or not. The question of Lock-outs had been fully discussed yesterday. He thought there had been sufficient said on the question, and that it should come to a vote.

Mr. ODGER replied: As to the remark of Mr. Sweeting, as to the efforts to bring men into the societies, he could only say they had done it over and over again. Another remark was, that societies were responsible for bad pay in their own want of exertion. He begged him to reflect that some men were born into a bad trade, and how had they the same chance of receiving good wages that those had who were born into a well paid trade? To him the idea was perfectly preposterous. If an iron shipbuilder or a carpenter had a son, and brought him up to the business, was it any credit to the man that he was well paid; while, suppose a child was born into an unfortunate trade, was he responsible for it? Not for a moment. Efforts of a greater character were required to raise up the ill paid trades than were required to sustain the wages of those that were well paid. He had been misunderstood with regard to his meaning by many speakers. He never meant anything that was not consistent with the words of the resolution. The words applied simply to Lock-outs, and his remarks as to giving prior consideration to those poor trades pertained to that, and that only. Whatever that support, it was not yet exactly determined in what form it was to come; but in whatever form it came, if there should be two applications, one from a poor and another from a well paid trade, the better paid trade being better off, and probably better capable under existing circumstances of looking after itself, he would merely give first consideration to the most severe case; not to say that they ignored the other case—nothing of the sort; but as they must consider one first, to consider the worst extreme case first. This was perfectly just. With regard to the contributions paid, he believed that in his trade they paid as much contributions to real trade purposes as any society in England; but they must understand that the bulk of those societies that paid large contributions had large benefits which were not for trade purposes at all. If he paid one shilling a week to a society, and absorbed sixpence out of that for out-of-work money, that was not a legitimate trade purpose. He gave the men who paid it credit for paying it; but it was purely a benefit purpose. If, however, he paid threepence a week for legitimate trade purposes, he paid as much to sustain his trade as another man who paid one shilling, if ninepence out of it went for benefit purposes. Therefore he wished them to dissect the contributions, and ascertain what they went for, and see if those who paid much received benefits in the shape of incidental funds, or out-of-work funds, or large sick perquisites, or anything of that sort, for if these were allowed to obtain in opposition to legitimate trade funds, they might call Odd Fellows' or Foresters' societies, trade societies.

The second amendment was then put and lost, three hands only being held up for it.

The first amendment was lost, fourteen hands being held up for it.

The original resolution was then carried *nem. con.*

ENTRANCE FEE, ANNUAL SUBSCRIPTIONS, &c.

MR. DRONFIELD moved the third resolution, which was as follows:—

"That all Trades' Societies connecting themselves with this Alliance shall be required to pay an entrance fee of one halfpenny per member; and that an annual subscription of one penny be contributed by each member to meet the requirements thereof, such subscription to be paid half-yearly."

He hoped the meeting would adopt this resolution, believing with the sub-committee, who drew it up, that it was the best under the circumstances. They had said all through that they thought the organisation should not attempt too much. If they wanted to make it a grand national affair, they must make it as cheap and simple as possible. They wanted to embrace the whole of the trades, not sectional trades, for if they were to succeed, they must have this, and nothing else; and if they wanted to make it a success, they must keep it at a low figure. The present organisation in Sheffield had been a greater success than any other that preceded it, because its contributions were not too high, and it did not aim at too much. He wanted this Association to do the same, and more if possible, to embrace the larger trades, and devise the best and simplest means of arriving at it. The calculation had been made of what would be produced supposing all the trades there represented were comprised. The question was, had they a fair right to expect that the number of men represented at the Conference would join the Association. He thought they had a right to expect as many, and probably more. He hoped they would reach 200,000, and with that number a contribution of one halfpenny per man would amount to £416 13s. 4d., a penny contribution £833 6s. 8d., and so on. Of course, this was only for working expenses, paying the executive and officers, printing rules and reports, and all expenses such as delegations, &c. He did not want to press the subject further, but to call their attention to the amount of money that would thus be realised from 200,000 members.

MR. W. LEIGH seconded the motion. If the entrance fee were large, men would not pay it. Some would be in favour of getting a large sum for the purpose of raising a fund; but he knew that there was an objection to accumulate funds in localities where the contributors did not reside. It was therefore better to have as small an amount as was necessary for the working expenses. The smaller the amount the better for the organisation.

MR. S. MEAKIN, Basford, spoke, but with such indistinctness that he could not be reported.

MR. HUMPHREYS, understanding that this provision was merely for incidental expenses, wished to ask how the principal expenses of the Association were to be borne. He saw no provision for them.

The CHAIRMAN: We will enlarge on that further on.

MR. WILLIAMS said his own opinion was, that the sum proposed in the resolution was a very reasonable one, and if only well taken care of, and judiciously expended, it would hurt no one, but do many a vast amount of good. In Liverpool they paid a penny per month per member for all the trades in connection with their protective institution. He liked the word "protection." He had not the least hesitation in taking up the resolution and sending it round the United Kingdom for the approbation of the lodges he represented.

MR. SHORROCKS proposed to add the words "in advance" to the resolution.

To this the mover and seconder agreed, and the words were added.

MR. BURNS, London, thought they ought to make a bold stand at once, by endeavouring by every means in their power to raise capital sufficient to prevent Lock outs for the future. He could not see what a halfpenny subscription would do, but to pay the expenses of the committee of management, and so on. He wanted something further than this, and so did the body of men he represented. They wanted something like this. Take the Sheffield Lock-out. Had there been sufficient to meet it, and sent down, he felt convinced it would have prevented it altogether. Take the builders' Lock-out. Had they had the means at once promptly to have met it, capital would have thought twice before locking out once. It was a question of starvation for them,—could they starve the men sufficiently. They knew what the men had in hand as well as they did themselves, and could calculate how long it would last, and saw when the Strike must cease. If the men had the means, Lock-outs would be prevented, and the masters must come to terms. He believed that their only course was not to go childishly to work. Trades' Unions knew what they had gone through, and wanted to prevent it in future;

and if a bold step were taken, it would be prevented. Of course, he was not anxious to go and pay a great amount of money; but he thought that if a levy of 5s. per member were made throughout the trades, payable at a small sum, 3d. or 6d. per week, it would not take long to raise a fund, and the committee should have full power over it, to support the trades that required help. His experience of many years told him that a plan of this description would be preventive, and it was with that view that he made the following proposition:—"That we recommend to the various societies represented at this Conference that a levy of 5s. per member be made, in order to raise a fund to support any society belonging to this organisation, whose members may be locked out by their employers, either for requiring an advance or opposing a reduction of wages, or an encroachment on their rights or the privileges or customs of their trade, the same to be paid at 3d. or 6d. per week, as the members of the society may decide, to be paid to the proper officers of the society, who shall forward the same to the secretary of the organisation, with the number of members, and amount sent."

Mr. B. ORAN, London, seconded the motion. If it were carried out, they might depend upon it they would not have the trouble of using the fund. The brickmakers were as badly paid a people as any, and if they got this fund, they would be in such a position that no masters dare lock them out; they would be afraid of them.

Mr. WALTER SPENCER, Nottingham, quite agreed with the Secretary as to going slowly and surely at the onset; but he thought a halfpenny and a penny were too little; and as a further amendment, to carry out the object, he proposed—"That all Trade Societies connected with this Alliance shall be required to pay an entrance fee of one penny per member, instead of one halfpenny, and that an annual subscription of twopence be contributed by each member, to meet the requirements thereof, instead of one penny, such subscription to be paid half-yearly in advance."

Mr. MARSDEN, Sheffield, seconded the amendment.

Mr. W. H. WOOD was very sorry there was so much difference of opinion on the question of admission, and he should like, first of all, to ask Mr. Burns whether he would guarantee that his constituency would pay 5s. for an entrance fee on the number of members he was stated to represent?

Mr. BURNS: Yes, they will.

Mr. WOOD was sorry to say that, even with the small number he represented, he should not be placed in the same happy position. He should therefore like to lay before his constituency a reasonable proposition, and such as would meet with their acceptation. Such an amendment as Mr. Burns had proposed would act as a strong deterrent to prevemt associations from joining this proposed Alliance. In fact, if they took it as a general rule, it was twelve and a half per cent. upon the capital of the trades associations throughout the country. The great bulk of the trades' organizations, the largest and most powerful, could only command £2 per man in their trade funds, and if they took five shillings from each individual member, they took one-eighth portion of the property, of the wealth of the organizations; so that he could see it was totally impossible for the great bulk of the Trades' Unions of the country to accept such a proposition. He should therefore support the motion of the committee. They must learn to creep before they ran; and they would perceive, from a subsequent resolution of the committee, that in the event of their wanting extraordinary assistance to meet extraordinary claims, the committee had made a suitable provision, and they would entirely depend on the members composing the Association as to what extent they should go. He thought the delegates had not treated the special committee in the way they ought to have done. The last resolution passed, after exhausting a couple of hours in its discussion, was carried almost unanimously. These resolutions had been calmly and dispassionately deliberated upon in a small assembly, far more qualified to inquire into the claims of a resolution than a larger one, and it would be treating the committee with a little more respect if members would come prepared to inquire thoroughly into the questions submitted.

The CHAIRMAN said the committee all felt extremely obliged to the delegates for giving their ideas, and would be perfectly repaid if they gave the meeting satisfaction in giving the delegates every opportunity of expressing those ideas.

Mr. MEMMOTT thought Mr. Burns was too sanguine. He believed that if every member represented at that Conference had the well-being of the working classes at heart as much as Mr. Burns had, a great amount of money would at once be raised for the defence; but while he thought Mr. Burns's proposition was too strong, he also thought that the proposition of the committee was a little too weak. He considered that they might very well adopt the amendment doubling the sums named by the committee. This was a very small amount, and would raise a large sum of money. He was much obliged to the committee for the care they had exercised, and the time they had occupied in framing the resolutions.

Mr. T. SMITH quite endorsed the resolution. The scale was a very just one, and if they put a higher figure on, it would debar many from joining. They were indebted to the committee for the able manner in which they had brought out the resolutions. He wanted all societies throughout the country to be members, and to benefit each other. His trade suggested to him that the scale should be low, and then they would go hand and heart with the Conference. His trade would subscribe the amount, and if they found they wanted more means at the next Conference they would then be able to get it.

Mr. G. TROUP said the Working Men's Association, at a meeting held in St. Martin's Hall, had an intention of taking such steps as the committee had done, and for which he felt grateful to them. He hoped, however, he should not be considered to be disrespectful to the committee if he said that if the meeting adopted the resolution they had better never have met. They would be exposing themselves to the laughter of their opponents. Did they think that the employers of the country cared anything for their halfpennies and pennies? Suppose that 120,000 persons joined the Association—and he told them that they had not 120,000 persons represented there that day, or not much more—the payment would amount to £750 a year, equivalent to the cost of the Sheffield Lock-out for three days. If the committee proposed this to meet the working expenses let them say so, and put it into the resolution, and let them propose means for attaining the objects desired in another resolution, and not bring the delegates there like a mountain to produce mice. As to halfpennies and pennies, it was just impossible that the object they had in view could be accomplished in that way at all; they would be 140 Mrs. Partingtons keeping back the Atlantic with a broom. He had been engaged endeavouring to understand the dispute between wages and profits for twenty-five years, and what he thought they were met together for was, simply to extend the area of the combat. He had no objection to tell them further, that, unless they went further than the proposition, they would be beaten to a certainty. He rejoiced in the meeting as a step to the final result which he expected to see, if he lived long enough, of the working classes endeavouring to get capital of their own, and getting profits and wages themselves. He wished to tell them, that when they had found a proper association to resist Lock-outs, he should regard it chiefly of value as a means of promoting co-operation. He did not think, with one large Association, the capitalists would have a superior power to theirs; but they would not give themselves the trouble to attend to an Association on the halfpenny and penny scale. He could name capitalists who could sign a cheque to pay such sums as they were likely to raise that way for a thousand years. It was quite absurd to think of accomplishing the matter in that way. He had been quite delighted at the tone of the meeting: in many respects it was the best meeting of representative men he had ever been at. He liked its temper and conciliatory character, but he was not altogether sure that he liked its penitence. There had been many speeches like calling past sins to remembrance. Some of the speakers seemed to sing penitential psalms, as if they had committed great sins, and would commit them no more. He thought many of the working-classes should know what were the benefits and profits derived from the misery —and there was a great deal of misery—and suffering in past struggles. He wished to bring this to their remembrance before he sat down. They talked of the expense of Strikes. There was one trade represented there—8000 of them in England—and he could count to-morrow the cost to them, and the result had been an increase of £187,200 every year in the wages paid. The builders in London had a very memorable Strike; but it affected the wages by 2s. 6d. a week.

The CHAIRMAN: You cannot talk about Strikes now. Keep to the question.

MR. HUMPHREYS observed that the final gain to the building operatives amounted to £652,000 a year in London, and the gain to them in point of time was fifteen days a year. He stated this, because he wished to say one word more. He thought many Strikes were founded on good principles, had a good design and object, and had brought out a very glorious result; and it ought not to be forgotten in these latter times of conciliation that this was the case.

The CHAIRMAN: I say you are not speaking to the question.

MR. HUMPHREYS said he wished to propose an amendment, namely—"That the societies connecting themselves with this Association be required to pay at the rate of one penny weekly, or one shilling per quarter, to a Lock-out fund." This would produce, from 240,000 members, £52,000, and he thought this would not be felt to be a heavy burden, if they really were in earnest in making themselves independent of the Lock-outs. If they did not think safety, guardianship, or protection from Lock-outs was worth a penny a week, or if they could not persuade the working-men that it was worth that, they did not fear Lock-outs very much, or value safety very highly.

The CHAIRMAN explained that these speeches really related to the next resolution, and that the proposed levy was not for the purpose of opposing Lock-outs, but for working expenses.

MR. REANEY rose to speak, but was silenced by cries of "vote."

MR. THOMAS BANKS concurred in the proposition made by the committee, because, at the commencement of a new era like the present, they might depend on it there would be a difficulty in many localities in getting the resolutions adopted by their constituents. Mr. Troup had made a very serious allegation against some of the delegates present. They had put down on paper something like 200,000 members represented, but he said they could not depend on more than 120,000. If this was a fact, a levy, no matter for what object, could not be depended upon either. Mr. Troup's criticism on some speakers had been better unsaid. Whether he (the speaker) was a psalm singer or not, he was sent there to do his duty, and he would do it, irrespective of what Mr. Troup or anybody else might say. In trade matters he had had some experience, and that was the reason why he agreed with the resolution of the committee, to begin with a very low levy, and one easily adopted in all the localities. They wanted the Association forming; it was not yet formed. Though the levy was small, it would be enough at any time to check capital in any future Lock-outs, if they were thoroughly united. The trades in the various towns were bound in honour to agree to what was transacted there. If there had been such an Association a short time ago, the result at Sheffield would have been very different. Let them get the thin edge of the wedge in, and capital would be deterred from locking out in future.

MR. M. LAWRENCE wished distinctly to state that time was being wasted. How long would they be if they discussed every question at a quarter of an hour each? They would get to the end of the week, members would want to go home, and the business would not be properly finished. There were errors committed in the report of the numbers represented, owing to different sections of societies being represented, and the total number being put down to each member; but he thought the extreme number to be deducted on this account was 20,000. One great error had run all through the discussion. When the preamble was proposed, members went and discussed the first and second proposition, and on every proposition they discussed the one following. The committee had made a better provision for the support of a Lock-out than any that had come from the meeting yet; these were all for increased pay, and that would only drive men away, and make the numbers of the Association less. For every man prepared to pay 6d. a week, they would get 20,000 ready to pay the committee's terms. As Mr. Odger had said, when a Strike or Lock-out was on, the men knew what money was wanted for, and that it went to its object, and its object alone. The amount now asked for was only as a preliminary to carry the Association on. Was it possible that working men could vest funds without having the protection of the law? It was this want of legal protection they were cursed with; it made members distrustful of giving large contributions, but they knew members were always ready to contribute in time of need. If they had 200,000 men in the Union, and placed a case of Lock-out before them, it would take a very small levy, and their sympathies would come out. He believed that instead of 200,000 members, they would have half a million before the end of the year, if they went on a broad basis and a fair free principle, and that the trades' delegates would be able to assemble and represent as large a constituency as the House of Commons. It was a waste of time to discuss the raising of money for Lock-outs now; let them take it on the next resolution.

Mr. GEORGE SHIPTON hoped to represent 50,000 men by that time twelve months. He thought the resolution as it stood embodied everything they could expect to go forth to their constituencies. He was prepared to do his best with the body he represented for the resolution as it stood, and he had every confidence that when the time came for a levy, the men would be only too proud and happy to pay it.

Mr. J. BRIGGS, Accrington, said he would guarantee that if they once got this to meet the low paid trades, his 6000 would be 30,000.

Mr. W. BROADHEAD, as one of the committee, rose for the purpose of supporting the committee's proposition. He endorsed what had been so fully expressed, that they would make the Association a larger success by making the contribution low. The observation of one speaker would not only apply to the trades represented, but to the other trades throughout the country. The Committee put the resolution forth as nothing further than working expenses. He had had sufficient experience to know that the most certain means of getting aid was not to go for a large sum. If the resolution was passed, it would give such universal satisfaction throughout the country, that the Association would increase year by year, as other societies had done that had adopted a similar principle, until, in time, it would be able to meet any difficulties that might come upon it. He felt that they would make a very great mistake if the resolution was not adopted.

MR. ALLEN said it was a question whether, after all, they should not have a fund to fall back upon in case of a Lock-out. He was aware that sometimes levies were longer in coming in than they could wish them to be, and the men locked out might be in great difficulties before the money came to hand. He would suggest to the committee the advisability of looking over the resolution again, and adding that a fund of so much be left in their hands, to stand by when called upon until the returns from the various societies were got in.

The CHAIRMAN: That is the next resolution. (Cries of "Vote, vote."

MR. DRONFIELD then replied. He said he avoided saying much at first, in the hope that the delegates would see the necessity of passing the motion at once. All through, they had gone on the assumption that the payments should be as light as possible. and he was astonished that any one should propose 5s. levies. How could they get the Amalgamated Engineers to pay for each of their 30,000 members? The brickmakers might do it, but the trades generally would not. It was a waste of time to discuss questions that could not be carried out. It was a much more reasonable proposition to double the payment proposed by the committee, and for his own part he could adopt Mr. Spencer's suggestion, but having moved the committee's resolution, he intended to adhere to it. They did not want the members, in addition, to have this levy to pay; he contended that it should come from the trade fund. Executives, as a rule, would not pay even a small levy from the general fund, and the members thought it hard they should have to pay a double rate. He believed if any town in the kingdom would pay, it was Sheffield; but many would not. The committee went on the assumption that the sum asked for was for preliminary expenses, and for granting assistance while the trades paid their levies. The trades would grant money if the case was a deserving one.

The amendments were then put to the meeting and lost, the number of hands held up for each being respectively 2, 18, and 9. The original resolution was then carried *nem. dis.*

MR. NEWTON proposed the next resolution, and said that a good deal of the speaking upon it had been done already:—

"That in the event of the ordinary subscription failing to meet the requirements of the Alliance, the votes of the trades forming the same be taken as to the desirability of increasing the contributions to meet all necessary contingencies."

The resolution was by no means so full and explicit as the law based upon it would be. It was unnecessary to be so tremendously precise. He believed a considerable saving would be effected off the steady income of the committee; but Lock-outs, as a rule, were gigantic affairs, and required much money to meet them. Now, supposing they had 200,000 members, and they took the average of wages as low as £1 a week, a levy of 2½ per cent. would realize £5000 a week. It had been very properly said that in the present state of the law, it was not advisable to collect a large amount of money and leave it in the hands of persons to meet contingencies, and from past experience many trade societies were extremely jealous of having their money put to anybody's credit but their own. If a society was solvent, the money was better in their own hands than in the hands of any central body. The resolution did not contain a description of the *modus operandi* by which the money should be got in, but he thought the plan suggested by Mr. Ross, yesterday, would be a very good one. The central committee, or executive, whenever a case came to be considered, should be informed of the whole facts, the number of persons likely to require support, and the amount of money required. With regard to the proportions of payment, they could scarcely expect the men who earned fifteen shillings a week to pay as much as those who earned thirty shillings. Before any appeal was made, it would be necessary for all trades to report their wages to the executive, so that the information would be at hand, and the circumstances of the parties Locked-out, so that the executive could see at a glance what was necessary, strike the per centage, and apply it.

The motion was seconded by Mr. H. MITCHELL.

Mr. SHORROCKS very briefly supported it.

Mr. ODGER stated distinctly and advisedly that the resolution would do no good whatever. What were these contingencies? Were they contingent upon the money referred to in the third resolution not being sufficient to meet the working expenses? He could not see that it was anything else.

The CHAIRMAN: We say we shall not want half of that.

Mr. ODGER said the plain meaning of the resolution was, that if the committee wanted more money for the working expenses this resolution afforded it, and nothing else. He proposed that the resolution be handed back to the committee for further consideration, and that they prepare another resolution upon the subscription necessary to meet Lock-outs. He hoped the committee would not think for one moment that he was attaching any blame to them whatever—he thought they had an extraordinary piece of work last night, after the public meeting, to go and prepare these resolutions, and every one must be grateful to them for completing the work as far as they had gone. But here was an inconsistency when compared with the third resolution, and with the declaration that this resolution was meant to serve in obtaining funds to resist Lock-outs.

Mr. G. HOUSELEY, Sheffield, seconded the motion.

Mr. LEIGH thought that the third and fourth resolutions should have been thrown together, and the fifth ought to be—

"That in the event of a Lock-out, the executive shall, on ascertaining the requirements, call for such an amount from each trade in proportion to their numbers, as will meet the necessities of the case." He thought the resolution should go back to the committee for consideration, or some would think that the whole amount they were to pay was the halfpenny for entrance money, and a penny per annum.

THE CHAIRMAN: We will withdraw the resolution, and carry out the wish of Mr. Odger.

MR. ROSS moved as an amendment:—"That in the event of a national levy being required to meet any contingency that may arise, the votes of the trades be immediately taken thereon; such levy to be equal to the then existing emergency, and to be an equitable per centage in proportion to the wages received by each trade connected with the Alliance; and the support received by any trade in difficulty to be also in proportion to the wages received previous to such dispute arising."

It was now five o'clock, and the amendment not being seconded, and the resolution withdrawn for reconsidertion, the Conference adjourned.

FOURTH DAY.

The minutes of Thursday's deliberations were passed as read.

The resolutions of the special committee to be submitted today being likely to cause considerable discussion, it was resolved that the standing orders regulating time be altered from fifteen minutes to five minutes for each speaker, allowing movers of resolutions ten minutes to reply.

Credentials were given in from Mr. W. Crossley, of the Power Loom Weavers' Society, Blackburn, representing 5000 members, also from Mr. John Kane, of the Amalgamated Ironworkers, Gateshead, numbering 5000.

MR. G. NEWTON moved the first of the resolutions of the special committee. He said: The arrangement as to the time allowed to each speaker reminds me of an inverted pyramid; you have no time to explain your motion, and a lot of time to reply to other people's remarks, which might have been prevented. Certain suggestions were agreed to yesterday with reference to the scale of payment, and it was proposed that the fund should be organized on the per centage basis. Previous to going to their business, the committee were informed by several delegates that the principle of per centage was totally inadmissable, and some, who spoke from practical experience, said it was utterly unworkable, giving rise to dishonest practices in the returns of members, and the circumstances in which they were situated. The committee felt this difficulty, and being desirous of drawing up such resolutions only as were likely to meet with the approval of the majority of the Conference, they have drawn up the following, which I move:—

"That any surplus funds over the expenses of management be

reserved for the main objects of the Society, viz.: assistance in cases of Lock-outs, and in the event of such reserved funds being insufficient to meet the requirements of cases voted worthy of support by the Judicial Council, they shall declare what amount should be paid to the same, and submit it to the trades for approval or rejection." The proposition does not contain any specific proposal as to the amount to be subscribed. The wisdom of leaving out any such specific proposition will be apparent, when we consider that the circumstances of the Lock-out must, in every case, regulate the amount that may be required from the different societies. If you say that there shall not be more than a given sum asked for from the societies, then you circumscribe yourselves, and place it beyond your power to render that assistance which may be necessary under the circumstances. It becomes a matter of grave importance that the judicial committee shall be men of sound judgment, in whose probity the bulk of the trades have the greatest confidence, and who will be likely to secure that co-operation which is necessary to success. It may be well, I think, here to explain that the *modus operandi* which the committee have in view is this, that in any case of Lock-out occurring, immediate application be made to the executive committee, who shall convene the Council, and take the whole matter into consideration, and decide whether the case is one worthy of support, and if it is, to draw up a statement of the requirements of the parties, and send it forth to the societies connected with the Association to afford the assistance required. (Hear, hear.)

MR. T. P. BAILEY seconded the proposition.

MR. P. SHORROCKS proposed the following as an amendment:—"That any surplus fund over the expenses of management be reserved for the main objects of the society, namely, assistance in case of Lock-outs, and that, for the purpose of meeting any extraordinary expenditure, the Judicial Council shall propose a levy equally upon all members comprising the Association, subject to the decision of the majority of the societies comprising the amalgamation, all such levies to be paid weekly." This may call forth an amendment for assessment according to the wages earned; but I think that, considering how many trades are constituted, this would be altogether unfair. Although I represent Manchester, I am connected with the Executive Council of the Amalgamation, and there are towns that pay from 3½d. per hour upwards. Different wages are paid because of the difference in cloth. It would be next to impossible to average the wages in my trade.

MR. HIGGINBOTHAM seconded the amendment. The system of poundage on earnings was a very unfair one, resulting in unjustly taxing the industrious and frugal man, and lightening the burden of the idle and shuffling.

The VICE-CHAIRMAN: The amendment is perfectly unnecessary, and the object aimed at by it may be accomplished by adopting the original proposition. Every man who really wishes for equality of payments and receipts should vote for the proposition, and that will effect his purpose; but the meeting ought not to be hampered with special resolutions without regard to the matter. Let us try to confine the discussion to the real principle, and not make mere verbal alterations. We can all write the same meaning in different words, and that is all they are trying to do.

MR. HARDY, MR. RISHWORTH, and MR. WILLIAMS having addressed the meeting,

MR. ALLEN proposed the insertion of the following words:—"That all members of this Association be placed on the basis of equality, both as to payments and receipts."

MR. E. GREEN thought that the original proposition was all that reasonable men could require. Nothing could be more plain and definite.

MR. BOND moved that a fixed sum of 2s. a year be paid by each member.

MR. WEST was afraid that the last clause of the resolution placed too great a restriction on the power of the executive. There ought to be power in the hands of the executive to call at once for support from the various societies, so that prompt action might be taken. He would move that the resolution should be "That any surplus fund, over the expenses of management, be reserved for the main objects of the society, and, in the event of a Lock-out taking place, the Council shall have power to make a levy upon all societies in connection with the Alliance, of not more than 6d. per member, each call to be paid immediately."

MR. W. SPENCER seconded the amendment.

MR. DAVIES supported the original motion.

MR. W. H. WOOD: The speakers generally seem to anticipate a Lock-out. My idea is, that when we become strongly organised, Lock-outs will cease to exist. Lock outs can only take place where there is a weak organisation. The existence of a powerful organisation will prevent Lock-outs altogether, and the provision which you made, yesterday, for the formation of a power to draw up the rules and settle the ordinary expenses of the Alliance, will be all that will be required for the present. When we have a Lock-out to deal with, then it will be the duty of the Council to ascertain the requirements of the Lock-outs, and to place before the members the amount required to afford the requisite assistance. Our object in not naming the amount is to simplify the matter in the first instance. It would be a total impossibility to exact subscriptions according to the rate of wages. It is impossible to obtain correct returns, and, therefore, we must go upon the common footing for the obtaining of subscriptions from all the members.

MR. ODGER: The resolution does not appear to me to be workable. In the first place, the judicial council considers what sum will be required to support the Lock-outs. That, according to the proposition, will have to be submitted to the trades for approval or rejection. Our object is to afford prompt support in case of Lock-outs; but how can it be prompt if the trades reject the sum submitted? (Hear, hear.) In that case the executive will have to consider the matter again, and send out another proposition. That would be almost as bad as the existing state of things. I would move as an amendment. "That any surplus fund over the expenses of management be reserved for the main object of the society, namely, assistance in cases of Lock-out, and that a subscription of one penny a month be paid, these funds to be held by the societies to assist cases approved of by them." A penny a month, surely, will not be an extravagant amount. If we cannot pay a farthing a week to resist these terrible Lock-outs that cost societies thousands of pounds, it is not worth our while meeting here to-day.

MR. T. SWEETING seconded the amendment, and

MR. J. CARLISLE supported it, though in his estimation it scarcely went far enough.

MR. THOMPSON said there must be money at the disposal of the executive, and Mr. Odger's amendment would provide all that was required at present.

MR. JONES also spoke in support of the amendment.

MR. AUSTIN suggested that the amendment should be in the following form:—"That any surplus funds over the expenses of management be reserved for the main objects of the society, viz., assistance in cases of Lock-out; and in the event of such reserved fund being insufficient, they shall declare what amount shall be paid to the same by the various trades in connection with the Alliance."

MR. STANSFIELD moved—"That the Conference strongly recommends to the members of all societies which may join this Alliance the desirableness of their setting apart a sum equal to one penny per month, to be held by each society under its own control, but at the service of this Alliance, to meet the difficulties contemplated."

After considerable discussion on the foregoing motion, amendments, and suggestions,

MR. NEWTON replied. He said, nothing has given me greater pleasure than to listen to the debate. Sound argument and earnestness have characterized the various speakers, and there has been indicated a strong desire to arrive at the best mode of conducting the operations of the Association, which is a very good guarantee I think that good results will follow. The various propositions which have been submitted, and which led to considerable confusion, don't differ very materially from this submitted to the Conference by the committee. I cannot accept Mr. Shorrock's amendment, because it refers to "extraordinary expenses." I suppose he refers to the expenses of Lock-outs, and these are not extraordinary but the ordinary expenses of the Association. The committee feel unable to accept the proposition that the levy shall be on the per centage principle, their idea being that it should be equal upon all. (Hear, hear.) Mr. Odger proposes that in place of a special call by the judicial council, there shall be a regular payment of one penny per month. If that were adopted, the operations of the council would be confined to a penny a month, because there is no provision for an increase of the amount. It is, perhaps, as well that this matter, having been so extensively debated, should be left to the trades themselves for final determination. There must be some mistake about the proposition to subscribe according to the average wage. No such statement was made, but the intention is that the trades themselves shall report to the executive committee what is their average wage, and upon that they are to be assessed. There have been many arguments in favour of this, but the committee are assured that it would be repugnant to the feelings of the great body of the members. We have heard the absurd dogma, that "the best way to secure peace is to be prepared for war." This has been denounced by the wisest heads, and it is our duty to give a flat denial to such a doctrine. The best way to preserve peace is to conduct yourselves peaceably, and to menace nobody with your preparations. Was it the fleet of Great Britain or her fortifications that guard our shores, or the standing army, that prevented the threatened invasion from the other side of the Channel? Was it not the outburst of popular enthusiasm that brought out the volunteer corps which prevented and totally snuffed out the idea? (Applause.) The creation of a reserve force similar to that, which is no menace to anybody at all, will be sufficient to deter those who would impose upon us. The adoption of Mr. Austin's proposition would, no doubt, strengthen the hands of the judicial council and the executive very much indeed, but I fear it would be objectionable to many trades. I think no harm would come to our proposition if we accepted the suggestion of Mr. Banks that the last few words be taken out. If it be the desire of the meeting that they should be struck out, I am instructed to say that the committee will offer no objection. (Hear, hear.) It is argued that unless a large reserve fund be kept in the hands of the executive committee, it will be impossible to render effective aid in the time of disaster, and that the circumlocutory proceedings will take so much time, that suffering may occur before the fund can be forthcoming. I am not aware that any Lock-out has taken place in a day, or a week, or a fortnight, and it will be the business of the committee so to take time by the forelock that all these matters will be looked into, and the whole reserve force set in motion and ready to act the moment it is required.

On being put to the vote, all the amendments were lost, the original proposition being carried by a large majority.

THE GOVERNMENT OF THE ALLIANCE.

Mr. G. NEWTON: I bring forward at once the remaining business with reference to our future organization. It has reference to the Government of the Association. I think we have taken all care to explain the general scope of the functions of the three constituent elements of government, and it does not seem to me that the remaining propositions require very much discussion. I move—

"That for the Government of this Alliance there shall be an Annual Conference, Deliberative and Legislative, with power of reviewing the proceedings of the past year; also, a Judicial Council to consider all Lock-outs in Trades in connection with the Alliance which may be submitted to the Executive, and determine what action should be taken in reference thereto, together with an Executive Committee to carry out the resolutions of the Annual Conference and of the Judicial Committee, and to superintend the affairs of the Society generally."

Mr. A. RIDDELL seconded the proposition, and after some conversation it was adopted.

APPOINTMENT OF A COMMITTEE.

Mr. NEWTON moved—

"That a committee be appointed by this Conference to draw up a code of rules for the management of the Association; and that the committee call a Conference as early in 1867 as possible, for the purpose of finally adopting such rules, and inaugurating the Alliance."

Mr. RIDDELL seconded the motion.

It was decided that the committee should consist of seven persons, and with that alteration the proposition was adopted.

A conversation took place as to the advisableness of at once appointing the committee, and it was ultimately resolved—"That the subject of the appointment of seven members of the committee from seven distinct localities be referred to the special committee, such number not to include the Chairman, Treasurer, and Secretary."

It was resolved, on the motion of Mr. M. LAWRENCE, seconded by Mr. G. ROSS—"That the seat of government be held in Sheffield until the first meeting of the Conference, and that the executive be appointed from seven industrial districts of the country." It was further resolved—

"That Mr. C. Bagshaw, of Sheffield, be appointed to act as Chairman."

"That Mr. W. Dronfield, of Sheffield, be appointed Secretary."

"That Mr. W. Broadhead, of Sheffield, be elected Treasurer."

Mr. W. HARRY moved—

"That this Conference do most earnestly recommend to the various societies the necessity of considering the following:—That should an endeavour be made by any Trade Society joining this Alliance to increase their wages, or to resist an attempt on the part of the employers to reduce them, they shall, subject to the approval of the executive committee, be entitled to all the moral and pecuniary support of this Association."

When they reached home, they would ask themselves what position they occupied. If they left things as they were, in the event of a great and mighty Strike taking place in any part of England, the Alliance would not be able to use one single halfpenny out of the funds subscribed. At present they were pledged only to support Lock-outs. The majority of trades suffered from Strikes which were brought about by attempts on the part of employers to reduce wages.

Mr. AUSTIN seconded the proposition.

Mr. TODD, Mr. KANE, and Mr. SHIPLEY supported.

MR. ROSS suggested that it should be withdrawn. They had met to consider Lock-outs alone.

The proposition was adopted.

It was resolved, on the motion of MR. ODGER, seconded by MR. FOX—

"That this Conference, although having declared its opinion in favour of paying to the support of the Association by levy, deems it judicious to send the amendment regarding fixed contributions to the societies for their consideration."

The special committee having given in their financial statement, which amounted to £35 4s. 1d., it was resolved that the amount be obtained by a levy of one shilling per hundred members on the various societies represented, which, it was calculated, would cover all the expenses incurred up to publishing the full report of the proceedings.

RESOLVED,—"That the Special Committee, together with the Door-keeper, receive the sum of £1 each, and that the Assistant Secretary receive in addition for his services, £1 extra."

"Also, that the sum of £1 be granted to each of the six Members of the Committee who undertook the promotion of this Conference."

"Levies to be paid at once by all prepared. Those who are not prepared, to pay within one month."

FIFTH DAY.

The minutes of Friday's proceedings were confirmed.

In accordance with the resolution of yesterday's Conference, the special committee recommended the following seven delegates to form the committee to draw up rules, &c., for the government of the Alliance:—

MR. J. ALLEN, Boiler Makers and Iron Ship Builders	9000	members.
„ THOS. BANKS, Preston Associated Trades	1682	„
„ M. LAWRENCE, London Operative Tailors	7000	„
„ JOHN NORMANSELL, South Yorkshire Miners	4000	„
„ GEO. ROSS, Associated Carpenters and Joiners of Scotland	5000	„
„ C. WILLIAMS, Operative Plasterers' Association	5000	„
„ W. H. WOOD, Provincial Typographical Association	2500	„
CHAIRMAN, TREASURER, and SECRETARY, Sheffield	5000	„
Number of members represented..	39,182	

MR. GEO. NEWTON moved the adoption of the list. The matter had been considered in various aspects, with a view to the proper industrial division of the country. The small number of seven totally precluded the possibility of having the committee organised on a scale that might be called just in respect to the representation of the parties present. The Chairman, Secretary, and Treasurer represented 5000 members, so that, taking the division of the parties over the country, the committee will be found to be pretty fair. Messrs. Allen, Banks, and Williams were all located in Lancashire, that great hive of industry in the country. There were about two millions of people there, all pursuing industrial occupations, and no other part of the country could claim such a number. Mr. Lawrence would represent London; Mr. Ross would represent that part of Her Majesty's dominions called Scotland; and Mr. Wood would represent the important city of Manchester. The Conference would regret to hear that Mr. Banks wished his name to be struck out; but the sound judgment he had shown was a sufficient warrant for the retention of his name.

Mr. J. HIGGINBOTHAM seconded the proposition.

Mr. MITCHELL moved that Mr. Humphreys be substituted for Mr. Allen.

Mr. FOX seconded the amendment.

Mr. FINLINSON moved the substitution of Mr. Kane's name for that of Mr. Williams.

Mr. ROSS wished his name to be struck out, and that of Mr. Newton to be put in its place.

Mr. JONES moved that Mr. Thomas Sweeting be elected in the place of Mr. Allen.

After considerable discussion,

Mr. NEWTON replied. He observed: All that had been said in opposition to the list was stated by him at the outset. There were only seven members, and it was perfectly impossible for all the trades to be represented. They wanted men who had a knowledge of the working of societies and trades' disputes, and who were capable of drawing up rules to meet the necessities of the case, men who had given their attention for years to the matter, who were thoroughly known, and not likely to allow any interest to swerve them in the framing of rules; men who had the objects of the Conference at heart, and who would do all they could to make the Association as wide spread as possible.

All the amendments were lost, and the list proposed by Mr. Newton was adopted.

It was resolved, on the motion of Mr. CARLISLE, seconded by Mr. LAWRENCE, "That all Societies be permitted to send in suggestions to the committee respecting the formation of rules for the government of the Alliance."

THE NEXT CONFERENCE.

Mr. G. NEWTON moved, and Mr. JONES seconded, "that the next Conference be held in Manchester."

Mr. G. SHIPTON moved, and Mr. CROSSLEY seconded, "that it be held in London."

Mr. CARLISLE moved—"That the next Conference of this Alliance commence its sittings in Manchester, on the first Monday in September next, and that no person be allowed to have either voice or vote in any question brought before the

Alliance who is not a *bona fide* representative of a trade society."

After some discussion,

MR. NEWTON, in reply, said: It is undeniable that London is the first city of the empire, and that there are a great many workmen there; and it is also undeniable that comparatively few are in union. When we go there, let us go with a compact, strong organisation. (Hear, hear.) There are very few great associations that hold their first or their second meetings in London. They generally allow a year or two to elapse, and finding their strength, they go there as a reality. The London people have a wonderful faculty for discovering that a great many things are not real; let this Association be in such a position that it will be impossible to conceive such an idea, and if weight is wanted to induce the non-unionists to join, a solid compact Association will be more likely to have that influence than one newly fledged. The time suggested for holding the Conference is out of all reason. I don't know where our iron friends have learned the knack of making resolutions, but they must have a wonderful facility. I could not do it in the time if I had it all to myself, without any discussion at all.

The amendments were lost, and the original proposition adopted.

COURTS OF CONCILIATION OR ARBITRATION.

MR. W. DRONFIELD: I have very great pleasure in introducing the next business to the Conference, inasmuch as it will give me an opportunity of expressing not only a few opinions of my own, and the action of the Trades' Societies I represent, but also of making a few remarks on what has been done by a very important body—the National Association for the Promotion of Social Science. The bill that was before the House of Lords was introduced by Lord St. Leonards in May, 1865. That bill, so far as I have been able to understand it, and so far as I have been able to gather from the discussions that have taken place upon it, was a very simple and unpretending bill; but such as it was, Lord St. Leonards felt compelled to withdraw it for the present session, owing either to the opposition of employers, or to their not supporting it. Lord St. Leonards found, on the contrary, that the working-men generally were in favour of the establishment of some such courts; and I think it a somewhat significant fact, that should not be lost sight of by the representatives of labour, that the employers, as a rule, are opposed to the establishment of courts that would have a tendency to put a stop to strikes, whilst the men desire to get such courts established. To my mind, that gives the lie direct to those who charge us with being interested in strikes and lock-outs. (Hear, hear.) The bill of Lord St. Leonards provided that the courts should be of a permissive nature. In 1860, the Association of Organised Trades of Sheffield took an active part in promoting the passing of Mr. Mackinnon's bill, which passed the House of Commons, but was rejected in the House of Lords mainly, I am told, through the influence of the landed proprietors and the opposition of the manufacturers, as expressed through the Chambers of Commerce. At the meeting of the Social Science Congress, held in Sheffield in October last, a paper was read, entitled "What are the best means of establishing a system of authoritative arbitration between employers and employed in cases of strikes and lock-outs?" That question, unfortunately, when it came on for discussion, turned out to be a very secondary affair indeed, the writer devoting the greater portion of his paper to arguments against Trades' Societies. Mr. Thomas Hughes, the member for Lambeth, had been put down for a special paper on that subject, but, unfortunately, he had not been informed that he was to read such a paper, and when the question came before the department (Economy and Trade), Mr. Hughes was not present, and did not hear the paper I have alluded to, or the one by myself. Professor Fawcett took a lively part in the discussion. He gave some very sound advice, and I believe that he is in favour of the establishment of Courts of Arbitration; but he wishes chiefly to promote a better feeling between employers and employed. Mr. Hughes declared that he was not aware he had been put down to read a paper on the subject, and expressed himself to the following effect:—"Of course, there is nobody who will deny—at least I suppose so—that arbitration is in itself a far better thing than war. If you can find any tribunal that will be satisfactory to the two sides—to the masters and to the men—and if you can manage to give to it some power of enforcing its awards on such matters as are referred to it—if you can find such a tribunal, it would be a great advantage, for no man in his senses, or, at any rate no man in his senses who had not some personal interest to lead him wrong, would deny that it was far better that trade, questions should be submitted to such a tribunal than that they should be fought out to the end, between masters and men, by the only means by which it is possible they can be fought at present, by strikes and lock-outs." I hope this Conference will fully endorse the sentiments of Mr. Hughes on that point. (Hear, hear.) Professor Fawcett said, "You may attempt to establish arbitration, where the decision of the court will be compulsory; but then the difficulty is, to get the men to be bound by that decision." As far as I know, the men are quite willing to go in favour, and have gone in favour, of the establishment of Courts of Arbitration. (Hear, hear.) It so happens that the gentleman who read the paper on the other side of the question was decidedly opposed to the establishment of Courts of Arbitration, contending that, if established, they would be ineffective for the purposes aimed at, and would not be a success in other ways. I do not intend to read you much more on the question, but would call your attention to one significant fact—that whilst the paper I read on the question of Trades' Societies, which was really in defence of them, is omitted from the authorized report, the paper on the opposite side (the writer of which is opposed to Courts of Conciliation, and also to Trades' Unions,) occupies several pages of the report in question, a great portion of which is devoted to arguments against Trades' Societies. An important discussion followed the reading of the papers, in which several of the representatives of the trades of Sheffield took part, yet not a word they said is recorded in the official volume of the Society. Do you think it is fair on the part of gentlemen professing to seek truth, that, when they issue their report, they should give a one-sided version? ("No.") It is a most disgraceful thing, and I hope the trades of the country will not forget the matter. (Hear, hear.) Hitherto, I have had a very strong impression that it was the duty of the working classes to take part in such proceedings as those of the Social Science Association, but if we cannot get justice done to us—if we cannot get our views represented—if, when we express ourselves, either by writing papers, or in attempting to reply to the attacks made upon us, we are ignored in the official documents of the Society, it is no use our spending our time and our money, or the money of our constituents, in going to such places. (Hear, hear.) As to Courts of Conciliation, I think the Delegates would do well to urge upon their members to take every opportunity that presents itself, to bring the question fairly and fully before the House of Commons, through the medium of the representatives of the various constituencies in Parliament. We had a long discussion with our Borough Members on this subject, and I hope you will give your attention to it, and your support to any bill for the purpose of establishing Courts of Arbitration, fully believing, as I said in my paper, that the establishment of such courts would tend to lessen the number of strikes in the country, a thing which we, as the representatives of labour, are most deeply interested in; and I hope we shall not be deterred by trifles, but use every legitimate means of obtaining the establishment of a court that will give us a fair representation, and a medium for the settlement of those unfortunate disputes that are so frequently taking place. (Applause.) In conclusion I beg to move—

"That this Conference highly approves of the establishment of Courts of Conciliation or Arbitration, and exceedingly regrets to find that, owing to the opposition of the employers, Lord St. Leonards has withdrawn his bill for the present session; and the Delegates present hereby pledge themselves to use every exertion to obtain the establishment of such courts."

Mr. G. ODGER, in seconding the motion said: There is a little misapprehension existing with regard to the actual state of the feeling that exists with the masters on this question. Lord St. Leonards, in a letter to me, gives us good hopes that Conciliation or Arbitration Courts will ultimately be established. It has been represented that Lord St. Leonards said in his letter that the employers were not favourable to such a bill. That is not true. What he said was, that the employers were not favourable to the bill then before the house, which is a marked difference. It is only fair that the employers should have credit for what they do that is creditable. I am not sorry that the employers opposed that bill, and for this reason, that it was a very small one, merely giving power for these Courts of Conciliation to decide disputes upon wages already determined, but in no sense whatever to interfere with wages not determined. That you will admit was very simple, not at all embodying the great principle we desired to see taken up by these Courts when established. In his letter to me, Lord St. Leonards says, after speaking about the want of support for his bill, "I have ascertained that both masters and operatives are anxious for a Court which could settle future wages." Now you see here is the great principle admitted, which is far preferable to the previous bill, and I have no doubt his Lordship will be disposed to take the matter up again in order to effect a beneficial settlement of it. He further goes on to particularize what he thinks would be the mischievous effects of attempting to establish such Courts. He thinks that when there was a great demand for labour the men would make an effort to raise their wages, and if they were not able to do so, he fears they would violate the law, and that much mishief would follow. He suggests that they might be established for a limited time, so that if the men found they were not so conducive to their interests as they were led to hope, they would be enabled when the time came to withdraw them. He also suggests twelve months duration, and that in the event of their being established, no trade should be compelled to have them, except on the application of the working-men themselves, and the mode of proceeding would be as follows:—Notice of the application would have to be inserted in the papers circulating in the town in which the trade existed; voting papers would then have to be sent to the various workshops, so that the workmen might say whether they would have such Courts or no, and then a public meeting would have to be held, at which the result of the votes would be given. In the old bill, the Courts once established would be permanently established, and we could not get rid of them; but by this they could be tried, and if they were liked, retained, if not, abandoned. I believe Lord St. Leonards to be so earnest in the matter that he will be disposed to frame another bill of a more comprehensive character, with every probability of success. Some of the leading members of the House of Commons looked upon the bill with suspicion, and thought it likely to produce dissatisfaction and discord instead of harmony and good will. With these views, they did right not to push the matter too far; but finding that the Courts are to be established on a principle not committing the men to them for more than a given period, it is very likely that those members who did not look upon them with satisfaction before, will now be disposed to give their entire concurrence. I have no doubt this Conference, met to consider the great difficulty of Lock-outs, will have its influence, and that the masters, finding that the mishievous element is becoming more mischievous on account of its dimensions, will see the necessity of something of this sort being done. I believe, also, that by getting these Courts established, this Alliance will be able to watch over every interest connected with its members, and that instead of discord we shall be instrumental in promoting peace and goodwill.

The proposition was adopted without discussion.

It was resloved—

"That the remaining resolutions on the business paper be simply proposed and seconded, no opening remarks being made by either mover or seconder."

This was adopted in order to close the proceedings of the Conference by one o'clock.

The following resolutions were unanimously adopted without discussion:—

Moved by Mr. G. NEWTON, and seconded by Mr. J. CARLISLE—

"That this Conference regards the statutes which empower Justices of the Peace to send workmen to prison with hard labour, for breach of contract of service, as a grievous injustice, and wholly unsuited to the present state of society, and resolve to use their best endeavours to procure the repeal or amendment of the same by subscriptions, petitions to Parliament, and such other means as may be requisite."

Moved by Mr. W. H. WOOD, and seconded by Mr. KANE—

"That this Conference recommend every trade represented to encourage, and, if possible, adopt the principles of Co-operation, with a view of enabling working-men to obtain a more equal share of the benefits resulting from their labours."

Moved by Mr. J. CONSTANTINE, and seconded by Mr. W. HARRY—

"That this Conference, fully appreciating the efforts made by the International Association, to unite in one common bond of brotherhood the working-men of all countries, most earnestly recommend to the various societies here represented the advisability of becoming affiliated to that body, believing that it is essential to the progress and prosperity of the entire working community."

Moved by Mr. G. ROSS, and seconded by Mr. G. FOX—

"That the cordial thanks of this Conference be tendered to the local daily press, for the full and impartial reports of the proceedings of the Conference."

Moved by Mr. G. TROUP, and seconded by Mr. J. CARLISLE—

"That this Conference deeply regrets the omission of a fair report pf the arguments and statements of the Members and Friends of Trades' Societies in the Transactions of the Social Science Congress, recently issued."

A resolution, proposed by Mr. G. ODGER, and seconded by Mr. W. HARRY, was passed, to the effect that the Conference urge all Trade Societies to be cautious in making demands for increased wages, until the commercial panic has passed away, and the war on the Continent has somewhat abated.

A further resolution was also adopted, moved by Mr. G. ODGER, seconded by Mr. R. BAILEY, recommending the trades of the country to assist the Sheffield File Trade in paying off the heavy debt incurred during their late Lock-out.

Votes of thanks to the Chairman, Vice-Chairman, Secretary, Assistant-Secretary, and other officers of the Conference, brought the proceedings to a close.

LIST OF DELEGATES, TRADES REPRESENTED, &c.

SHEFFIELD.

Association of Organised Trades (5000 members): Charles Bagshaw, Spring street; William Dronfield, Princess street, Moor; George Austin, Oaks' green, Attercliffe; Henry Cutts, 11, Cambridge street; George Fox, 32, St. Thomas street; Joseph Mallinson, Stannington; including the following Trades belonging to the Association, with their special Delegates :—

Bone Scale Cutters: John Gill, Compass Inn, Orchard street;

British Plate Filers: J. Sanderson, Dove & Rainbow, Hartshead;

Cabinet Case Makers: Mr. Jackson, 95, Arundel street;

Common Scythe Makers: Peter Bingham, Albion Inn, Little Sheffield;

Copperplate and Lithographic Printers: A. McDougall, Red House Inn, Fargate;

Edge Tool Forgers: James Reaney, Henry Mitchell, Grapes Inn, Trippet lane;

Edge Tool and Wool Shear Grinders: Samuel Stacey, J. Higginbotham, J. Turner, Grapes Inn, Trippet lane;

File Grinders: Joseph Rolley, James Rippon, Moseley Arms, Westbar;

File Hardeners: E. Memmott, J. Smith, Three Tuns Inn, Orchard street;

Filesmiths: T. Holland, E. Grayson, H. Howe, I. Chatterton, W. L. Todd, G. Martin, R. Brownhill, 11, Cambridge street;

Fork Makers: H. Richardson, 68, New Hereford street;

Joiners' Tool and Brace Bit Makers: James Cann, J. Parker, Royal George Hotel, Carver street;

Railway Spring Makers: J. Hardy, Robert Marsden, J. Rishworth, White Lion Inn, Wicker;

Razor Blade Grinders: Henry Womack, Thomas Gledhill, Star Inn, Gibraltar street;

Razor Scale Pressers: J. Vickers, T. Shaw, Star Inn, Gibraltar street;

Saw Grinders: Wm. Broadhead, W. Peace, Royal George Hotel, Carver street;

Scissor Forgers: Joseph Thompson, Star Inn, Gibraltar street;

Scissor Grinders: Joe Gale, 135, Milton street; R. Holmshaw, Star Inn, Gibraltar street;

Surgical Instrument Makers: Mr. Widdowson, Boot and Shoe Inn, Pinstone street;

Work Board Hands, Scissor Trade: Wm. Simpson, Star Inn, Gibraltar street.

Bricklayers: G. Housley, Spring Place, Marshall street, or Black Lion, Snig hill 300
Brickmakers. W. Owen, E. Elms, Nags' Head Inn, Old Haymarket 200
Cordwainers: W. Bishop, Dog and Partridge, Trippet lane.. 50
Operative Millers' Friendly Society: John Allen, Stannington
Pen and Pocket Blade Forgers: W. J. Holden, — Harrington, Black Lion, Snig hill 280
Pen and Pocket Blade Grinders, John Broomhead, Samuel Spooner, Black Lion, Snig hill 560
Saw Makers: Thomas Smith, Moseley Arms, Westbar.... 400
Spring Knife Cutlers: Peter Hall, Moseley Arms, Westbar; Edwin Green, 28, Brunswick road 1800
Stove Grate and Fender Grinders: E. Platts, King William Inn, Russell street.............................. 40
Table Blade Forgers: Joshua Walker, B. Furniss, Burns' Tavern, Townhead street 500
Table Blade Grinders: Jonathan Wragg, Robert Lowe, Thomas Roberts, Moseley Arms, Westbar 450
United Carpenters' and Joiners' No. 1 Lodge: Thos. Hinchcliffe, 110, Wellington street, or White Lion, Barker Pool ....
Wire Drawers: H. Bond, 23, Bramber street, Occupation road 86

OTHER PARTS OF THE COUNTRY.

Alliance Cabinet Makers' Society, London: Mr. Mayle, Alliance Hall, Old street road.............................. 1030
Amalgamated Carpenters and Joiners: W. Harry, 7, Richard place, First street, Chelsea, (London General Office, 8, Northumberland street, Strand, W.C.)................. 7000
Amalgamated Iron Workers: John Kane, 36, Ellison street, Gateshead 5000
Amalgamated Journeymen Tailors' Association, Manchester: A. Stansfield, Old Green Man, New Church street, Sheffield 8000
Amalgamated Trades of Preston: W. Lindsey, Wellington Inn, Glovers' Court; Thomas Banks, 4, Edgar street; and Alfred Bailey, 130 North road 1682
Associated Carpenters and Joiners of Scotland: George Ross, 9, Scotia street, Glasgow 5000
Associated Trades: John Davies and Wm. Leigh, Hyde.. 1400
Association of Organised Trades, Derby: William Brown, London road, and John Sharpe, 15, Quarm street.......... 700
Barrow-in-Furness Brickmakers: J. Hunter 133
Bleachers' and Trimmers' Society: Samuel Meakin, Lincoln Terrace, Old Basford, near Nottingham 160
Boiler Makers and Iron Ship Builders: J. Allen, 5, Camden street, London road, Liverpool; P. Jones, Birkenhead; Thos. Sweeting, London; A. Rogers, Wallsend, near Newcastle-on-Tyne; A. Riddell, Glasgow.......................... 9000
Brass Founders: Richard Yates, Manchester............ 160

75

Brickmakers' Society, London: William Burns, 12, North street, Whitechapel road, and B. Oran 1000
Bristol Associated Trades: John Kingdom, St. John's Bridge, Bristol 1265
Carpenters and Joiners: Joseph Head, Bowling Green Hotel, Greengate, Barrow-in-Furness,; Charles Peach, Bull's Head, Queen street, Derby 404
Carpenters' General Union. John Teasdill, 3, Fleece square, Meadow lane, Leeds 65
Coopers' National Association: Thomas Metcalf, Victoria Crescent, Horninglow road, Burton-on-Trent; James Crawford, 88, Rose street, South side, Glasgow; and Andrew Nicoll, 239, Attercliffe road, Sheffield 3000
Deal Cabinet Makers, London: W. H. Wire, 28, Northampton street, Essex road, Islington 400
Derbyshire and Nottinghamshire Miners: William Ball, Wood house, near Sheffield................................ 2000
East Lancashire Power Loom Weavers: Thomas Birtwistle, 11, Free Trade Terrace, and James Briggs, Accrington.. 6000
Engineers, Machinists, and Millwrights: John Constantine, 8, Brunswick street, Halifax 290
Flat Pressers: R. Joynson, 13, Marsh street, Hanley 300
Flat Pressers' Society: Henry Pope, Burslem............ 80
Friendly Society of Ironfounders of England, Ireland, and Wales: W. H. Markham, (London General Office, 200, New Kent road, S.E.).... 10,669
Glass Bottle Makers of Yorkshire Trade Protection Society: W. Lindsay, High Town, Castleford.................. 700
Halifax Trades' Council: W. H. Hey and Joseph Atkins, Moorfields, Skircoats 700
Hollow Ware Pressers: J. J. Pope, 94, Well street, Hanley, Staffordshire 700
Hose, Shirt, and Drawers' Society, Leicester: T. P. Bailey, Red Cow Inn, 142, Belgrave gate 1500
International Association of Working-men: George Odger and M. Lawrence, London 12,000
Iron Workers: James Carlisle, Mill street, Brierley hill, Staffordshire 3500
Joiners' Society: Richard Bowles, 151, Upper Kent street, Leicester 240
Lancashire Cotton Spinners: Thomas Maudsley, 57, Jersey street, Ancoats lane, Manchester 4000
London Trades' Council: George Odger, 115, St. Martin's lane 60,000
Manchester Tailors' Society: P. Shorrocks, 6, New York street, Ardwick.. 800
Manchester Typographical Society: W. H. Wood, 29, Water street, Manchester.
Masters' and Workmen's Act Executive: G. Newton, 125, Castle street, Glasgow.
Nottingham Joiners' Society: William Hicking, 7, Curzon street 300

76

Nottingham Operative Lace Makers: Wm. Willbond, Outgoing lane, Alfreton road, Radford, Nottingham............ 1000
Operative Potters' Society: George Newton, 125, Castle street, Glasgow.. 500
Operative Tailors' Association, London: Matthew Lawrence, 70 Davies street, Berkeley square, London, W 7000
Organised Trades of Nottingham: T. West, Trinity street 1100
Ovenmen and Kilnmen, Potteries: W. Gilmore, 22, Navigation street, Burslem, and Joseph Adams, Railway Terrace, East Vale, Longton, Staffordshire 1000
Painters' and Plumbers' Society: John Rigg, 38, Clive street, Barrow-in Furness.. 30
Painters' General Council, London: George Shipton, The Merlins' Cave, Margaret street, Cavendish square 4000
Painters' Society, Nottingham: Walter Spencer, 5, Lark Dale Terrace, Goldsmith street 108
Plasterers' National Association: Charles Williams, 35, Doulby street, London road, Liverpool 5000
Power Loom Weavers: W. Crossley, Back lane, Blackburn 5000
Provincial Typographical Association: Henry Roberts, 30, Bretherton terrace, Peru street, Salford, Manchester 2000
Rotherham and Masbro' Stove Grate Fitters: Joseph Norburn, Dalton Parva, near Rotherham 20
Self-Acting Spinners and Minders: C. Shepley, Bolton .. 400
Small Wire Drawers: John Shaw, Halifax.................. 146
South Yorkshire Miners: John Normansell and Philip Casey, Nelson street, Barnsley 4000
Tailors, Shoemakers, Bricklayers, and Plasterers: William Finlinson and Samuel Heydock, Victoria Inn, Church street, Barrow-in-Furness 225
Thick Iron and Steel Wire Drawers: Ely Dyson, Manchester 100
United Operative Brickmakers of Nottingham and District: C. Oakland.. 110
United Trades' Protection Association: J. Lewis, Liverpool 3000
United Trades of Warrington: John Ashton 642
Warrington Filesmiths: Mr. Clare 700
West Yorkshire Miners' Association: John Dixon, Drighlington, near Leeds 2500
Wolverhampton Trades' Council: Joseph Humphreys, Chapel street, Dudley road 509
Working-men's Association, London: G. Troup, *Bee Hive* Office, Bolt Court, Fleet street, E.C.......................... 400

CHARLES BAGSHAW, CHAIRMAN, Sheffield.
WILLIAM BROADHEAD, TREASURER, Sheffield.
WILLIAM DRONFIELD, SECRETARY, Sheffield.

☞ *All Communications to be addressed to the Secretary, Princess street, Moor, Sheffield, and Post-Office Orders made payable to the Treasurer.*

LOXLEY BROTHERS, PRINTERS, FARGATE, SHEFFIELD.

THE TAILORS' STRIKE AND THE SYSTEM OF PICKETING.

The Decision of the Judge at the Central Criminal Court, August 21st.

The following defendants who had been convicted of conspiring to injure and molest by means of picketing were then placed at the bar. They were George Druitt, Matthew Lawrence, John Adamson, Samuel Partridge, Alfred Bayley, William Glinden, Zolu Maurice, Robert Knox, Edward Ormerod, Alfred Shermer, Thomas Geary, Thomas Flood, Christopher Winterton, Robert Drüler, George Templing, and Robert Newbury.

Baron Bramwell then left the Court for a short time, and upon his return he addressed the defendants, and said that they had been found guilty of the offence of conspiracy, and it was now his duty to pass sentence or to say what should be done in consequence of that verdict. Before he did so, he asked them to listen to a few observations he should make to them as an impartial man, one who was neither an employer of labour nor a receiver of wages, and to whom the only difference that it could possibly make would be that he would have to pay a few shillings more for a suit of clothes, and he thought they would believe that this was a circumstance not likely to bias his opinion. He would tell them in the first instance that he was perfectly satisfied that they had broken the law—if it was possible to carry out this system of picketing without annoyance or molestation to others, it was not in itself possibly illegal, but it was impossible to do so. He was perfectly satisfied that it was utterly impossible to carry out effectually the system of picketing without molestation and annoyance, and without a breach of the law. The jury had, in his opinion, returned a true verdict, and he was perfectly satisfied with it; and if the defendants themselves only considered the question fairly and dispassionately, he thought that they must see that reason and justice were against them in the course they had taken. In the first place, what right had they to dictate to any other man in what way he should employ his talents and his labour? The men of Sheffield seemed to consider that they had a right to put compulsion upon others in their trade, and though he did not for a moment suppose that the parties concerned in the present proceedings would think for a moment of going to such fearful extremities as the persons to whom he had referred, still it was an exemplification of the results that were likely to follow from combinations of such a character. What right had they to demand that all their fellow workmen should throw their labour into one common stock? They might depend upon it that such a proceeding was wrong. If all the journeymen tailors were to be allowed to combine and carry out this object, the shoemakers and the men belonging to every other trade might do the same, and so might all the masters, and what would be the consequence? Did they think for a moment that if this was done it would be for the benefit of society at large? Certainly not. The aggregate of the happiness of mankind was created by each man being left to his own discretion, and to do what he pleased in reference to his own affairs. For more than forty years some of the best men in the country—men whom, he was assured, they (the defendants) admired—had been exerting themselves in removing restrictions from every branch of trade and commerce, and there was nothing like class legislation at the present time, and no set of men engaged in any branch of trade or commerce had any exclusive rights; and yet the defendants were actually trying to legislate among themselves in an entirely contrary direction. They were endeavouring to put restraints upon trade, and to re-establish those corporate guilds which might have been found of use in ancient times, but which certainly were not so in the present times. They must be aware that even among those public men who were their best friends, they had no sympathy; and no man of eminence or distinction, or any known friend to the working classes, had said a word in their favour, or had expressed an opinion that they were doing right in the course they had taken. He had consulted with the learned Recorder, and also with the experienced alderman who sat by his side (Alderman Lusk), as to the course that should be taken. They had done wrong—they had broken the law, but it was only fair to them to say that it was possible that they might have acted under an erroneous impression as to what the law really was. Ignorance of the law was of course no excuse, but it was a reason why the punishment that was inflicted should be lighter. He had therefore come to the conclusion that the right way to deal with them was not to pass any sentence at all, but to take their word that the act should not be repeated. This, however, would not apply to one of the defendants—Knox—who appeared to have been guilty of some acts of personal annoyance and violence, but, with regard to the remaining defendants, he should take their word that what they had done, and which they now knew was a breach of the law, should not be repeated. With regard to the remaining defendant, Knox, he had heard that such acts as he had been guilty of were entirely repudiated by those with whom he was supposed to be acting; and he hoped the observations of his own counsel, and the stinging terms of reprobation they had made use of, with reference to such conduct, would have due effect upon him. He felt called upon, however, to treat him differently to the other defendants, and the sentence upon him would be that of imprisonment and hard labour for three months.

The remaining defendants then entered into their own recognisance, in the sum of £100, to appear and receive judgment if they should be called upon to do so, and they were then discharged.—*Star*, August 24th.

THE

AMALGAMATED SOCIETY

OF

CARPENTERS AND JOINERS.

BY

EDWARD SPENCER BEESLY,

Professor of History in the University College, London.

REPRINTED BY THE GENERAL SECRETARY FROM THE "FORTNIGHTLY REVIEW," BY PERMISSION OF THE EDITOR.

WITH APPENDICES.

London:

PRINTED BY J. KENNY, 40, PARKER STREET, LITTLE QUEEN ST., W.C.

1867.

AMALGAMATED SOCIETY OF CARPENTERS AND JOINERS.

I DO not propose in the present paper to deal with the abstract economical questions connected with Trades' Unions. The elucidation of those abstract questions is indeed far from complete, and the public mind has much to learn, and much more to unlearn, with regard to them. But what seems most wanting at the present stage of the controversy is, a clearer conception of Trades' Unions, not as they may be or as they ought to be, but as they are. There are few educated people who have not acquainted themselves more or less with the line of reasoning by which it is sought to demonstrate the folly and futility of combinations among workmen. That reasoning is largely based on hypotheses, and the delusive semblance of scientific exactness which it wears has led the educated public to overlook the very hasty and one-sided observations on which those hypotheses are raised, and to dispense with anything like a careful verification. Mr. Harrison's acute and searching criticism of the vulgar economic doctrines has already led to a sensible modification in the supercilious self-confident tone of the anonymous writers who are in the habit of lecturing workmen so loftily on their infatuated attachment to their unions. But the general public has not the patience to follow these discussions. Its attitude towards the unions is determined, partly of course, by its blind reliance on the authority of the pseudo-economic school, but far more by the gross and systematic misrepresentations propagated by a press devoted to middle-class interests. So long as unions are believed to encourage or tolerate tyranny, outrage, and idleness, and to discourage industry, providence, and independence, they will continue to be regarded with a dislike bordering on horror.

How are these misconceptions to be removed? They are strengthened from time to time by statements appearing in newspapers, many absolutely false, many exaggerated or uncandidly coloured, a few too true. The question is, whether the facts, or alleged facts, so adduced, are fair specimens of the practical working of unionism, and the natural result of the system, or whether they are—when authentic—as exceptional as the appearance of clergymen in police courts. To collect and examine these cases during the last five or ten years, with anything like completeness, is of course impossible. Any number of isolated examples, pointing in a different direction, might be objected to as exceptional, and would be of no avail to rebut the sweeping charges made. I have thought it well therefore to take a single society, to describe its origin, growth, and present state, to give

2

a faithful account of its rules, to describe and illustrate its organisation and policy. Although most people are forward enough to express strong opinions about unionism, the ignorance on the subject is, if I may judge from my own observation, profound. It might have been supposed that employers, at all events, who are constantly coming into collision with their workmen, would be rather better informed. But so repugnant is it to their pride and obstinacy to seem to recognise the societies, that they deliberately prefer ignorance where knowledge is within their reach. The information contained in this paper will, I am satisfied, be to most readers entirely novel, and, since Trades' Unions are to be the topic of the session, ought not to be without interest.

I have selected for description the Amalgamated Society of Carpenters and Joiners, because it exemplifies all the modern improvements in organisation which are giving a new signification and importance to the Unionism of the present day, and because I happen to have watched its development more closely than that of other societies. I have before me its rules, forming a small book of fifty pages, a complete set of its reports, annual, quarterly, and monthly, and specimens of all forms used in transacting its business. I have known its general secretary, Mr. Applegarth, for several years, and have always found him trustworthy. He has been often fiercely attacked by the *Beehive* newspaper, the organ of Mr. George Potter, and if any story to his discredit had been going about, I could not have failed to hear it. The best proof of his good character and his ability is the confidence of his society, which has five times elected him its general secretary. I must apologise to him for making these remarks, which he might justly consider more offensive than complimentary. But in the eyes of the educated public, there is a heavy presumption against the character of a "paid secretary of a union;" and it will be evident that I have derived much of my information from him.

The Amalgamated Society of Carpenters arose out of the great builders' lock-out of 1859-60. At the time of that struggle the masons alone of the building trades were organised into a single society, extending throughout England, and providing not only for trade purposes, but for the ordinary benefits. Its head-quarters were then at Bristol. I remember thinking that its machinery was too cumbrous, that there was in it too much of check and countercheck for the exigencies of a campaign. Still, its solidity was remarkably exemplified. The London masons locked out were supported regularly and punctually by their society, and could have continued the struggle for an indefinite time; but the other trades, split up into numerous local societies, were soon reduced to extremities.

The lock-out of 1859-60 proved that unions established for trade purposes only, or confined to a particular locality, must inevitably

3

fail whenever a serious crisis occurred. Impressed with this experience, some members of the carpenter's trade determined to found a union on the same principle as that of the Amalgamated Engineers, whose society, then in the tenth year of its existence, had increased and prospered to an extent hitherto unexampled. They sent out circulars, canvassed the various small societies throughout the country, and meeting with sufficient encouragement, started the new society in June, 1860. The main distinctive features as contrasted with the old societies were three:—(1) It consisted of branches scattered throughout the country; (2) it provided all the usual benefits offered by friendly societies; (3) it guaranteed a certain weekly payment to every member out of work *from whatever cause.*

The chief advantage of the first of these principles has been already explained. Nothing short of a lock-out extending over the whole country, and carried out by small as well as great masters, could cripple the funds of the society. But there are other merits less obvious, though not less real. The policy of this, and I believe of all amalgamated societies, is to promote a virtual equalisation of wages and working hours in different districts, regard being of course had to such circumstances as cost of living, housing, &c. A local union aims at raising the wages and shortening the hours of its own particular district. If it is successful, a rush of labour to that locality is the natural result. Thus the high remuneration of carpenters in London causes a perpetual influx from the rural districts. It may be said that this displacement would naturally produce a proportionate rise of wages in these districts. Here, however, as is so often the case in Political Economy, abstract reasoning only misleads. The country carpenter, or small builder, whose apprentice, having served his time, has made for London, simply takes a new apprentice from the weltering mass of agricultural labour lying at his door, or within a mile or two of it; and finds himself under no compulsion to offer higher wages for journeymen. I have watched the process in a country village. Again, an attempt by the London men to obtain better terms may be defeated by importing the low-paid artisans from rural districts. Thus, in 1860, the agents of the London builders scoured England and Scotland in search of men, and brought them up in droves to the metropolis. But when wages, in proportion to cost of living, shall have been equalised all over the country, passionate or obstinate employers will find it impossible to effect this reckless and mischievous displacement of labour.

The importance of the second principle will be best understood by looking at the character and working of the old-fashioned unions in which it is not adopted. The men combine purely for "trade purposes." The subscription is insignificant, sometimes only a penny a week. The members probably belong to the Oddfellows or Foresters

4

for the benefit purposes, and their financial tie to their union being so weak, they join it or leave it with equal carelessness. Nevertheless, small as the subscription is, a fund will in course of time be accumulated. There is nothing to do with this fund. There it is, eating its head off, so to speak. The men become impatient to use it; so a demand is made on the employers, irrespective perhaps of the circumstances of the trade. A strike follows. The members live on their fund for a few weeks, and when it is exhausted they give in. Such societies may be called strike-societies, for they exist for nothing else, and are of little or no use. Nevertheless there are many of them still existing. Some may perhaps have a somewhat higher contribution to meet certain contingencies not provided for by ordinary friendly societies, such as loss of tools; but practically all unions with small contributions are strike societies.

Now the improved union of modern times combines with "trade purposes" all the usual benefits. A member of the Amalgamated Carpenters losing his tools* by fire, water, or theft, receives their full value. If sick, he receives 12s. a week for twenty-six weeks, and after that, 6s. a week as long as illness continues. When he dies his wife receives £12 for his funeral. If his wife dies before him he receives £5 for her funeral,—£7 remaining for his own. If he sustains an accident incapacitating him permanently from following his trade, he receives £100 to set him up in some little business. When fifty years old, if through failing strength he is unable to make full time or wages at his trade, he is entitled to superannuation allowance—8s. a week, if he has been twenty-five years a member; 5s. if twelve years. Under certain circumstances assistance is given to emigration. The last and most important benefit is that of "donation," which will be explained presently.

To secure these benefits the contribution is, of course, proportionately heavy. There are two classes of members: those who are entitled to all benefits, and those who are entitled to all except sick benefit. The former pay 1s. a week, the latter 9d. All pay 3d. a quarter to a "contingent fund." The second category was established to suit those men who were already Oddfellows or Foresters. But it is much more economical to depend upon one society combining all benefits, than to contribute to a friendly society for sick and funeral benefit, and to a union for tool and accident benefit and trade purposes. The advantage of the former course lies in concentration of management and consequent lessening of working expenses. It may be asked whether it has been ascertained by proper calculation that the scale of contribution will enable the society to make good the benefits it promises. The Amalgamated

* A carpenter's tools are much more valuable than those of most trades. A full set costs about £20.

5

Carpenters simply copied the rules of the Amalgamated Engineers, who had submitted their calculations to an actuary, on starting their society in 1850, and had found them work well for ten years, although during that time they had passed through the ordeal of a gigantic strike and lock-out. But it must be admitted that when such an element as that last mentioned is included among possible contingencies, the most careful calculations may be baffled. The chances are evidently not such as an actuary can compute. If at any time the Executive Council should find that from labour disputes on an unprecedented scale, or any other cause, the funds have sunk so low as to make it doubtful whether all calls can be met, it would take steps to have an extraordinary levy made throughout the society, by which the funds might be raised to a proper level. No doubt a levy at a time of depression might press hard on the members. But such is their attachment and loyalty to the society, such a stake have they invested in it, so closely is their comfort and security bound up with it, that there is not the smallest doubt that they would meet the call rather than submit to the alternative—the break up of the society. On the other hand, it must be said that neither this society nor that of the engineers has ever approached such a crisis, and that the executive is always watchful to prevent too many labour contests in any one year.

The third principle—that which guarantees a weekly payment to every member out of work, from whatever cause—may be regarded as the backbone of this and similar societies. It does for the workman what the capitalist will do for him when society is better organised—secures him something like steadiness of income and employment. Any member who has been out of work more than three days, under circumstances satisfactory to his branch, is entitled to 10s. a week for the first twelve weeks, and to 6s. for twelve weeks more. He may not draw more than £9 12s. in this way in any one year. This is is called "being on donation." The donation, while large enough to keep a man from absolute want, is not so large as to be a premium upon idleness in a trade in which the wages vary from 21s. to 36s. a week. Moreover, it is the business of the secretary of every branch to know if any work is to be had in the neighbourhood, and to send to it members on donation. Any member knowing where work is to be had and not informing his branch secretary, is fined: while, on the other hand, he or any other person reporting work, and so taking a member off donation, receives 6d. Similarly the branch secretary has to inform the general secretary in London whether employment is brisk or slack in his district. If the Manchester secretary informs the general secretary that he can place a dozen carpenters in his town, the latter glances over his reports, and writes (say) to Birmingham and Worcester, where work is slack, and there are consequently several members on donation, and desires them to send so many men to Manchester. Any man declining to go would have his donation stopped. It is better both for him and for the society that he should be earning 30s. at Manchester than idling on 10s. donation at Birmingham or Worcester. He receives his railway fare from his branch secretary, unless the Manchester employer pays it—which is often the case, for the employers or their foremen generally apply to the branch secretaries to find them men. If a man so required to migrate alleges any reasonable excuse, such as illness in his family, the branch secretary will pass him over. In case of dispute, he can appeal first to a meeting of his branch, and secondly to the Executive Council in London: for nothing is left in this society—or, indeed, in any that I am acquainted with—to the arbitrary discretion of that bugbear of employers and journalists, the "paid secretary." Experience proves that this system works admirably. Secretaries and private members are alike vigilant to keep down donation. The sense of comfort and security it inspires cannot be estimated by those who, having never known what it is to depend for their bread on payment by the day or the hour, draw foolish comparisons between the precarious intermittent wages of an artisan and the steady salary of a bank clerk or curate. On the other hand, employers and the public are gainers by the prompt transference of labour to the districts in which it is needed. All these arrangements would be organised by the capitalists if they understood their social function. They have had centuries to learn their duty. But they have not learnt it or done it, and in modern times they have digested their negligence, incompetence, and selfishness into a theory. The despised and ignorant workman, the drunken venal being known to Mr. Lowe, has therefore been obliged to take the supervision of industry on a large scale into his own hands, and to do for the capitalists what they are too selfish and mutually distrustful to do for themselves.

The next thing to be described is the organisation and government of the society, local and central. Each branch is a completely organised body in itself, choosing its own officers, collecting, holding, and disbursing its own funds, and generally managing its own business. A branch may not consist of less than seven or more than three hundred members. There are at present one hundred and ninety branches, comprising 8,256 members. Presidents and committee-men are elected quarterly, stewards half-yearly, secretary, treasurer, referee, and trustees annually. There is a general meeting every fortnight to transact business. On election nights, or at specially summoned meetings, members absent without written excuse are fined 3d. The committee sits when summoned, and its minutes must be read and confirmed at the next general meeting of the branch. The treasurer is, if possible, a householder, and gives security to his branch. He can only hold so much money as the branch each fortnight authorises, the rest being banked by the trustees. He can only disburse money on written order from the secretary. There are many other elaborate checks on malversation.

The central authority of the society is vested in a general council consisting of sixteen members and a president, six of whom, with a chairman, are elected by the London branches, and the rest by the country branches. Half the council retires every six months. But as the country council-men could not attend frequent meetings in London, the ordinary management is vested in the six London members, who are called the Executive Council, and a chairman elected annually by the London branches. Most general societies show considerable jealously of London, and provide that the seat of government shall be changed from time to time. The head-quarters of the masons, I believe, are seldom or never in London; but the Amalgamated Carpenters are sensible of the advantages of a real centralisation. Although the reports show that disputes have frequently arisen between the Executive and various branches, I do not find that as Londoners the former have given any dissatisfaction.* It is the duty of the Executive Council to see that the branches conform to rules, to decide appeals, to authorise the opening or closing of branches, to order one branch to transmit money to another (as will be explained presently) to lake legal proceedings on behalf of the society when necessary, and, lastly, to authorise strikes. The Executive communicates with the branches through the general secretary, the only officer of the society whose whole time is devoted to its service. He is elected annually by universal suffrage of the whole society. His office is the head-quarters of the society, and his business is to act as secretary to the Executive, to transact all correspondence with branches, to compile and issue the annual, quarterly, and monthly reports, to arrange the transmission of members on donation from one place to another, and to keep a register containing the name of every member and such information about him as is necessary. In short, the working of the society as a whole depends on his industry and activity. He has a salary of £2 10s. a week.

Though each branch is a complete union in itself, the funds belong to the society as a whole. Consequently, if Plymouth, owing to slack trade or a strike, or any other cause, has exhausted all its own funds, the Executive directs Exeter or Lambeth, whose coffers are full, to remit the requisite sum to Plymouth; and at the end of the

* A delegate meeting was found necessary last year for a general revision of the rules. It was held in Manchester, many of the London branches magnanimously voting for the provincial town. Bristol, however, expresses its "disgust and intense dissatisfaction at the innovation."

8

year a general "equalisation" of funds is arranged by the general secretary throughout the society; branches more or less rich remitting to branches more or less exhausted, until each branch has the same amount per member to begin the new year with. Thus no large fund is ever deposited at head-quarters. The general office draws what it wants for its own working expenses each month from any branch that is most convenient. If a branch squanders its funds in a manner not authorised by the rules, the Executive compels it to make good the deficiency by a levy on its members. If a branch gets its accounts into confusion, or gives any cause for suspicion, the Executive can order a special audit, or send a deputation to investigate. If a branch disobeys rules or resists lawful requisitions of the Executive, that body can expel it.

Every precaution has been taken to prevent arbitrary action on the part of officials. The Executive can decide where the rules are silent: but it cannot alter or suspend a rule. The rules may be altered by a meeting of delegates specially deputed for the purpose by their districts. Two such delegate meetings have been held since the foundation of the society. In cases of great emergency, the *General* Council can take a general vote of the society through the branches as to the alteration or suspension of a rule, without calling a delegate meeting, which is a slow and costly proceeding. If any branch, by a majority of two-thirds, disputes a decision of the Executive, it has an appeal to the society. The case on each side is printed and circulated through the branches, and the vote of the majority decides. Similarly if any individual is aggrieved by an official, he may appeal first to the branch committee, then to the general meeting of his branch, and lastly to the Executive in London, whose decision is final. From a dozen to twenty such appeals reach the Executive in a year.

The fortnightly meetings of the branches are, as a rule, well attended. They are necessarily held in public-houses, but from the large majority of them drink is excluded, the members preferring to pay a fixed rent for a room. When society business is concluded, and the meeting has been formally closed, it is not unusual for the members to remain sitting, with the same chairman, to discuss social and political questions.* To belong to such a society is an education far more efficacious for all political and practical purposes than any training gone through by the majority of the wealthier classes. The business of the society is conducted by some 1,600 officers. All these must accustom themselves to keep accounts according to a uniform pattern, and to perform their other duties with method and regularity. Office is a burden, to decline which when it comes to a member's turn involves a fine. Thus all the members acquire a familiarity with the management which makes usurpation or maladministration very improbable. Of course this large governmental

* In the Shoemakers' and Tailors' Societies this is still more common.

9

staff is not kept up without expense. The nation, or those who rule it, may be content to entrust parliamentary and judicial functions to unpaid volunteers; but carpenters cannot afford to have their business bungled or neglected. They pay their officers to do it, and so acquire the right to fine them heavily for neglecting it. The pay is indeed of the smallest. Branch secretaries receive from 6s. 3d. to £2 11s. a quarter, according to number of members; treasurers, from 15s. to £3 5s. a year; sick stewards, who visit sick members twice a week, 6d. a week for each; all who attend a committee, 6d., &c. The working expenses of the society are certainly not light. They have been reduced in the last five years from 38 to 19 per cent., which, I am assured, is below the usual rate of friendly societies. It must be remembered that the collection and disbursement of money is just as troublesome when the sums are small as when they are large. In the former case, therefore, the working expenses will necessarily bear a larger proportion to the gross receipts and expenditure. Moreover, many of the branches are at present very small. Thus Middleton has only seven members, yet it pays as much for room rent as Bradford with 125 members.

I will now explain the action of this society in case of a strike. On this point it has been found impossible to have a code of rules. For such is the monstrous state of our law as made by an upper and middle class Parliament, that any society which can be proved to contemplate spending any portion of its funds in supporting a strike is denied the protection of the tribunals. An officer who has robbed it in the most impudent manner, when brought before a magistrate, has only to show that its rules operate "in restraint of trade," and he will at once be discharged. Thus the English plutocracy, while not daring any longer to make combination of workmen criminal, endeavours, with unexampled baseness, to cripple their action by encouraging fraud and embezzlement. Such being the state of the law, the society of which I am giving an account is obliged to be very circumspect. The only rule which applies to a strike is the following:—

"Any member or members leaving his or their employment under circumstances satisfactory to the branch or Executive Council shall be entitled to the sum of 15s. per week." (Rule xviii. clause 7).*

By wording the rule thus vaguely, and interpreting it according to a well understood usage, the society has hitherto escaped outlawry. It has on several occasions successfully invoked the protection of the tribunals. Last August the secretary of the Birkenhead 1st Branch was convicted of embezzlement and forgery at the Chester assizes. Baron Pigott sentenced him to five years' penal servitude, observing

* The *Builders' Trade Circular* of January 24, thinks it honest to quote this rule in inverted commas, as follows: "Any member withdrawn from his employment (to maintain the existing privileges of the trade)" &c.

that "it appeared to him that the prisoner was one of those who thought that because he was entrusted with the business affairs of a society composed of working men, he could rob them with impunity; but it was due to the public as well as to the society which had been robbed of funds subscribed for such laudable objects, that proof should be given to the contrary." Whether, after the late judgment of the Court of Queen's Bench, in the case of Hornby *v.* Close, the Amalgamated Carpenters will continue to enjoy the protection of the law, remains to be seen. The rules of the Boiler-makers' Society, on which that case turned, were indeed much less circumspect than those of the carpenters, for they contained the damning word "strike," and fettered the discretion of the Executive by certain explicit provisions. But I fear there would be no great difficulty in proving that the carpenters' rule quoted above points to the "restraint of trade." Although it empowers a branch to decide on the circumstances under which a member has left his employment, yet the Executive has for some years claimed the exclusive right of determining all cases where more than one or two individuals are involved; in other words, where it is a question of striking; and this interpretation of the rule was formally endorsed by the delegate meeting at Manchester last year. Whenever, therefore, a branch desires to resist any innovation on the part of its employers, or to make a demand on them, it is required, before taking the smallest step, to submit its case to the Executive in London. The wisdom of this provision is manifest. The Worcester or Chester branches might be misled by a little local prosperity into advancing pretensions not justified by the general state of trade, or of the finances of the society. In times past many a strike has been thus ignorantly brought on. But the Executive Council have before them complete and circumstantial tables, compiled every month from the branch returns, and showing the exact state of trade all over the country. The following is a specimen from the report for November, 1865.

Branch.	Number of Members	State of Trade.	Members on Donation. 10s.	Members on Donation. 6s.	Members on Sick Benefit 12s	Members on Sick Benefit 6s.	Superannuation.
Sheffield	160	Good			1		
Oldham	30	Good. Setting on					
Birmingham	168	Bad	6		2		
Colchester	9	Very good					
Aldershot	7	Men wanted					
Torquay	80	Unsettled					
Poplar	31	Dull					
Droylsden	25	Improving					
Sunderland	60	Moderate					
Coventry	30	Steady					
Scarborough	94	Slack	1		2		
Plymouth	167	Strike and Lock-out					
Dartford	23	Setting on occasionally					

11

And so on through the 127 branches of which the society then consisted. To look at labour on this large scale, with a view to provident arrangements for its employment, is the province of masters rather than men, did the former but understand and recognise their responsibility to society. I will not assert that the London master builders do not provide themselves with similar statistics, but I very much doubt whether any one of them troubles himself about anything else than obtaining the men necessary for the works he has in hand. If the President of the Board of Trade wanted such information at an hour's notice, I wonder whether it would occur to him that the only place where he could get it is in a little office up two pair of stairs in Northumberland Street, Strand.

The answer of an Executive to an application depends mainly on these returns. The branch anxious to fight may be told that its demand is unreasonable, or that its grievance is not one the society can recognise, or that other branches are more aggrieved and must have their turn first, or that its case is a good one, but that the society has a heavy strike on hand or impending, somewhere else, which may press heavily on the funds. If on the other hand the case appears good, and the society has its hands free, the branch is authorised to address a letter to the employers. When the men are not resisting an innovation, but are preferring a new claim, the Executive requires the branch to give at least three months' notice; for the society steadily discountenances all attempts to take the employers by surprise. Its aim is to introduce everywhere a "code of working rules" signed by the representatives of masters and men, and not to be altered without three or six months' notice on either side. This system is in force in Manchester and many other towns.* If the concession demanded is refused, the branch strikes on the expiration of the notice, and the men out of work receive fifteen shillings a week from the society. But arrangements will have been made beforehand for drafting off all the men who can be easily moved to other localities where hands are wanted, so that the burden on the funds may be lessened. A fairer application of the strictest rules of the political economists could not be desired. When the strike is at an end, many of the men drafted off, especially those who

* The *Times* of February 7 says, "In many trades it is unsafe, indeed it is hardly possible to take a contract, for an edict of the union may at any moment suspend the work. Of course this is a natural proceeding. The master's difficulty is the workman's opportunity; and if men have a right to strike, they have a right to choose their own time for striking." That the *Times* should concede this "right" to workmen is only natural. The plutocracy of which it is the organ would be only too glad to eliminate all moral considerations from economical questions. But whatever workmen may have done, or may still be doing, in their struggle for life with capitalists steeped in the anti-social doctrines of the *Times*, they have a higher ideal; and in proportion as they become solidly organised, and are in a position to deal calmly as well as firmly with their employers, they aim at moralising the relations of labour and capital, as is shown by the policy of the Amalgamated Carpenters described above.

12

have left families behind them, will find their way back to their old abode. A branch striking without the approval of the Executive, or remaining on strike after the Executive has called on it to give in, has no right to use the funds of the society for that purpose. Last year the Plymouth branch having disbursed £20 out of its treasury to men on strike after the Executive had required then to go to work, was compelled to refund the money by a levy on its members, Branches generally take a pride in carrying a strike through on the funds in their own treasury, without drawing on other branches, though if the struggle is protracted the latter course must be adopted.

Two or three illustrations of the policy of this society as to strikes will probably be interesting.

In December, 1864, the Birmingham and Midland master builders announced that they would not for the future take on any man who could not produce a "discharge note" from his last employer. The men very naturally fired up at this tyrannical innovation, and struck work till the notice should be withdrawn. The Executive of the Amalgamated Society disapproved of the strike, considering that the men should have taken no notice of the announcement, and should simply have declined to receive the discharge note when it was tendered. If all or most of the carpenters of the district had been members of the society no difficulty could have arisen. Any member declining the discharge note, and so failing to get fresh employment from Midland masters, would have been quietly removed by the society to some other part of England. By this masterly inactivity the scheme of the employers would have become ignominiously inoperative. But only 211 of the Birmingham carpenters belonged to this society; and as the rest of the men rushed hotly into a strike, the Executive had no choice but to let the society men do the same, and pay them the usual strike pay. The resolution of the Executive was couched in the following terms:—"That, while sanctioning the Birmingham strike, we at the same time feel convinced that the members have acted injudiciously in striking before the time expired for the discharge note to come into operation." However, prudent or imprudent, the action of the Birmingham men was attended with immediate victory: the discharge note was unconditionally withdrawn. The strike, however, was not immediately brought to a close, the men being stimulated by unwise advisers to make further demands. Therefore the general secretary of the Amalgamated Carpenters went down to Birmingham, without waiting for directions from his Executive, and urged the men to be satisfied with their victory and resume work. By taking this step he drew down upon himself much vituperation, but the Executive approved his action by a majority of five to two; and in the following summer he was re-elected to his office, after a sharp contest, by a majority of 1,814 votes to 612.

13

In 1865 the London carpenters struck for an advance of a halfpenny an hour. Here, again, the Amalgamated Executive objected to the strike on the ground that the general interests of the trade would be better served by endeavouring to raise wages in the west of England, where they ruled very low. There would have been no difficulty in getting the amalgamated men to acquiesce in this view, because they are accustomed to a large and far-sighted policy. But only 2,000 of the London carpenters belong to this society, and the strike was directed by a temporary committee elected from the whole trade, in a hit-or-miss way, at disorderly meetings called by no recognised body. It is easy to imagine the vexation of unionists when they see an agitation of this kind, which has been got up by self-appointed agitators working chiefly upon the non-unionist mass, ascribed by ignorant and reckless journalists to the "dictates of the union," or the "tyranny of paid secretaries." The Executive of the Amalgamated Society had again no choice but either to let its members do as others did, or play into the hands of the employers. It of course adopted the former alternative, and the advance was gained.

From the attitude of the Executive on these occasions of successful strikes, it may be gathered what policy is pursued in less hopeful cases. It is time the public ceased to swallow the nonsense that is talked about "paid leaders of unions." It is for the interest of masters, men, and society at large that the workmen should be represented, not by noisy irresponsible agitators, seeking to fill their pockets in the scuffle, but by men who have been elected in calmer moments, with proper formalities, by a well-defined constituency, and whose great anxiety is to husband the funds of their society, and to carry it through a struggle, when unavoidable, as rapidly and economically as possible.

The Amalgamated Society does not attempt to lay down any rule about work, wages, or hours. Every branch negotiates these matters for itself. The Executive would permit a strike against the introduction of piece-work only where it had been forbidden in a code of working rules mutually agreed on between masters and men, as at Scarborough. In London piece-work is perfectly recognised. Similarly where a code has limited the number of apprentices, resistance to an alteration might be authorised. But no branch is ever urged from head-quarters to enter on a conflict; for there are always far more applications for permission to strike than can be granted. The power of the Executive, in short, may be summed up as the right of veto on strikes. Subordination to this power is what the *Saturday Review* calls "submission to the Vehm-gericht of Unionism."

If the space conceded to me permitted, I could add copious illustrations of the action of the society. Two more must suffice. Recently a clerk of works at Dorking, being in want of men, and knowing

14

there was a lock-out at Manchester, applied to the secretary of the Dorking branch to obtain him some men, promising to pay railway expenses. The men were fetched, but as some of them left after a short time, the clerk of the works represented to the Executive that he had been a loser by paying the travelling expenses; upon which the money was refunded to him. At Sheffield, last May, the masters' association sent for a deputation of the Amalgamated Society (which has its largest branch in that town), and voluntarily offered an advance of 2s. a week on condition that a code of working rules was established. The men closed with one offer as gladly as the other. The code regulates hours of work and wages during winter and summer, and time for walking, or lodging-money, when work is beyond a certain distance; provides that extra rates shall be paid for over-time, and requires four months' notice of alteration on either side. It is signed by the president and secretary of the master's association, and by the secretaries of the amalgamated society, the general union, and a local society. Every workman has a copy of it. At the same time a committee of masters and men was appointed to settle all disputes, an arrangement which has hitherto worked very satisfactorily. Times are altered since the London master builders in 1859 passed the following resolution:—"That no deputations of trade unions, committees, or other bodies, with reference to any objects referred to in Article 3 [the regulations of any establishment, the hours or terms of labour, the contracts or agreements of employers or employed, or the qualification or terms of service], be received by any member of this association on any account whatever; it being still open to any workman, individually, to apply on such subject to his employer." I doubt if such a resolution will ever be passed again by any body of employers. The affectation of not recognising unions and their officers has been found to be as stupid as it is insolent. When the agricultural labourers begin to combine, as I venture to predict they will within the next five years, it is to be hoped that the landlords and farmers will be wise in time.

The Amalgamated Society has now been nearly seven years in existence. The subjoined table will show how rapid has been its progress.

Year.	Branches	Members.	Surplus.
			£ s. d.
1860	20	618	321 7 2½
1861	32	650	593 12 0½
1862	38	949	849 8 10
1863	53	1718	2042 11 3
1864	81	3279	4566 10 0½
1865	134	5670	8320 13 7
1866	190	8256	Accounts not yet made up.

15

The Amalgamated Society, however, is not the largest in the trade. The "General Union" is a much older society, dating from 1827 (three years after the repeal of the combination laws), and extending, as its name imports, throughout the kingdom. This society in August last had 140 "lodges," and 9,490 members, being an increase of 40 lodges, and 2,504 members on the previous year. It was formerly a society merely for "trade purposes," or, at most, for some small benefits, and each of its lodges had its own code of rules. Since 1860 it has been gradually assimilating its constitution to that of the younger society. But its regular contributions and benefits are still on a smaller scale; it has not yet adopted the principle of equalisation; strikes are authorised or prohibited by the cumbrous process of a vote of the whole society; while a special fund for trade purposes and certain benefits is deposited at the seat of government, and permanently kept up to £500 by means of levies. Thus a member of the General Union, in addition to his regular contributions, pays an uncertain sum for trade and other purposes. The Amalgamated Society has never been obliged to resort to a levy, and the responsibility of permitting strikes lies, as we have seen, with the Executive. I have no doubt that the latter is the wisest plan. The necessity of framing rules on this delicate subject is avoided, and power and responsibility rest with competent men who are thoroughly acquainted with the state of the trade throughout the kingdom.

There are many other societies of carpenters, but all comparatively insignificant. In 1860 there were some forty of them in London alone. I cannot ascertain how many there are now; but they are being gradually eaten up by the two great societies which, notwithstanding past jealousies, will themselves, no doubt, sooner or later coalesce. At present, however, the non-unionists must vastly outnumber the unionists. There are probably 13,000 or 14,000 carpenters in London alone; while the General Union and Amalgamated Society together number less than 17,000 members for the whole kingdom. But combination on sound principles dates, in this trade, from the formation of the latter society in 1860. Both are now increasing at a prodigious rate. To all appearance they will include a majority of the trade in six or seven years; and those not included will be chiefly men who are too old for admission, incompetent workmen, and men who have been expelled.

The limited space at my disposal unfortunately prevents me from giving many other interesting particulars respecting this admirable society. I can fancy, however, that some will think my sketch must be very incomplete, because it has not thrown any light upon the violence and intimidation commonly supposed to be essential to a trade organisation. I simply affirm that nothing of the kind exists in this society. If any one can impugn this assertion, let him do so.

16

The Executive has often announced that it would prosecute any member guilty of violence as promptly as it would prosecute a master. Last year it offered £25 out of its "Contingent Fund" for the discovery of the perpetrator of the Sheffield outrage, although it was not a carpenters' matter.

Since the foregoing part of this paper was written, the fears expressed respecting the effect of the decision in the case of Hornby *v.* Close have proved only too well founded. That decision was given on January 16. The *Builders' Trade Circular*, of January 24, exultingly pointed out that the principle thus established would apply to the Amalgamated Society. A month had not elapsed, when the society had occasion to prosecute an officer of the Hull branch. The magistrate looked at the rules, pronounced them to be framed with a view to strikes, and dismissed the case. Just as these remarks are going to press, a similar case, followed by a similar decision, is reported from Bradford. Thus eight thousand carpenters, the flower of their trade, find themselves proclaimed fair game for swindlers. The excitement throughout the society may be imagined. No wonder that out of the 2,000 London members, some 1,500 marched with Mr. Beales to Islington on the 11th of February. The society took no part in the December demonstration. For my part, I am not sorry that they should learn what their liberty amounts to under a Parliament chosen by the upper and middle classes.

EDWARD SPENCER BEESLY.

ADDENDUM.—The income of the society in 1865 was thus expended:—Donations, £347 6s. 10½d.; tool benefit, £114 4s. 11½d.; sick benefit, £1,369 15s. 5d.; funeral benefit, £257 10s.; trade purposes, £1,941 8s. 1½d.; accident benefit, £200; grants to distressed members from Contingent Fund, £121 10s.; grants to other trades from ditto, £69; property, £234 9s. 10d.; working expenses, £2,078 6s. 3d. Balance, £3,754 3s. 6½d.

17

APPENDICES.

APPENDIX A.

I find that the statement on page 14 about Sheffield is not quite accurate. The following extract from Mr. Applegarth's evidence before the Trades' Union Commission contains a fuller account of the facts:—

190. (*Mr. Harrison.*) You have spoken of the customs of the trade. Are there many towns in which a code of rules agreed upon between the employer and the employed is in force?—(*Mr. Applegarth.*) Yes.

191. Does the society desire to extend that, or do you consider that a beneficial state of things?—We are doing all we can to extend it; we believe that nothing can be more advantageous than for the masters and men to meet and agree upon certain conditions, and we think that it should be a matter left to themselves.

192. Have you any evidence that that has acted beneficially upon the state of the trade, that any town, for instance, has been the better for it since it has been enforced?—Yes. There is a very remarkable instance at Sheffield. Our society has been in existence since March, 1862, in that place. They were receiving then 26s. per week, and working 58½ hours for that, and receiving the same rate of wages in winter as in summer. After our branch had increased there, the men began to think that they were not so well paid as men in other towns, and they applied to the masters for more. It was not settled without a dispute, they had a strike of some few weeks, and then they got what they desired, including a code of working rules. Since that they have had many meetings with their employers, and the employers were well satisfied with the conduct of the men, and expressed themselves so. It happened that the masons and other branches of the building trade had asked for the Saturday half-holiday, and our men thought it is not fair that we should work longer than they do on the Saturday, so they went to the employers, simply instructed by a resolution of our branch, to ask them if they would be prepared to grant the half-holiday. The employers said, "Yes, we will," and a short time afterwards granted it. They said, "what is more; we have a suggestion to make to you carpenters."

18

We are well satisfied with the manner in which you have met us during the last few years on trade affairs, but there is one thing that we should like to suggest to you, and that is, that we should have a board composed of an equal number of masters and men, and that any little grievance on the part of either one or the other should be submitted to this board so composed, thus preventing our getting into a large dispute; "we will agree to appoint a number of masters if you will appoint an equal number of men." The men went back to those who had sent them, and they gladly accepted that proposal, and there has been no such thing as a strike since that board of arbitration was established; there is not a town of England where, taking into consideration the working hours and the wages and price of provisions, the men work on better conditions than they do in Sheffield.

193. (*Chairman.*) Do you mean that the men are paid higher and that provisions are cheaper?—I mean that taking into consideration the amount of wages received, the number of hours they work for that amount of wages, and the price of provisions, there are no men in the trade better off. I think that the manner in which the employers met the men may be attributed to this fact. When I left Sheffield some five years ago they were receiving 26s. per week the year round. The employers felt this was unfair, as the time worked was so much shorter during the winter, and they said to the men, "You ask for more wages, but you have an advantage in getting the same rate in the winter as in the summer." The men said, "We will agree to 2s. less in the winter time than we receive in the summer." This showed a desire on the part of the men to deal fairly with their employers, and they have benefited by so doing, for at one of the last interviews they held with their employers, it was agreed, on receiving proper notice, to give the men an additional 2s. per week if the state of trade would warrant it; this the men had not asked for.

APPENDIX B.

I have extracted from the last Yearly Report the following tables, which will doubtless prove interesting. Table II. shows the full amount paid by the society to its members since its formation, and for what purposes the sums were paid. Table V. shows the advantages gained in reduced working hours and increased wages.

TABLE II.—Amounts paid each year for the various benefits since the formation of the Society, in June, 1860, and the balance at the end of each year.

Year.	No. of Members.	Donation.		Tool.		Sick.		Funeral.		Accident		Trade Privileges.		Benevolent Grants.		Grants & Loans to other Trades.		Balance at the end of each Year.
		Yearly Amount	Per Member.	Yearly Amnt.	Per Member	Yearly. Amount	Per Member.	Yearly Amnt.	Per Member	Yearly Amnt.	Per Member	Yearly Amnt.	Per Member.	Yearly Amnt.	Per Member	Yearly Amnt.	Per Member	
		£	£ s. d.	£	s. d.	£	£ s. d.	£	s. d.	£	s. d.	£	s. d.	£	s. d.	£	s. d.	£ s. d.
1860	618	25	0 0 9¾	10	0 4	11	0 0 4¼	17	0 6½	......	......	......	....	......	......	......	......	321 3 2½
1861	650	247	0 7 7	40	1 2¾	92	0 2 10	23	0 8½	......	......	......	....	......	......	......	......	593 12 0½
1862	947	267	0 5 8½	39	0 9¾	249	0 5 3	73	1 6½	......	......	7	0 1¾	......	......	......	......	894 8 10
1863	1718	435	0 5 0¾	101	1 2	571	0 6 7¾	95	1 1¼	30	0 4¼	10	0 1½	......	......	......	......	2042 11 3
1864	3729	211	0 1 3½	88	0 6½	1031	0 6 3½	212	1 3½	......	......	762	4 7¾	......	......	......	......	4566 10 0½
1865	5670	347	0 1 2¾	114	0 4¾	1370	0 4 10	257	0 10¾	200	0 8½	1941	6 10½	121	0 5¼	69	0 3	8320 13 7
1866	8002	1335	0 3 4	362	0 11	2246	0 5 7¼	423	1 0¾	360	0 10¾	2525	6 3¾	271	0 8¼	109	0 3¼	13052 4 3½
		2867	1 5 0¼	754	5 4¾	5570	1 11 9¾	1100	7 1¾	590	1 11½	5245	18 1¼	392	1 1¼	178	0 6¼	

SUMMARY OF THE ABOVE:—

Donation	benefit	£2867	being	£1 5 0¼	per Member for 6½ years.	
Tool	"	754	"	0 5 4¾	"	"
Sick	"	5570	"	1 11 9¾	"	"
Funeral	"	1100	"	0 7 1¾	"	"
Accident	"	590	"	0 1 11½	"	"
Trade Privileges		5245	"	0 18 1¼	"	"
Benevolent Grants		392	"	0 1 1½	"	2 years *
		16518				
Grants and Loans to other Trades		178	"	0 0 6¼	"	"
		16696				

* This fund was established in 1864.

20

TABLE V.

Advantages gained in Wages and Working Hours during 1866.

TOWNS.	No. of members in each town.	Amount per member per week gained. s.	d.	Weekly amount gained by the members in each town. £	s	d.	TOWNS.	No of members in each town.	Amount per member per week gained. s.	d.	Weekly amount gained by the members in each town £	s.	d.
Ballymena	17	5	2	4	7	10	Wood Green	21	2	4	2	9	0
Maidstone	57	4	2	11	17	6	Ealing	24	2	4	2	16	0
Leek	16	3	10	3	1	4	Greenwich	80	2	4	9	6	8
Halifax	117	3	9	21	18	9	Forest Hill	25	2	4	2	18	4
Newport	10	3	9	1	17	6	Barrow-in-Furness	24	2	0	2	8	0
Sowerby Bridge	11	3	8	2	0	4	Exeter	57	2	0	5	14	0
Droylsden	29	3	7	5	3	11	Heywood	40	2	0	4	0	0
Luton	14	3	7	2	10	2	Hull	258	2	0	25	16	0
Birmingham	265	3	6	46	7	6	Lymm	10	2	0	1	0	0
Brentford	9	3	6	1	11	6	Newcastle-on-Tyne	97	2	0	9	14	0
Hereford	63	3	5	10	15	3	Oldham	42	2	0	4	4	0
Richmond	12	3	2	1	18	0	Royton	16	2	0	1	12	0
York	147	3	2	23	5	6	Saltburn-by-the-Sea	16	2	0	1	12	0
Ashford	8	3	0	1	4	0	Wellington	6	2	0	0	12	0
Aigburth	15	3	0	2	5	0	Dublin	126	2	0	12	12	0
Bath	47	3	0	7	1	0	Lisburn	15	2	0	1	10	0
Birkenhead	114	3	0	17	2	0	Londonderry	18	2	0	1	16	0
Bootle	12	3	0	1	16	0	Portadown	13	2	0	1	6	0
Chatham	12	3	0	1	16	0	Scarbro'	200	1	11	19	3	4
Devonport	41	3	0	6	3	0	Tunbridge Wells	16	1	9	1	8	0
Doncaster	35	3	0	5	5	0	Hastings	42	1	8	3	10	0
Dorking	51	3	0	7	13	0	Brighton	39	1	6	2	18	6
Folkestone	17	3	0	2	11	0	Rochester	24	1	6	1	16	0
Harrogate	10	3	0	1	10	0	Lynn	18	1	5	1	5	6
Surbiton	24	3	0	3	12	0	Darlington	86	1	3	5	7	6
Liverpool	298	3	0	44	14	0	Barnet	20	1	2	1	3	4
Liscard	25	3	0	3	15	0	Crewe	7	1	2	0	8	2
Malvern	76	3	0	11	8	0	Twickenham	21	1	2	1	4	6
Middleton	12	3	0	1	16	0	Woolwich	66	1	2	3	17	0
Portsmouth	30	3	0	4	10	0	Lowestoft	10	1	2	0	11	8
Rotherham	8	3	0	1	4	0	Sheffield	252	1	1	13	13	0
Stafford	30	3	0	4	10	0	Bradford	172	1	0	8	12	0
Seaforth	16	3	0	2	8	0	Colchester	20	1	0	1	0	0
Southampton	14	3	0	2	2	0	Cambridge	27	1	0	1	7	0
Torquay	132	3	0	19	16	0	Shipley	35	1	0	1	15	0
Whitby	23	3	0	3	9	0	Stockton	61	1	0	3	1	0
Stone	9	2	11	1	6	3	Worcester	124	1	0	6	4	0
Wolverhampton	29	2	8	3	17	4	Burton-on-Trent	12	0	9	0	9	0
Manchester	458	2	7	59	3	2	Wallsend	26	0	9	0	19	6
Bridlington Quay	17	2	5	2	1	1	Hartlepool	6	0	6	0	3	0
Rugby	18	2	5	2	3	6	Jarrow	12	0	6	0	6	0
London	1808	2	4	210	18	8	Belfast	80	0	3	1	0	0
Croydon	39	2	4	4	11	0							
								6509			750	15	1

The advantages gained in increased wages and reduced working hours, which in the preceding table has been reduced to a money value, are as follows:—In 48 towns an advance of wages has been obtained, in 11 towns a reduction of working hours, while 26 towns have obtained both. In all, in 85 towns the members have received advantages in wages varying from 3d to 5s. 2d. per member per week, with reductions of working hours, varying from half-an-hour to 5½ hours. These advantages will be better understood from the following summary:—

No of Members.	Average amount per member per week gained.	Average amount per year per member gained.	Total amount of gain per week.	Total amount of gain per year
6509	2s. 3¾d.	£6 0s. 3d	£750 15s 1d.	£39,038 0 0

Deduct cost to the Society	£2,600	
Amount of wages lost by members during suspension of work	2,500	
		5,100 0 0
Leaving a net gain of		£33,938 0 0

For the first year, and £39,038 per year after.

REPORT

OF THE VARIOUS PROCEEDINGS TAKEN BY THE

LONDON TRADES' COUNCIL

AND THE

CONFERENCE OF AMALGAMATED TRADES,

IN REFERENCE TO

THE ROYAL COMMISSION

ON

TRADES' UNIONS,

AND OTHER SUBJECTS IN CONNECTION THEREWITH.

LONDON:

J. KENNY, PRINTER, 40 PARKER STREET, LITTLE QUEEN STREET, W.C.

1867.

CONTENTS.

	PAGE
ADDRESS TO THE TRADES' UNIONISTS OF THE UNITED KINGDOM ..	5
Report of Deputation on Sheffield and Nottingham Outrages.—Explanation of the term "rattening."—Interviews with Mr. Wilson, Mr. Taylor, Mr. Fearnehough, and the Organised Trades of Sheffield.—Conclusions arrived at.	
THE SHEFFIELD TOWN COUNCIL AND DEPUTATION TO THE HOME SECRETARY	9
Correspondence between the Town Council and Mr. Applegarth.	
INTERVIEW OF THE TRADES' DELEGATES WITH MR. ROEBUCK ..	11
REPORT OF THE INTERVIEW WITH THE HOME SECRETARY IN REFERENCE TO THE ROYAL COMMISSION	12
Speeches of Mr. Roebuck, Mr. Dronfield, Mr. Austin, Mr. Applegarth, and Mr. Odger.—Mr. Walpole's reply.	
REPORT OF THE DEPUTATION OF THE SHFFFIELD TOWN COUNCIL TO THE HOME SECRETARY	19
IMPORTANT DECISION OF THE COURT OF QUEEN'S BENCH IN REFERENCE TO TRADES' SOCIETIES' FUNDS	22
Decisions of the Lord Chief Justice, Mr. Justice Blackburn, Mr. Justice Mellor, and Mr. Justice Lush.	
APPOINTMENT OF A PERMANENT CONFERENCE	26
INTERVIEW WITH THE HOME SECRETARY	27
TRADES' UNIONS' PROTECTION BILL	30
TRADES' UNION COMMISSION—DEPUTATION TO THE HOME SECRETARY ..	31
GREAT MEETING AT EXETER HALL	34
Petition and Resolutions adopted.	
TRADES' SOCIETIES' FUNDS IN POST OFFICE SAVING'S BANKS	36
Interview with Mr. Gladstone.	
THE CONFERENCE AND THE ROYAL COMMISSION	38
GREAT MEETING OF LONDON TRADES IN EXETER HALL	39
Letter from Mr. J. M. Ludlow, and speeches of Messrs. Allan (engineer), Guile (ironfounder), Odger (shoemaker), Howell (Secretary of the Reform League), Harry (carpenter), Dunning (bookbinder), Applegarth (carpenter), Professor Beesly, and Mr. E. Beales.	
DELEGATE MEETING OF LONDON TRADES' COUNCIL.	45
Resolutions passed in which this Conference concur.	
CONCLUSION	45

5

ADDRESS TO THE TRADE UNIONISTS OF THE UNITED KINGDOM.

FELLOW WORKMEN,

The following pages contain an account of the proceedings of the London Trades' Council, and other Trade Societies in London, in reference to the Sheffield and Nottingham outrages, and the decision of the Judges in the Court of Queen's Bench in the case of *Hornby* v. *Close;* which declared that the Friendly Society of Boiler Makers, though its rules were deposited with Tidd Pratt, pursuant to 18 and 19 Victoria, cap. 63, s. 44, had not the protection for their funds which is implied in that section of the Act. The first step taken was the appointment of a deputation to visit Nottingham and Sheffield, with a view of ascertaining, if possible, if any, and to what extent, connection existed between the trade societies and the outrages. The deputation consisted of Mr. Danter, specially appointed by the Council of the Amalgamated Engineers, of which society he was at the time President, and Mr. G. Odgers, Secretary of the London Trades' Council. We think it necessary to state that this deputation was appointed on the 20th of October, 1866, and proceeded immediately on their mission, and presented the subjoined report some time before the Royal Commission to enquire into the internal working of trades' unions had been asked for:—

SHEFFIELD AND NOTTINGHAM OUTRAGES.

Report of deputation appointed by the Trades' Council of London, and the Council of the Amalgamated Engineers, to visit Nottingham and Sheffield, with a view to ascertain whether the outrages recently committed in those towns were in any way traceable to the Trade Societies:—

"Fellow Workmen,—Soon after our arrival in Nottingham, we visited Mr. Dalton, the proprietor of the eating house where the outrage was committed. We found that the reports in the local and other newspapers were not in the least degree exaggerated regarding the brutal treatment Mrs. Dalton had been subjected to. She had been struck on the nose by some cowardly scoundrel with such force as to break the bone in two or three places, and to blacken both her eyes, whilst in the laudable act of protecting an aged man called Thurkettle, and others who were lodging in her house, from the dastardly attack of three ruffians who had forced their way into an inner room, and assaulted every one there in a most fiendish manner. The lodgers referred to were non-unionists, who had been brought into the town to work by the 'Master Builders' Association,' to fill the places of unionists who were on strike. As the men arrested were members of the Bricklayers'

6

Labourers' Society, a strong suspicion existed that the society had induced the outrage. We then had a long interview with the Bricklayers' and Labourers' committees, who not only disclaimed any knowledge of the affair, but added, that nothing could have done their cause more harm, either morally or pecuniarily, than the outrage had done. A general meeting of the two societies was called, to enable us to put what questions we thought proper to the members. It is our bounden duty to state that nothing transpired to give us the slighest suspicion of the societies being implicated in the outrage, for which the members expressed their extreme regret, and have started a subscription amongst the societies for the sufferers. We also obtained an interview with the Master Builders, who received us in a courteous manner, and who have offered a reward of £10 for the apprehension of the instigators of the outrage. We then proceeded to Sheffield, and had several interviews with both unionists and non-unionists. We have thought it advisable to give a few of the statements made by workmen on rattening. Mr. John Wilson, who for years past, both in public and private, has been one of the most determined opponents of the doings of the Sheffield Trades' Unions, said that he had been a member of the trade society about three years; he was not so now. The trade was the Pen Bladegrinders. He had been a collector and one of the committee part of the time, and was sure that rattening was encouraged by some of the societies; that it was a disgrace to those who countenanced it, and that it damaged trade societies in the estimation of those who otherwise might be disposed to think favourbly of them. He had known a great number of men who had been rattened, and could prove that they were compelled to pay large sums of money before they could get their things returned to them. He believed the principles of trades' unions to be sound, and should have no objection to join a well-regulated trade society. He would never rest satisfied until the rattening, which he must admit had very much diminished of late years, should be entirely done away with.

"To those who are not acquainted with the term 'rattening,' we beg to offer this explanation :—

"When a workman has given offence to the society connected with his trade, either by working against the rules, or not paying his contributions, it has been the custom, in Sheffield, with a few of the societies, say three or four, to take away, at night, the bands or nuts from his wheels, to prevent him working until the contributions are paid or the rules complied with. We should state also that several respectable working men informed us that rattening had been admitted by some of the employers, and tacitly countenanced by others, as the best mode of making the men comply with the rules and regulations of their societies.

"We were introduced by Mr. Wilson to Mr. Thomas Cawthorn, who said he was 64 years of age, was formerly a member of a trade society, but was not so now; had himself taken workmen's bands away to make them comply with the rules, but had never received any pay for anything of the sort. It was a regular custom for bands to be taken; some of the men agreed to have their bands taken if their contributions were not paid by a certain date.

"Mr. Taylor was a member of the saw-grinder's society six years. In

1862 he lost the whole of his bands and tools whilst in the employ of Mr. George Taylor. The reason for this was—his lad was asked by the employer to carry some work to him (Taylor) which was contrary to rule. A man who worked in the shop informed the society, and then his bands were taken away. He went on working. On the following Monday morning, the secretary, Mr. Broadhead, called for his contributions, which he (Taylor) refused to pay until he got his bands returned. This the secretary would not agree to. Three weeks afterwards he lost his nuts. The master became uneasy about the affair, had a watchman put on, and requested an amicable settlement with the society. Taylor refused at first to move in the matter; but the master, having seen Mr. Broadhead, informed him that he must pay the money. He then went to Broadhead, who told him to pay £2 down; his contributions he could pay by instalments. This he complied with, and asked for a receipt, but was told that it was against the rules. The tools he was to have back the next morning. The next day the proprietor of the wheel brought him a note, saying he would find his nuts and bands in the hen roost where he settled his grievance. He went to Broadhead, who said, 'Have you got your things?' Taylor said, 'No; they are in your hen roost.' Broadhead and he then went to the hen roost, and there they found the things. He then returned to his work. Three weeks afterwards a general meeting of the society was called. He attended, and was very much abused and threatened for having placed certain letters which he had received from Broadhead in the hands of Mr. Wilson. He left the room, iniending to have no more to do with the society, although he continued to pay to it some time afterwards. A man who had worked with him, and had been taken away from his work in consequence of the bother, called for some tools he had left behind. Taylor told him he could still work there if he liked. The man said he could not unless the society was settled with. To this Taylor objected, and the man said, 'If you don't mind you will have your head blown off.' Taylor then said, 'If any one interferes with me or mine I will blow Broadhead's brains out.' He had not received any harm from them since.

"We have had four interviews with Mr. Fearnehough, the man whose house was blown up. The conversation on each occasion lasted for some time; the substance of it amounts to this: He was a saw-grinder by trade, but had never been a very regular member of the society, although, taking one time with another, he had belonged to it about ten years. He would have been more with the society but could not agree with the officials. They did not act fair. He was always having rows with the secretary. Broadhead did not like him. At last he got behind with his contributions; then they took his bands away, while he was working at Butcher's wheel. He did not trouble the society about it, but went and bought new bands. This was on the 25th of last November. The new bands cost him £5. He went on working till the 1st of January, when he told Green, a fellow-workman and a member of the society, that he should like to join the society again. Green spoke to Broadhead about it, and the result was a meeting of Broadhead, Green, and himself, at Green's house. He spoke to Broadhead in a friendly manner, and asked what he owed to the books of the society. Broadhead said about £13 or £14. Fearnehough then wished to have the £5 which he had paid for his bands taken out of it, but was told that he must pay £6 down and the rest in instalments, and when the whole of the money was paid he could have his old bands returned. He was sure that he did not owe the money, as he had only been three months out of the society. It was put on by Broadhead, who was at the bottom of it all. Since then he had done all in his power to 'beat' Broadhead. He knew Broadhead had vengeance against him, and he had sworn vengeance against Broadhead, and, if it cost him his life, he would beat him before he had done with him. He was not surprised at being blown up, he expected it long ago. He liked the trade society, and would join it again if Broadhead was out of the way.

"We felt it to be our duty to visit the Council of the 'Organized Trades' of Sheffield, for the purpose of putting such questions to them as we thought consistent with the responsible position in which we were placed. The result of which is, that Mr. Broadhead unequivocally admits that rattening, to the extent of taking away bands and tools, and detaining them until fines and contributions are paid, has been the custom of some of the societies ever since they have had an existence. He states that it has very much diminished during the time that he has been secretary, and that he has checked it as much as he possibly could; and feels certain that the denunciation of of it by the trade societies will prevent any further repetition of it in any way whatever. At the same time he repudiates, in the most emphatic matter, any connection, of the most remote character, with destruction of property, or injury to the person of any human being.

"In conclusion we say, that we have not been able to trace any complicity in the Hereford-street outrage to the trade societies, or to any one member of them, and hope that the guilty person or persons will speedily be brought to justice. We feel it to be our duty also to caution the men of Sheffield against lending their countenance to the abominable practice of rattening, which is calculated to demoralise those who are concerned in it, and to bring disgrace upon all trades' combinations.

"Robert Danter, President Amalgamated Engineers.
"George Odger, Secretary London Trades' Council."

The report of the deputation should be sufficient to convince any one who may have any doubt on his mind as to the opinion of the London Trade Societies regarding outrages, that not only were those societies free entirely from them, but that they condemned, in unmistakeable terms, the system of rattening, which they found existing in Sheffield. Not only did the Societies do this, but they opened a correspondence with the Executive of the Organised Trades of Sheffield, as to the best means to be adopted for the purpose of bringing the perpetrators of these horrible crimes to justice. We found that that Executive had met, within three days of the

9

perpetration of the Hereford Street outrage, expressed their abhorrence of the deed, resolved on calling a Delegate Meeting of the whole of the trades at which a reward of £100 was offered for the apprehension of its perpetrators. This sum was afterwards supplemented to £1000 from the Government, and £25 from the Amalgamated Society of Carpenters and Joiners, and other sums, amounting in all to £1,140. From this time commenced an active communication between the Sheffield and London Trades, which has continued up to the present time, and still continues. On the 29th of October the Sheffield Town Council resolved on sending a deputation to the Home Secretary to urge for the appointment of a commission of enquiry in reference to the Hereford Street outrage, and the following correspondence will show the reader the precise means which were employed to bring about the enquiry:—

THE SHEFFIELD TOWN COUNCIL AND DEPUTATION TO THE HOME SECRETARY.

At a meeting of the Sheffield Town Council, held in the Council Hall, the Town Clerk read the following letter from the Home Office:—

"Whitehall, 6th November, 1866.

"Sir,—I am requested by Mr. Secretary Walpole to acknowledge the receipt of your letter of the 3rd inst., and to inform you that he will hold himself in readiness to receive the deputation from Sheffield on Tuesday, the 13th inst., at Two o'clock.

"I am, Sir, your obedient servant,

"The Mayor of Sheffield." "BELMORE.

Ald. Saunders wished to point out that the resolution passed at the last meeting of the Council referred simply to an enquiry "into the cause or causes of the explosion in Hereford Street." Now, he saw by the advertisement with regard to the deputation that they were to solicit the Secretary of State to inquire also into the working of the general system of trades' unions. He wanted to know whether the deputation from this Council were going to exceed their powers by asking for any such general inquiry, because that would be a question open to very grave discussion—(hear, hear)—and it was not justified by the resolution of the Council. The Mayor: The deputation of this Council must confine itself entirely within the resolution. (Hear, hear.) It must not go beyond that, and I believe it will not. (Hear, hear.) The Town Clerk read the following letter:—

"Amalgamated Society of Carpenters and Joiners, Northumberland-st., Strand, W.C.

"London, November 6th, 1866.

"Sir,—I am directed to respectfully inform you that Mr. R. Danter, Chairman of the Amalgamated Society of Engineers, Mr. G. Odger, Secretary of the London

10

Trades' Council, in conjunction with myself, have been authorised by the Executive of the Organised Trades of Sheffield to make the necessary arrangements for a deputation of representatives of trades from Sheffield and London, to accompany the deputation appointed by the Sheffield Town Council to wait on the Home Secretary, and to support the applicatipn for a commission of inquiry, in accordance with the resolution of the Sheffield Town Council, passed on the 29th ult.

I am, Sir, yours respectfully,

"R. APPLEGARTH,

"General Secretary of the above Society.

"To the Town Clerk, Sheffield.

"P.S.—We shall feel obliged if you will kindly inform us of the time named by the Home Secretary at your earliest convenience."

To that letter the following reply was received:—

"Sheffield, 8th November, 1866.

"Sir,—By direction of the deputation appointed by the Town Council to wait on the Home Secretary in reference to the Hereford Street outrage, I beg to forward to you a copy of a resolution passed by them at a meeting held this day.

"I am, Sir, your obedient Servant,

"JOHN YEOMANS, Town Clerk.

"R. Applegarth, Esq."

At a meeting of the deputation appointed by the Town Council to wait on the Home Secretary in reference to the Hereford Street outrage, held at the Council Hall, Norfolk Street, Sheffield, on the 8th day of November, 1866, the Mayor having laid before the deputation a letter received by him from Mr. R. Applegarth, dated London, November 6th, 1866,—Resolved—

"That in reply to that letter Mr. Applegarth be informed that the Home Secretary has fixed Tuesday, the 13th instant, at Two o'clock, to receive a deputation from the Town Council, the Cutlers' Company, the Chamber of Commerce, and the New Hereford Street Outrage Committee, and that it would not be right or proper to take any other deputation with them to that interview, or, in fact, any other persons except the Members of Parliament who may be asked to introduce them. This deputation submit that it is for the Organised Trades to apply to the Home Secretary."

The following letter was sent to the Town Clerk, in reply to his announcement that a deputation of the trades' societies could not accompany the employers' deputation to the Home Office:—

"November 9th, 1866.

"Sir,—I am instructed to acknowledge the receipt of your letter, and copy of resolution, and to assure the deputation that the representatives of the trades are aware of the fact that the proper course to obtain an interview with the Home Secretary is to apply to him for that purpose. But seeing that application had been made for an interview, they did not think that any formal objection would be made to a deputation from the trades attending with the deputation from the Sheffield Town Council, as their object is ours. We thought that the one interview would serve the purpose, and a commission be more likely to be granted when applied for by employers and workmen at the same time.

"I am, Sir, yours respectfully,

"R. APPLEGARTH.

"John Yeomans, Esq., town clerk, Sheffield."

11

The advice of Mr. Thomas Hughes, M.P., was sought in regard to the matter. The hon. gentleman recommended the representatives of the trades to endeavour, by all means, to obtain an interview with Mr. Walpole, in conjunction with the manufacturers, if possible; if not, as a separate body. The desire was general amongst the leaders of the great trade societies in London that a Royal Commission should be issued, in order that, if possible, the truth might be sifted from the mass of conflicting statements that were being put forth on both sides. The delegates resolved to lose no time in the matter, and at once communicated with the Home Secretary in the following terms:—

8, Northumberland-street, Strand,
November 13th, 1866.

"Sir,—I am instructed to respectfully request you to grant an interview to representatives of trade societies, who are deputed to urge the appointment of a Commission to inquire respecting the late diabolical outrage in New Hereford Street, Sheffield, as proposed by the Sheffield Town Council. We think it right also to inform you that the representatives of the trade societies, desirous of avoiding any unnecessary intrusion on your time, applied through the Town Clerk to the deputation appointed by the Sheffield Town Council to accompany them at the time named by you to receive them. This they declined, for what reason we are at a loss to know—the more so as we considered the Commission of Inquiry asked for to be on public grounds, and in the interests of truth and justice. We have not been able to communicate with you as early as we could have wished, as our deputation is composed of trades' representatives of Sheffield and London; and for that reason we have had to confer and decide as to the course we should pursue in an emergency for which we were not prepared.

"On behalf of the deputation, I am, Sir, yours obediently,
"R. APPLEGARTH.

"The Right Hon. S. Walpole."

The following reply was promptly received:—

WHITEHALL,
14th November, 1866.

SIR,

I am directed by Mr. Secretary Walpole to acknowledge the receipt of your letter of the 13th instant, and to inform you that he will be happy to see the deputation, on whose behalf you have applied, at the Home Office on Saturday next, at two o'clock.

I am, Sir,
Your obedient servant,
BELMORE.

Mr. R. Applegarth,
8, Northumberland-street, Strand, W.C.

INTERVIEW OF THE TRADES' DELEGATES WITH MR. ROEBUCK.

On Thursday, November 12th, the Sheffield delegates, Messrs. Dronfield and Austin, accompanied by Mr. Odger and Mr. Applegarth, had an interview with Mr. Roebuck, at his residence in Ashley-place, to apprise him that the Home Secretary had fixed the time for the interview.

The delegation took the opportunity to explain to the hon. member the action taken by the Organised Trades in the Acorn-street and the New Hereford-street outrages. On both these questions Mr. Roebuck, upon receiving information of the course that had been taken, expressed an opinion that the executive had acted very properly, and had taken the most efficient means to free themselves from any suspicion of complicity with the incendiaries. The hon. member entered fully into an explanation of his views in respect to the proposed Commission, similar to that which we are given to understand he made to the Home Secretary when he accompanied the employers' deputation. He pointed out that such a Commission as he had mentioned in a letter to Mr. Dronfield could only be issued by the authority of Parliament. A most comprehensive investigation was, in his opinion, required, and in order to carry that out the commissioners must have full powers; whereas, if a commission were to be issued with the limited powers with which the Home Secretary could invest it, the attendance of witnesses and the production of books and papers, &c, would be purely voluntary, and no indemnity could be given. He did not think for a moment that the Home Secretary would grant such a limited commission, but that he would bring the whole subject before Parliament. The hon. member further remarked that in his view such a commission was necessary in the interests of the working classes themselves, to relieve them from the suspicions that attached to them. For his own part, he felt considerably relieved by the information which had been afforded to him by the deputation. He alluded to the part he had taken in 1824, in the repeal of the Combination Laws, with the late Mr. Hume, and the predictions that were then made of reckless and lawless conduct on the part of the working classes. Those predictions had been to a great extent falsified; but he regretted that such outrages as those which had occurred at Sheffield and other places continued to afford grounds for unfavourable comment. The deputation furnished the hon. member with a large amount of information, and he said that he should not fail to make use of it.

REPORT OF THE INTERVIEW WITH THE HOME SECRETARY IN REFERENCE TO THE ROYAL COMMISSION.

On Saturday, November 17th, the representatives of the London and the Sheffield Trade Societies had an interview with Mr. Walpole, the Home Secretary, on the subject of the proposed Commission of Inquiry into the incendiary explosion in New Hereford-street. The deputation were extremely desirous that the fullest publicity should be given to everything that took place at that interview, and they requested several reporters to accompany them. The representatives of the press were met, however, by "an express order" from the Home Secretary that they should not be admitted. Mr. Roebuck informed the reporter that personally he had not the least objection to the publication of the proceedings, and Mr. Applegarth and Mr. Dronfield sent in a special request that the reporters might be admitted; but the "Minister" was possessed with the idea that the presence of reporters would be injurious, and he refused to withdraw the order.

Mr. Roebuck had taken a lively interest in this deputation, and he acceded to their request that he should accompany them to the Home Office with great readiness. Mr. Hadfield was equally anxious to assist them; and the hon. member came up specially from Manchester on Friday night in order to attend with the deputation. Although deeply engaged just now with the preparations for the forthcoming reform banquet, the hon. member said that this was too important an occasion for him to hesitate for a moment whether he should come up to the metropolis or not.

The deputation consisted of Mr. William Dronfield, Secretary to the Sheffield Association of Organised Trades, and Mr. Austin, Secretary of the Railway Spring-makers' Society, as the representatives of the Sheffield trades' societies; Mr. R. Applegarth, General Secretary of the Amalgamated Society of Carpenters and Joiners; Mr. Powell, Assistant-Secretary of the Amalgamated Society of Engineers; Mr. Odger, Secretary of the London Trades' Council; Mr. Coulson, Secretary of the Operative Bricklayers' Society; Mr. Lawrence and Mr. Mackay, of the West-End Tailors' Society; and Mr. Cope, of the West-End Boot-Closers' Society.

Mr. J. A. Roebuck, M.P., introduced the deputation to Mr. Walpole. The hon. member said that on Tuesday he had the honour of introducing a deputation of employers; and he now had the honour of introducing a deputation of workmen. On this, as on the previous occasion, he had felt that the subject was too important for him to speak upon it without due preparation. He had read a paper on the subject on Tuesday, and he had now prepared another, which, with the permission of the Home Secretary, he would read. The purport of his paper was, that trades' unions were desirous of the fullest possible investigation into these outrages, they having the fullest confidence that the charges of complicity which had been brought against them could not be sustained. The unions courted the fullest investigation into their conduct, because they believed that these outrages not only injured them, but were detrimental to the town of Sheffield and the country at large; and in order to remedy the evil, they were present to request that a Commission of Inquiry might be instituted, which should trace those crimes to their true source. Since the interview on Tuesday, the hon. member said he had given the subject his most serious consideration, and he had come to the conclusion that it would be in all respects desirable to limit the period over which the inquiries of the proposed Commission should extend. He thought that the time should be limited to ten years. Charges might be brought against the unions by employers which might concern workmen who were dead, or had left the country; and on that ground he thought such a limitation of time as he had suggested would be desirable. He wished the Home Secretary distinctly to understand that this suggestion was purely his own, and that he alone was responsible for it.

14

In conclusion, Mr. Roebuck alluded to a point of great practical importance in the working of the Commission. The workmen, he said, had to labour hard for their daily bread; and whilst they were desirous to have the most comprehensive assistance, and would be willing to give every assistance in their power, yet they were poor men, and attendance upon such an inquiry would entail great expense, and probably that would be the means of deterring many from giving all the information in their power. Under these circumstances, and seeing that the question was one of national interest, he thought that the nation ought to defray the expenses of the witnesses. The suggestion seemed to him to be so reasonable and fair that he could not doubt of its acceptance by the Government. Mr. Roebuck then introduced

Mr. Dronfield, who said he had the honour to appear with his friend Mr. Austin as the representatives of the Association of Organised Trades of Sheffield. They were appointed to wait upon Mr. Walpole for a twofold object; in the first place to explain to him the action which that association had taken in reference not only to the late diabolical crime committed in New Hereford Street, but the Acorn Street explosion; and, secondly, to state what their views were in regard to the proposed commission. In the outset, he wished it to be distinctly understood—and the feeling was shared by all his friends present with him—that he most heartily detested and abhorred such crimes, and protested against trades' unions being stigmatised as having in any way connived at or countenanced them. They did not for a moment contend that their trades' societies were perfection. Like all other societies, they were capable of improvement, and if it could be pointed out where a change was requisite, they would be only too glad to adopt any practicable suggestion. Shortly after the incendiary explosion in Acorn Street, in 1861, the delegates of the organised trades met and passed resolutions strongly condemnatory of the outrage, and also authorising the executive to issue a document setting forth their views on the question. That document was very widely circulated; a copy was sent to the editor of every newspaper in the kingdom as far as their number could be ascertained. He handed a copy to Mr. Walpole, along with copies of every other document that had been issued upon the subject by his association. At a public meeting that was held in the Mechanics' Hall, a committee was formed to consider the best means that could be adopted for the suppression of these crimes. That committee was composed indifferently of unionists, non-unionists, and manufacturers; and at its first meeting representatives of twenty-two trades were present, representing the principal trades of the town. The gentleman who was appointed as secretary (Mr. John Fisher) shortly afterwards left the town in order to enter the ministry; and the committee appointed him (Mr. Dronfield) as Mr. Fisher's successor. Mr. Edwin Round, then a member of the Council, was appointed chairman; and the committee held numerous meetings, at one of which the chairman and others were appointed a deputation to wait on Mr. John Brown, who was then Mayor, and such other members of the Council as were manufacturers, as might be thought desirable. From the reports which that deputation gave, and the correspondence which he had with Mr. Brown, it appeared that the manufacturers were not willing to co-operate with the committee, and after an existence of some nine or ten months, the committee collapsed, after passing this resolution:—

"The committee appointed at the public meeting held on the 19th March last, to take steps for the suppression of outrages, having frequently met in conference with a number of trades' representatives and after a correspondence with the Mayor (John Brown, Esq), as well as a personal delegation with his Worship and various manufacturers, regret that their labours have proved so ineffectual. That under the circumstances in which the committee are now placed, they can see no other course left than to settle their few liabilities, and to dissolve forthwith."

Speaking of the Acorn Street outrage, he thought it only proper to call Mr. Walpole's attention to the fact that after Thompson (the man charged with the offence) was acquitted at the assizes, he was expelled from the Fender-grinder's Society, and he (Mr. Dronfield) was informed that Thompson was now employed by the firm the man worked for whose house it was alleged Thompson had blown up. Mr. Dronfield also referred to the third annual report of the Organised Trades, in which there was an address to Lord Palmerston upon the Acorn Street murder, explanatory of the action taken by the Organised Trades in reference thereto. He then referred to the late unfortunate and diabolical outrage in New Hereford Street. That occurred on the 8th of October. On the following day the chairman of the Association of Organised Trades waited upon him, and they resolved to call a special executive meeting in order to take steps in reference to the crime. At that meeting, which was held on the 11th, the following resolution was unanimously passed:—

"That this executive cannot but regret the recent outrage in New Hereford Street, and hereby enters its strong protest and utter detestation at the perpetration of such diabolical acts, by whomsoever committed, not only on account of the heinousness of such offences, but also because they tend materially to damage the trades' society of the town and country generally, and to retard the progress of trades' union principles; and are prepared to recommend the delegates, at a special meeting to be called as early as possible, to offer a reward for the discovery of the villain or villains who committed the said outrage."

It would be seen by a comparison of dates, that the representatives of the workmen had actually taken steps before the manufacturers themselves, inasmuch as their meeting was held on the day preceding the meeting in the Council Hall. A special delegate meeting was held in the following week; resolutions strongly condemnatory of the outrage were passed; the executive were authorised to offer a reward of £100 (to be raised by levy upon the members); Mr. Broadhead, the secretary of the Saw Grinders' Union, offered £5, the union itself offered £10, and the Amalgamated Society of Carpenters and Joiners, through Mr. Applegarth, added the further sum of £25, making, in all, a total of £140 as the reward offered by the trades for the discovery of the criminal. Hundreds of placards conveying this offer had been posted in Sheffield, and sent to the neighbouring towns; and it was important to observe that the executive did not request that information should be given to them, but to the Chief-Constable of Sheffield. From this narrative he trusted that the Home Secretary would see that the trades had taken prompt action, and they hoped he would be convinced, as well as others, that they were earnest in their endeavour to bring the criminals to justice. They were desirous that a Commission should be appointed, in order to give them the fullest opportunity of removing the stigma now cast upon them. That stigma was detrimental not only to the trades themselves, bnt also to the town generally, in whose prosperity no body of men had a deeper interest than the working classes. In conclusion he explained that the deputation, having been refused permission to attend with the representatives of the Town Council and the outrage committee, for reasons they could not at all understand, had felt bound to request the honour of a separate interview. They were the more surprised at the refusal they had met with because the object aimed at by the Town Council was one of which they highly approved; and they felt that they were not asking too much to be allowed, as the representatives of a large body of working men, to form a part of that deputation. After thanking the Home Secretary for the attention with which he had heard his remarks, Mr. Dronfield gave way to

Mr. Austin, who began by expressing the detestation in which he held such crimes as that which formed the subject of this conversation, and by saying that he had been appointed, in conjunction with Mr. Dronfield, to ask for a commission of inquiry into the outrage, in the terms of the resolution passed by the Town Council. His principal reason for desiring that a commission might be granted was that the responsibility of these crimes had been cast upon the trades' unions, who, he felt, were not responsible for them. Instances could be adduced which would show that outrages in Sheffield were not always attributable, either directly or indirectly, to trades' unions. Mr. Austin concluded by asserting that he had adduced sufficient instances to prove that in regard to several of these violent outrages there was not the shadow of a pretence for accusing trades' unions.

Mr. R. Applegarth said: I regret extremely that I have not the opportunity of speaking before the gentlemen who were here on Tuesday last, because I believe that they may have made statements concerning us, and we may entertain opinions concerning them, which may be grounded in misapprehension; and if we had spoken in each others' presence, we might have found out our error, and they might have been convinced that some of their facts—or presumed facts—were not worthy of your consideration. I regret exceedingly that those gentlemen are absent. I also deeply regret that the press are excluded. I regret it the more because an accusation is being continually hurled at trades' unions that they are afraid of giving information of what they are doing, and that they are essentially secret organisations. To counteract that charge, our invariable rule is to make our proceedings as public as possible, and on all such occasions as this we ask reporters to accompany us. I think that no better guarantee of our good intentions could be adduced. Sir, it would be a waste of words to say that I am opposed to trade outrages. No honest man could say anything else; I am as anxious, with the union to which I belong, for an inquiry as any one can be. The following resolution was unanimously adopted by the Council, and ordered to be forwarded to Mr. Thomas Hughes, M.P. for Lambeth:—"That the Executive Council of the Amalgamated Society of Carpenters and Joiners regard with pleasure the notice of motion of Mr. B. Cochrane, 'that he would call the attention of the House to the dangerous increase of Trades' Unions, and move for papers,' and trust that nothing will occur to prevent the hon. gentleman persevering in his intentions, and that it may lead to a thorough searching inquiry into the operations of trades' unions. To such inquiry we should be willing to render every assistance in our power." Seeing that the inquiry is likely to take a wider scope than that suggested by the Town Council, and as Mr. Roebuck has intimated that he should desire to see a searching investigation, extending over the last ten years, by a Parliamentary Commission, and seeing, further, that these serious crimes are imputed to unionists, I think it is only fair that the representatives of the unions should claim to have a voice as to the scope of that inquiry, and the direction in which it should go. And I may here observe that it will be in the highest degree undesirable to conduct that inquiry as though the unions were on their trial "at the bar." The great bulk of the unionists repel with indignation the suspicion of complicity in such crimes, and if they are put, as it were, on their defence, and not asked to come forward as assistants in an inquiry to further the public good, they will refuse to do any such thing. If the object of the inquiry is not to attack trades' unionists solely, but to search out the causes of these disputes and outrages wherever they may be found, then I may venture to say that the unions will be happy to render all the assistance they can to find out who the offenders are. But don't let it begin with the assumption that the unionists are solely to blame. I have again, sir, to express my regret at the absence of the employers, because what I am about to say affects them very deeply. It has been the practice for many years to attribute to trades' unions all the heart burnings and disputes, and latterly these outrages, that arise; but I believe these are attributable to many diverse causes. In the first place, Sir, your own experience of the world will have taught you that there are men, as well as institutions, who still adhere to the same old style of doing business, and who persist in applying the rules that were applicable 20 years ago; and we don't claim that trades' unions are altogether an exception to this rule. We are free to admit that, on investigation, it may be found that there are some unions which have at the present time rules and regulations (commonly called "restrictions") that would be more applicable to the state of things that existed 20 years ago than to our present circumstances; but I believe that they will be exceedingly glad to alter those rules if upon due inquiry they should be convinced that they are not in accordance with the spirit of the age. This being admitted, it may be that these sad crimes are to some extent attributable to those causes; but while admitting this, we desire to make known our feeling on another matter that I cannot better explain than by giving an example. In one of the largest firms in the saw trade in Sheffield, a friend of mine was employed, and he signed an agreement to serve for twelve months. The terms of the agreement were, that he was to take 20 per cent. off the "statement" price,—that is, he was to deduct 20 per cent. from the list of prices at which the men of his trade were supposed to be working. This secret compact was drawn up by the master and signed by both sides. It was agreed that the man should have all the circular saws to hammer—the best class of work, at which he could make very good wages, even after deducting the 20 per cent. That agreement was kept for some time until another man came to work, and my friend found the best work leaving him, and it was being given to the new comer, who he suspected would of course be deducting a larger per-centage. At a meeting of the trade society, at which both these men were present, a complaint was made that some men in the town were dissatisfied because they were working at ten per cent. off the "statement." Some one cried out, "That is not much to complain of, when other firms are knocking off 30 or 40 per cent." As soon as that was heard, loud cries of "name" were raised, and the name was given. The meeting was a very stormy one; and when my friend went to work next morning, he was asked by the manager to point out on the "statement" where 40 per cent. was being deducted. He replied that he knew nothing about it. The other man came in and he was asked the same question. He instantly pointed out the items, and the manager struck him and knocked him down. The man, the friend of mine, who was taking off 20 per cent. complained when he found that the work for which he had bargained was being put into another man's hands.

Mr. Walpole: That would be a violation of the agreement, would it not?

Mr. Applegarth: In a moment you will see the value of that agreement. In answer to my friend's complaints, he was told, "Well, it amounts to this; you are not taking enough off," but he refused to take off any more. He refused, in the presence of his employer, to take out any more work, and his employer used personal violence to him in the office. He said he should seek to enforce the agreement, and the manager and the master both laughed, and coolly informed him that the document was worthless because it was not stamped. He charged them with having acted dishonestly, gave notice, and left the place. Now, Sir, I don't say that this is a general practice, but there are undoubtedly such cases constantly occurring. They engender, naturally, bad feeling between master and man, and are just as likely to lead to these sad outrages as any restrictive rules or regulations of trades' unions.

Mr. Odger thought that if the inquiry should extend over only five years the ends of justice would be met. He thought that the difficulty of obtaining witnesses to give evidence in reference to charges which may be ten years' old would be very great indeed. Mr. Lawrence and Mr. Mackay warmly supported the proposal to have a commission, and expressed their hope that it would be a thoroughly comprehensive and impartial one, conducted without reference to the opinions or the prepossessions of either one party or another.

Mr. Walpole said that, after his interview with the employers' deputation he named the matter to his colleagues in the Cabinet. They had had a very long conversation on the matter; and he thought he might venture to say,—although the deputation would not, of course, expect him to give a definite answer,—that in all probability the Government would be ready to grant a commission. If it were granted, both sides might rest assured that the inquiry would be thoroughly comprehensive and impartial towards both employers and workmen. He was glad that this deputation had waited upon him, and promised that the information they had given should have the fullest consideration on the part of the Government.

Mr. Hadfield: I am strongly of opinion that the inquiry should be thoroughly comprehensive and impartial. I have a gentleman in my mind now whom I should be glad to propose as a member of the commission. Would a Member of Parliament be competent to act?

Mr. Walpole: At the proper time I shall be glad to receive any suggestion from Mr. Hadfield.

Mr. Lawrence: I hope the commission will be issued as soon as possible.

Mr. Dronfield: I understand it cannot be issued until Parliament shall have met.

Mr. Roebuck: Precisely so It must come before Parliament, and no doubt there will be a long debate upon it.

Mr. Walpole signified that the matter must come before Parliament.

Mr. Roebuck then tendered the thanks of the deputation to the Home Secretary, and the interview terminated.

REPORT OF THE DEPUTATION OF THE SHEFFIELD TOWN COUNCIL TO THE HOME SECRETARY.

(*Obtained from a local paper; and is the only report of the doings of the Masters' Deputation with the Home Secretary.*)

The Mayor read the following report of the deputation appointed by the Council to wait upon the Home Secretary, praying for a Commission to inquire into the recent outrage:—

"The deputation appointed by the Town Council on the 29th October last, to wait on the Home Secretary, and confer with him on the necessity of granting a Commission of Inquiry in reference to the attempt made to blow up the dwelling-house of Thomas Fearnehough, in New Hereford Street, consisting of the Mayor (John Webster, Esq.), Ald. W. E. Laycock, Ald. John Brown (the Master Cutler), Ald. Robert Jackson (the President of the Chamber of Commerce), and Councillor Robert Thomas Eadon, met by appointment John Arthur Roebuck, Esq., M.P., and George Hadfield,

20

Esq., M.P., and deputations from the Cutlers' Company, the Chamber of Commerce, and the Sheffield Outrage Committee, at the Westminster Palace Hotel, on Tuesday, the 13th of November last, and proceeded from thence to the Home Office, and had an interview with the Right Hon. S. H. Walpole, the Home Secretary.

"The Mayor (John Webster Esq.) read the resolution passed by the Council on the 29th October last, and urged upon the Home Secretary the necessity of Government granting a Commission of Inquiry.

"The Home Secretary listened to the representations made by the Mayor and the deputations from the Cutlers' Company and the Chamber of Commerce, and accepted various resolutions; and then promised that the subject would receive the closest attention on the part of the Government, but until he had laid it before the Cabinet he could not be expected to give any more definite answer,

"J. WEBSTER, Mayor."

Ald. Saunders regretted, and he thought it was a regret which was generally felt, that there should be such an abstinence of explanation in regard to what the deputation did when before the Home Secretary. (Hear, hear.) He could not for one moment conceive—in fact, it was only just and reasonable—why the arguments which were placed before the Home Secretary by the deputation should not also be placed before the public. It now almost seemed as if the deputation were afraid of letting the public know what were their opinions or arguments and reasons for asking for investigation into the late outrage. The course which the deputation had taken was a very unwise one; because if what they said to the Home Secretary was true, it was only right that those who were impugned should know what their statements were—(hear, hear)—in order that they might be prepared to refute them. On the other hand, if what they said was not true, it should not have been said to the Home Secretary at all. This was the first time he had taken any part in a discussion with regard to the late outrage, and his object in now rising was to say that the secret and underhand way which was adopted by the deputation of going to the Home Secretary and making *ex parte* statements, was not honourable and straightforward. He should have liked for the deputation from such a public body as the corporation, to have gone with the working men's deputation, as the commission was asked, not for any individual or class, for the good of the town generally. He considered that a great example was set to the deputation from the Council and the other deputations that accompanied them by the deputation which went to the Home Secretary from the working men, as they boldly spoke out their sentiments, which were published to the world, and consequently the town knew what they had said upon a subject which so much affected it. (Hear, hear.) In that respect the Council and the manufacturers' deputations had been set an example. He personally and deeply regretted that a fuller and a much more general explanation of what was said before the Home Secretary had not been made known.

The Mayor thought that the remarks of Ald. Saunders were somewhat out of place. (Hear, hear.) The deputation which was appointed by the

Council went to the Home Secretary for one particular and express purpose—to lay before him the resolution passed by the Council, and to ask him to consider whether under all the circumstances it would not be advisable to comply with their request, and to issue a commission for a full inquiry into the late outrage. When, as their representative, he laid the matter before the Secretary of State, he made no accusations against any one; he impugned no man's conduct. He simply said this, that so far as such outrages were concerned, the arm of the law in Sheffield was paralysed, and that the police authorities could not make any discovery of the perpetrators, or the instigators, if there were any, of such crimes. (Hear, hear.) There was no doubt that a great crime was committed when Fearnehough's house was blown up, but he did not express an opinion to the Home Secretary as to by what man or by what class of men the deed was done. What he said was that a great and fearful outrage had been committed; that it was one which ought to be inquired into; and that the Town Council, through him, had asked that such an inquiry should take place. (Hear, hear.) He fully agreed with Mr. Roebuck and other gentlemen that the inquiry should be one by a Parliamentary Commission, an opinion which he expressed at the time the resolution appointing the deputation was passed. With regard to what had been said by Alderman Saunders about the deputation making *ex parte* statements, all he could say was that the deputation went to the Home Secretary for the express purpose of making *ex parte* statements—in fact, they could not do otherwise. They did not go to the Secretary of State to argue with him as to the persons who committed the crime, but all that they did, as he had before explained, was to ask the assistance of Government to find out the perpetrator, and, if possible, to adopt means for the prevention of such outrages for the future. (Hear, hear.) As to what was said by the members of the other deputations they had nothing whatever to do, but he could safely say that they made no charges against any one, and only expressed it as a matter of opinion, and not as a matter of fact, that the general impression in Sheffield was that such outrages proceeded from trades' unions. (Hear, hear.) The deputation from the Council did not even go so far as that; all that they requested was that a commission should be appointed. The statements which were made were purely *ex parte*, and they went to the Home Secretary to make them. Those who had set them an example, which it was said they should follow, did precisely the same, the only difference being that they published what they said, whereas the deputation from the Council did not. (Hear, hear.) He really did not think that what he said was deserving of publication, as he simply stated what was the request of the Town Council. In conclusion the Mayor moved that the report be entered on the minutes of the Council.

Mr. Ironside seconded the motion, at the same time remarking that he quite agreed with the remarks of Alderman Saunders, for wherever there was secresy there was corruption. For the good government of the town everything ought to be open and above-board. (Hear, hear.) The affair was a miserable one, and the sooner it was forgotten the better.

The motion was carried.

22

IMPORTANT DECISION OF THE COURT OF QUEEN'S BENCH IN REFERENCE TO TRADE SOCIETIES' FUNDS.

On the 16th of January the following important trial took place, and decision was given affecting materially the funds of Trades' Societies in the Court of Queen's Bench, before the Lord Chief Justice, Mr. Justice Blackburn, Mr. Justice Mellor, and Mr. Justice Lush.

HORNBY, APPELLANT, V. CLOSE, RESPONDENT.

This was an appeal from decision of magistrates at Bradford, dismissing an information under the Friendly Societies' Act against the treasurer of a society of working men, on the ground that the objects of the society, being partly those of a Trades' Union, were not within the Act. The information was for wrongfully withholding a sum of £24 belonging to the society. The society was the Leeds branch of "the United Society of Boilermakers and Iron Shipbuilders of Great Britain and Ireland," instituted in 1834 (as stated on the title-page of the rules), "for the purpose of mutual relief of its members when out of employment, the relief of their sick and burial of their dead, and other benevolent purposes, as inserted in their Rules," or (as the preface stated it) "for the relief of their sick, the burial of their dead, and support of their members when thrown out of employment through depression of trade, and for other benevolent purposes. The preface continued,—"There are but few among the working classes who are provident enough to realize out of their hard earnings a sufficiency to meet the demands of nature for even a few weeks when out of employment, and how much more serious is the consideration when a family is dependent on such individual? It is but too often in such cases that the domestic hearth, hitherto cherished as the greatest blessing and comfort of life, becomes a source of anxiety and care. To provide for the common wants of nature, to stay the cry of his beloved offspring for bread, the household goods on which his gentle partner and himself used to delight, one thing after another is sacrificed, and at length, when all is gone, then comes the heart-rending separation of a once happy family to seek food and shelter in a Union workhouse. To prevent these evils, or at least to obviate them in some degree, is one of the principal objects proposed to be attained by the society. These views can be carried into effect by uniting together while in employment and subscribing a small sum per week, which, under wholesome laws and economical management, will be capable of affording at least some assistance to those who occasionally have the misfortune of being put out of employment." Such, however, being the professed and perhaps the original objects of the society, the rules which, it is to be observed, had been revised and remodelled in 1862, contained several which were relied upon as showing that the present objects of the society were, in part, those of a Trades' Union Society—"Rule 28, Piece-work:—1. That in districts where members are compelled to work piece-work, and it be proved to the satisfaction of the executive committee that the firm is reducing the prices below the usual and reasonable prices, they shall allow the men resisting the reduction 7s. per week for two weeks; and, 2, That any member or members in a shop, either on piece-work or day work, where a dispute arises connected with our trade or society, no member or members shall be allowed to call at such shop or shops after being made acquainted with such dispute, or, for doing so, to be fined 10s. And that any member of this society encouraging any holder up or labourer to violate the rule by allowing him to practise with his tools, or otherwise instructing him in other branches of the trade contrary to these rules, shall, on proof thereof, be fined for the first offence 5s., &c., and for the third expelled the society." "Rule 29, Disputes on day Work and benefits:—Should a dispute arise in any shop, the members of that shop shall make it known to their branch, which, if it only affects the interests of two or three members, such branch to have power to settle it, and grant to members wishing to travel 12s. per week, &c. But should a general dispute arise in any shop which cannot be amicably settled by the branch, it shall be referred to the executive committee, who shall give their instructions on the subject. All members losing their employment through such disputes, after being sanctioned by the executive, shall receive the sum of 12s. per week so long as they shall remain out of employment; and this rule to be applied to all disputes except the settlement of piece-work prices." "Rule 42. Any member using his influence to obtain employment for a non-member shall be fined for such offence 10s." Such being some of the present rules of the society, the attorney for the party summoned contended that it was not a society within the Friendly Societies' Act, 18th and 19th of Victoria, cap. 62, and further that it was a society established for purposes illegal and contrary to public policy, as being a restraint of trade, and depriving the workman of the free exercise of his own will in the employment of his labour; and also in restraining him from getting employment, or continuing in employment, or obtaining employment for a non-member, and that the society was an organisation for or tending to the encouragement of strikes. The magistrates—the charge against the defendant being fully proved as laid in the information, were called upon to determine on the legal objections thus raised, and determining them in his favour, dismissed the information, from which decision this was the present appeal on the part of the society, their contention being that the society was established for some of the purposes mentioned in the Friendly Societies' Act; and that it was not established for any purposes which were illegal, and that, therefore, the magistrates should have convicted. To render the argument and the judgment intelligible it is necessary to state that the 44th section of the Act provides that a case of any Friendly Society established for any of the purposes mentioned in section 9 (which are those of an ordinary benefit or friendly society) or for any purpose which is not illegal, having rules certified by or deposited with the Registrar, disputes arising among the members may be summarily heard and determined by magistrates under the jurisdiction established by the Act.

Mr. Mellish (with whom was Mr. Macnamara) argued the case on behalf of the Society. His argument in substance was, first, that the objects of the Society were not illegal—(1) because the rules were not certainly a restraint of trade; (2) because even if a restraint of trade—though not enforceable, yet they were not illegal; (3) that they were not unlawful at common law; (4) that they were not rendered unlawful by anything in the Act. The Society, therefore, he insisted, was not illegal in the sense of being criminal; and, in the next place, it could not be denied that some of its objects came within the scope of the Act. The learned counsel cited and commented upon the great case of *Hilton* v. *Eckersley* (6 *Ellis and Blackburn's Reports*), which was affirmed in a Court of Error, and in which it was held that a bond given by a cotton-spinner, binding him to observe the regulations as to wages, and sanctioned by a majority of the Society, could not be enforced against him. In that case, however, he pointed out that Mr. Justice Earle dissented, and that Lord Campbell disclaimed holding that the combination was illegal in a criminal sense, and, though the judgment was upheld in a Court of Error, the Court expressly said—we do not say the combination was illegal in the sense of being criminal or punishable, and in this case the rules referred to were not those on which the information was founded. In the course of the argument

Mr. Justice Blackburn observed, that he could not help thinking that the trades' unions element was the main object of the Society; at all events, it was plainly a main object.

Mr. Justice Mellor observed that it could hardly be doubted that it was a main inducement to many of the men to joining it that it had this element in it.

The Lord Chief Justice said, at least the Society had the twofold object of the Friendly Society and the Trades' Union Society, and the rules as to the latter seemed to be in restraint of trade, and so not legal, even if not criminal.

Mr. Mellish urged that the main object, or, at all events, a main object, of the Society was still benevolent, and that the addition of these rules as to the terms of work did not so alter its character as to render it illegal. At the close of the argument,

The Lord Chief Justice at once said:—We cannot hesitate for a moment in saying that we thought the magistrates were right in holding that the Society did not come within the operation of the Friendly Societies' Act so as to give them jurisdiction. Supposing the main purpose of the Society were within the Act, as being benevolent, it would not merely by reason of one or two rules going beyond that object cease to be a Society within the Act. But here we find that the very purpose of the Society is not merely to carry out the objects of a benevolent society, but those of a trades' union. Under that term may be included every combination by which men bind themselves not to work except under certain conditions, and to support one another in the event of being out of employment, in conformity with the views of the majority. I am far from saying that a trades' union constituted for such purposes would bring the members within the criminal law, but the rules are certainly such as would operate in restraint of trade, and would, therefore, in that sense be unlawful, and on the same principle upon which a Court of Error held (confirming the decision of this Court) that a bond given by a master to observe rules which were in restraint of trade was so far unlawful that it could not be enforced at law, we think that these rules of a society of workmen having a like effect are in the same sense illegal. That is to say, if a civil action was brought on any contract or obligation arising out of the rules, they could not be recognized and enforced in such action. Therefore, upon the same principle, it was impossible to hold that the Society was within the Friendly Societies' Act; first, because the objects are not analogous to those of a friendly society; and next, because these rules, though they may not be illegal in the sense of being criminal, are so far illegal that they are not enforceable at law.

Mr. Justice Blackburn concurred. The magistrates, he said, had no jurisdiction unless the society was within the statute, and he thought it was not. The purposes of a trades' union were not at all analogous to those of a friendly society; and it was impossible upon any ordinary principles of construction to hold that sec. 44 applied to any society which was established for "a purpose not illegal." Such a literal construction could not be sustained; and it was obvious that the section only applied to societies the purposes of which were analogous to those of friendly societies. No doubt, a small degree of deviation from the purposes of a friendly society would not throw the society out of the scope of the Act. But, in the present case, it was manifest that the main object, or, at all events, a main object, of the society was to carry out the purposes of the trades' union. Further than that, he thought the rules referred to illegal in the sense in which the term was applied by the Court of Error in *Hilton* v. *Eckersley*, as against the masters—that is, illegal in the sense that they were not legally enforceable. The same principle must be applied to combinations of the men. He guarded himself carefully against being supposed to say whether the society was illegal in the sense of being criminal. He did not say that it was; he did not say that it was not. He left that point quite untouched. But he was clearly of opinion that the rules referred to were illegal in the sense that they could not be enforced; therefore the society did not come within the Act. Whether the illegality of some of the rules would taint the whole of them it was not here necessary to determine, because the objects illegal were not small and incidental, but constituted a principal, if not the whole, aim and object of the society.

Mr. Justice Mellor entirely concurred. He desired, he said, to express no opinion as to whether the rules referred to were illegal in the sense in which they would be criminal. But he thought that some of the substantial objects of the society were those of trades' unions, and were directed to the encouragement of strikes. He could not suppose that in the view of the members of the society the objects could practically be separated—they were too much mixed up together; nor could he doubt that numbers of men became members upon the footing of the illegal rules, and for the sake of the illegal objects. Those objects, the objects of a trades' union, were not analagous to those of a friendly society, because not really "benevolent," though some of the objects of the society no doubt were so. The objects of the society not being analogous to those of a friendly society, and not being legal in the sense of being enforceable, the society was not within the Act, and the magistrates, therefore, rightly determined that they had no jurisdiction to deal with the case.

Mr. Justice Lush also concurred. One main object of the society, he said, if not the main object, was to form a trades' union. That being so, the purpose of the society was not analagous to those described in the Friendly Societies' Act; and, further, its rules were illegal, in the sense that they were not enforceable in a court of law.

Judgment for the respondent, approving the decision of the Justices.

APPOINTMENT OF A PERMANENT CONFERENCE.

Immediately upon this decision being made public the delegates who had been attending to the Sheffield affair, recognizing the difficulties that were approaching Trades' Unions, thought necessary to call a Conference to further consider the aspect of affairs, and accordingly Delegates from the undermentioned Trade Societies assembled at the General Office of the Amalgamated Society of Engineers, 54, Stamford Street, at Two p.m., on the 28th of January, specially deputed by their respective bodies.

There were present T. Hughes, Esq., M.P. for Lambeth (by special invitation); Mr. W. Allan, Gen. Sec., Mr. R. Danter, Chairman, and Mr. D. Henderson, of the Council of the Amalgamated Society of Engineers; Mr. R. Applegarth, Gen. Sec., Mr. T. W. Hughes, Chairman, and Mr. J. D. Prior, of the Council of the Amalgamated Society of Carpenters and Joiners; Mr. D. Guile, Gen. Sec., Mr. J. Hemley, Chairman, and Mr. T. Powell, of the Council of the Iron-founders Society; Mr. G. Odger, Secretary of the London Trades' Council; Mr. E. Coulson, Gen. Sec. of the Operative Bricklayers' Society; and Mr. Thomas Spelling, Secretary of the Vellum Binders' Socieiy.

Mr. W. Allan having been appointed Chairman, and Mr. R. Applegarth Secretary to the Conference, a very animated discussion followed, during which an universal opinion was expressed that Trades' Unions, as at present constituted, paying vast sums of money for purposes truly laudable, and none of their objects being in opposition to law, ought to have the same protection as Friendly and other Societies; and that no means ought to be left untried to secure for the Trades' Unions of the country a Special Act for their especial protection. It was ultimately resolved:—

"That a Deputation, consisting of Messrs. Allan, Applegarth, Guile, Odger, and Coulson, be appointed to wait on the Home Secretary, to call his attention to the recent decision, and urge the enactment of a Law for the protection to which we feel convinced our Societies are entitled."

Mr. Applegarth was instructed to communicate with the Home Secretary to request him to grant an early interview, and to the Executive of the

Boiler Makers' Society to acquaint them with the proceedings of the Conference.

A vote of thanks was accorded to Mr. Hughes for his attendance; to the Chairman for his services during the Conference; and to the Executive of the Engineers for the use of their office; and the first sitting of the Conference terminated.

The Home Secretary having been communicated with, replied immediately, stating that he would be happy to meet the deputation on Friday, February 1st.

INTERVIEW WITH THE HOME SECRETARY.

The deputation to Mr. Walpole, by appointment, waited on the right hon. gentleman, for the purpose of representing to him, and through him to the Government, the position in which trades' unions have been placed by the decision of the Court of Queen's Bench in the case of the Bradford boiler-makers, and of impressing on the Government the importance of putting them in the same position with regard to protection as friendly societies in general. The deputation consisted of Mr. William Allan, representing the Amalgamated Engineers; Mr. D. Guile, the Ironfounders' Society; Mr. R. Applegarth, the Amalgamated Carpenters and Joiners; and Mr. George Odger, the Trades' Council. The deputation was introduced by

Mr. Thomas Hughes, M.P., who said he believed, and with great reason, that the decision of the Queen's Bench had placed trades' societies entirely outside the law. The deputation wished to lay their case before the right hon. gentleman, and ask whether the Government could not assist to place them in the position in which they understood they had been placed by the Acts of former Governments.

Mr. Allan said that so far back as 1855 they succeeded in getting two clauses inserted in the Friendly Societies' Act, by which they imagined they obtained legal protection against fraudulent trustees or other officials. It is now very questionable whether they had that protection; in fact, they were inclined to think they had not. The Court of Queen's Bench had decided that their rules placed them outside the pale of the law, and that they could not have the protection of the Friendly Societies' Act.

Mr. Walpole understood that the Lord Chief Justice decided that the society before him could only be looked upon as a society formed in restraint of trade, and could not be supposed to be legal.

Mr. Hughes said that in the rules of all the societies there were clauses that would probably be considered also as a restraint upon trade.

Mr. Allan said the Friendly Societies Act gave them certain privileges on depositing their rules with Mr. Tidd Pratt, provided they contained nothing contrary to law. In 1851 the society to which he belonged—in order that they might be perfectly satisfied as the legality of their rules, which went further then than they now did in regard to piece-work, overtime, length of hours, &c.—got the opinion of Sir Alexander Cockburn, at that time Attorney-General, who held that their rules were perfectly legal. If they were legal, the societies contended that they should have the protection of the Friendly Societies Act with regard to fraudulent trustees. On the faith of the protection they had deposited a large amount of money in the post office savings' bank. They had at first some difficulty about that, but they waited on Mr. Gladstone, who ultimately granted them the privilege of depositing their funds there. It was now questionable whether if any fraud took place they had any control over the persons who defrauded them. Mr. Gladstone decided that their rules were not illegal, and therefore they ought to have the benefit of the Friendly Societies' Act.

Mr. Walpole: What power had Mr. Gladstone to decide that? He only gave his opinion, I suppose?

Mr. Allan said he was Chancellor of the Exchequer, and representative of the Government.

Mr. Hughes said that there had been a direct sanction by the Government of the investment in the post office savings' banks.

Mr. Guile said the funds of the Amalgamated Engineers amounted to £140,000, and they were scattered in various parts of the country.

In reply to Mr. Walpole,

Mr. Guile replied that the rules were pretty nearly the same in all cases.

Mr. Hughes said those of the Bradford boiler-makers were rather stronger.

Mr. Allan said the rules on which Sir Alexander Cockburn gave his opinion were quite as strong, if not stronger, than those of the boiler-makers were ; and yet he said distinctly and clearly that they were not illegal, but perfectly sound according to law.

Mr. Walpole then read the judgment of the Lord Chief Justice, to the effect that he thought the society could not be brought within the Friendly Societies' Act—first, because the purposes for which the trades' unions were organised were not analogous to those of benefit societies, and because its rules, though not criminal, yet acting as a restraint of trade, were by the law of the land illegal. He also read the opinion of Sir Alexander Cockburn, which was to the effect that it was legal for men to combine to protect themselves, and to persuade others to act with them, provided no violence were done to person or property, and all threats, intimidation, and molestation were avoided, and that the rules were not objectionable in point of law, thoug they might have been prepared with greater accuracy. The right hon. gentleman then asked the deputation to point out what they specifically asked him to do.

Mr. Guile said the society which he represented instructed him to ask the Government to take their rules and put them into the hands of the law advisers of the Crown for examination. If they were of opinion that they were not legal, he was requested to ask that Government would provide such an Act as would place them within range of the law for all practical purposes. The speaker mentioned a case in which an agent of his society had £83 belonging to it, but which after the recent decision they could not claim. He also furnished a number of statistics to show them that the object of the societies was more benevolence than the advancement of the trades' unions. In eight years the society to which he belonged had expended on men out of work, but not through the operations of strikes, on the sick and disabled, on the aged and on funerals, no less than £138,503; whilst in the same period the total amount paid for strikes was £1,877. This was at the rate of 5s. 6d. per member to support strikes or disputes, while for benevolent purposes there had been paid £21 5s. 6¾d. per member.

Mr. Walpole said there was no doubt that would be perfectly legal. He did not know really how far the facts on the case as decided by the Court of Queen's Bench, were or were not identical for the purposes of the law with the case as regarded the deputation. He thought it so important a matter that the best thing the deputation could do was to put into writing, in the shape of a memorial, to be presented at that office, the exact thing they wanted him to do. That would enable him to consider and, if necessary, to consult with his colleagues as to how far they could fall in with the suggestions made to the Government. He saw a difficulty about consulting the law officers merely on the application of one society; for he should, if granting the facility in one case, have to afford it in others.

Mr. Hughes thought that the case of the societies was that they had been led into the present scrape by previous Governments; they had dropped their old system of protection by checking their officials, and they had been led to rely upon the protection of the Friendly Societies' Act. Now their funds were scattered all over England, and their affairs were exceedingly critical, because if their officers turned out rascals they had no power to recover a penny.

Mr. Walpole said the matter really depended on the ulterior question whether the societies were legal or not according to the judgment of the Court of Queen's Bench. The law officers could not decide such a question. Nothing but the court could decide whether the rules were contrary to law or not within the meaning or decision of the court. The law officers of the Crown could not give them any valuable opinion except as to what the decision of the Court of Queen's Bench would be.

Mr. Applegarth said the important question was, were trades' unions "as" trades' unions to be outside the pale of law. They had been of opinion that it was quite lawful for working men to combine for the protection of their interests, provided they did not bring themselves into collision with the law that provided against intimidation, &c. He would admit that if in any part of the country work was plentiful, men underpaid, and employers refused to raise their wages, they would draw the men from their employers, and pay them 15s. a week as strike pay; but they set their faces against all kinds of intimidation; and as for strikes, he would not say they dreaded them, but they did all they could to set their faces against them, and endeavoured to substitute for them a system of arbitration. He thought the suggestion made to them by the right hon. gentleman was a very good one; but he should wish that after they had put into writing what they desired, there should be some sort of inquiry to let the Government see what trades' unions really were. No doubt they did many sad and serious things, but they were not so bad as they were painted. Such an inquiry was necessary before any legislation was attempted. They required some specific law for their protection, for even now at Bradford and Hull they had cases of agents from whom they could not obtain large sums of money on account of the decision of the Court of Queen's Bench. The question was, what were they do do? Were they to be placed at the mercy of unscrupulous men, when the Government had led them to believe they had protection? A favourable solution of the difficulty would lead to their purging their rules if on impartial inquiry anything objectionable could be found in them; but with this exception, that they would not surrender the right to strike when a strike became a necessity, not even for a Bill.

Mr. Walpole said two distinct propositions had been laid before him. One was to deal with this particular case by some legislation immediately, and as to that he understood the deputation were prepared to lay a memorial before him as to the exact point on which they wished legislation. Secondly, with reference to the importance of a full investigation into the working of these societies, all he could say now was—and he believed Mr. Hughes would bear him out—that they would find in a very few days that he was not in the least unfriendly to such an object as that.

The deputation then thanked the right hon. gentleman and withdrew.

TRADES' UNIONS PROTECTION BILL.

Mr. Neate, who has written ably regarding trades' unions, at this time had given notice to the House of his intention to bring forward a Bill to secure legal protection for the funds of trades' societies. The conference waited on him several times with the view of assisting him to get the following Bill passed, as it embodied the views expressed at the interview with the Home Secretary.

The following is a copy of the Bill introduced into the House of Commons by Mr. Neate, to "Exempt Associations of Workmen from certain Disabilities for a limited Time :—Whereas doubts have arisen as to how far and in what cases associations of workmen not certified under the Act 18 and 19 Vic. cap 63, are entitled to any benefit under that Act, Be it enacted by the Queen's most excellent Majesty, by and with the advice and consent of the Lords spiritual, and Commons, in this present Parliament assembled, and by the authority of the same, as follows :—From and after the passing of this Act, until the end of the next session of Parliament, no such association of workmen shall forfeit any right or benefit to which they might otherwise have been entitled under the 44th section of the said Act, by reason of any rules or rule of such association of workmen being in restraint of trade, provided that snch rules or rule do not prescribe or recommend the performance of any act made subject to any statutory penalty."

It has been said by a few persons that the above Bill was not sufficient for the purpose, and that we ought to have asked for an Act which would give protection to the funds of trade societies, whether they have other than trade benefits attached to them or not. In answer to that, we say that the Bill was merely intended as a temporary measure, and we think it would have been extremely unwise on our part to have asked for complete legislation pending the enquiry of the Commission; but, small as it was, it was too big for the House of Commons, for they have thrown it out unceremoniously.

THE TRADES' UNIONS COMMISSION—DEPUTATION TO THE HOME SECRETARY.

The conference having read that another deputation had waited on the Home Secretary and expressed satisfaction with the present construction of the Royal Commission, thought necessary to have another interview with him on two especial points. The first was to have one working man appointed if possible, the other to secure the presence of a second representative of each trade to explain any difficult or complicated question which might arise. Mr. Samuelson, M.P., was consulted on the subject, and he went to the Home Secretary, who consented to again receive the deputation on February 15th. The following is what transpired:—

A deputation of representative working men waited on Mr. Walpole, on Friday, at the Home Office, in reference to the construction of the Royal Commission on Trades' Unions, which has been decided upon by Parliament. The deputation included Mr. J. Stuart Mill, M.P., Mr. Odger (shoemaker and secretary to the London Trades' Council), Mr. Danter (president of the Amalgamated Engineers' Society), Mr. Coulson (representing the bricklayers), and Mr. Guile (secretary to the Moulders' Society).

Mr. J. Stuart Mill, M.P., introduced the deputation. In doing so the hon. gentleman mentioned that Mr. T. Hughes, M.P., was to have accompanied the deputation, but he had been unavoidably prevented from attending.

Mr. Odger said that the deputation had to express their regret that they had to trouble Mr. Walpole on the question of the construction of the commission, but the belief that it would give general satisfaction, and indeed that it was essential that a working man should be placed on the commission was their excuse for doing so. Another thing they thought should occupy the attention of the commission was the advisability of having present during certain portions of the inquiry practical men who would be able to suggest the putting of questions respecting technical matters connected with workshops, the answering of which would help the members of the commission to arrive at proper conclusions. Otherwise, he believed the inquiry would be carried on under many difficulties, and would be concluded without the persons really most interested getting an opportunity of being heard. What he would suggest was, that, according as matters affecting particular trades came on for hearing, persons deputed by those trades should be allowed to be present to watch the proceedings, and suggest such questions as they considered advisable for eliciting the truth, and protecting the various trades from being misrepresented, through, it might be, a lack of proper information. Each of those persons need only be present while his particular trade society was being inquired into. He thought if some such plan as this was adopted by the commission it would be the means of giving satisfaction, and would show that the inquiry was being carried on fairly to the workmen.

Mr. Walpole: With regard to doing the working men any injustice, I do not think anything like that need be feared. As far as I could I tried to make the commission as fair and impartial as possible, and I think you may rest satisfied that those appointed on the commission will act as I am desirous they should. The last time I received a deputation from certain parties connected with trades' unions, I understood that if Mr. Harrison was appointed on the commission it would give entire satisfaction. I can only say, in reference to Mr. Harrison, that his works on the subject, and the great interest he has evinced, show you that you could not have any person so fully competent to assist in the inquiry, and I think you may rely on Mr. Harrison to bring all matters connected with trade unions forward which may be necessary.

Mr. Odger: Such is our opinion too.

Mr. Walpole: In consequence of the representations made to me by the other deputation, I wrote to the Queen, requesting the addition of Mr. Harrison's name to the commission. Her Majesty complied with the request, and the commission has been signed by the Queen, and confirmed by myself. Under these circumstances, I must say that it is not really in my power to make any other alteration in the construction of the commission. As regards the suggestions of having representatives of trades in attendance to explain matters of detail connected with the various trades, that is a matter that I could not well interfere with. The commission will have power to adopt such a suggestion, and I have no doubt that, if applied to, they will afford the representatives of the various trades every opportunity of getting a fair hearing. I think you may have full confidence in the gentlemen composing the commission. At the head of the commission is one who will be disposed to see that every part of the subject is thoroughly gone into, and that the views of the different trades are completely examined and weighed well before he will bring his judicial mind to bear on the subject—one who has never been supposed to arrive at any conclusion except when the justice of the case is clearly brought before him. I will write myself to Sir William Erle, the head of the commission, as to the possibility of having particular persons brought forward in the interest of particular trade societies, but, as I said before, the matter lies entirely with the commissioners themselves. The matter is really out of the hands of the Government, and I cannot again interfere.

Mr. Guile said, that all the working men wanted was that any possibility of their being held up to the country in a prejudicial light should be avoided, and to effectually guard against such a result he thought that, if they could not have a working man on the Commission, they ought at all events to be allowed opportunities of correcting mistakes or misrepresentations the moment they arose.

Mr. Walpole: I think it could be so arranged that Mr. Harrison could bring forward witnesses as he considered they were necessary, and I think you may rely upon Mr. Harrison to do so.

Mr. Odger said that the trades had perfect confidence in Mr. Harrison, and the appointment had given great satisfaction; but they felt that if a commission were appointed to inquire into the affairs of some of the universities, it would be stupid to entrust the inquiry to working men, who did not properly understand the matters to be investigated; and that, therefore, the present inquiry should not be left entirely to persons who had no practical acquaintanceship with the workings of trades' unions. Such was the opinion existing among a great many of the working men.

Mr. Walpole: I must say my experience of commissions is different from that. I have served on many of them, and, although knowing little or nothing of the subject beforehand, from the information brought before me I have been enabled to get a pretty good knowledge of the subject, and have thereby been enabled to come to what I considered a fair conclusion. I have always found commissions desirous of affording every means of eliciting reliable information, and I have no doubt the present one will do likewise.

Mr. Danter remarked that there were two masters on the Commission, and he thought one working man might at least be allowed on it.

Mr. Walpole: I believe there is one master on the Commission, but I am not aware that there are two. The one master I do know I believe will act with the utmost impartiality. As regards the suggestion to put an additional person on the Commission, I cannot make any further alteration.

Mr. Odger: We do not want an additional person added to the Commission. All we ask is to substitute a working man for one of the present members of the Commission.

Mr. Walpole: It cannot now be done.

Mr. J. S. Mill, M.P.: I have no doubt the Commission will examine every person that may be produced, and that any person the working classes wish to represent them will get a fair hearing; but if I understand the matter rightly the difficulty was not that witnesses will not get opportunities of giving all the evidence they consider desirable, but that some persons in the interests of the trades' unions, and properly understanding their working, should be present to answer any charge that may be made affecting the character of any one of the trades. I believe what is desired is, that some persons having practical acquaintance with trades' unions should be put in position to contradict anything that may be said, through, perhaps, ignorance, damaging to the character of these societies, or to put such questions as would have the effect of enabling the Commission to form a better and more impartial opinion than perhaps they otherwise could have done. Very likely Mr. Harrison may do it well, but Mr. Harrison with a working man may be able to do it better. If the Commission had the power to do what the trades' unions desired with regard to the attendance of persons to watch the interests of each trade as questions affecting that trade came up for inquiry, no doubt it would be better.

Mr. Walpole: I think the Commission can do so, but I should not like to interfere any further.

Mr. Mill then thanked Mr. Walpole for the courteousness of his reception, and the deputation withdrew.

GREAT MEETING AT EXETER HALL.

The next thing, and a very important one, was a meeting convened by the Amalgamated Engineers, in Exeter Hall, on Thursday, February 21st. This meeting was attended by representatives and members of the whole of the trades' societies, and was one of the greatest successes ever attempted by any association. We refrain from giving here a report of it, as in a short space we could not do it justice; but we beg to inform our readers that a *verbatim* report of the whole proceedings has been printed in the form of a pamphlet, and can be had at the Amalgamated Engineers' Office: 54, Stamford Street, London, S. We should advise all working men to possess themselves of this useful document. They are supplied at 1s. per doz. By post, 1s. 3d. per dozen.

At the meeting referred to, the following resolutions and petition were carried unanimously, and a copy of the petition and signatures, and another copy signed by the London Trades' Council, were forwarded to the Home Secretary.

First Resolution.—"That in the opinion of this Meeting the recent decision of the Court of Queen's Bench, in the case of *Hornby* v. *Close*, virtually destroys the protection which trade societies have enjoyed since the passing of the Friendly Societies' Act, and takes from them the privileges the Legislature intended to give in the passing of that measure. It therefore calls upon all trade societies to support the following Requisition to the Home Secretary in favor of Mr. Neate's Bill."

Second Resolution.—"That in the opinion of this Meeting no Commission of Inquiry into the doings of trade Societies can give satisfaction to the working classes unless they be represented on the Commission by members of their own order; and that, as the Royal Commission appointed to inquire into the organisations of Trades' Unions contains no such direct representation, we feel bound to express our disappointment and regret that the Commission has not been more equitably composed."

35

(COPY OF PETITION.)

To the Right Hon. Spencer Walpole, Her Majesty's Chief Secretary of State for the Home Department.

"SIR,

"During the discussion of the Friendly Societies' Act in the House of Commons, in the years 1854 and 1855, some of the representatives of the trades unions, thinking there was then an opportunity of getting legal protection for the funds of trades' societies, put themselves in communication with the late Mr. Sotheron Estcourt, Lord Goderich, and other members of Parliament who were most active and persevering in the laudable endeavour to secure as much as possible the contributions of working men against the defalcations of those who were entrusted with the management of their associations and funds.

"Previous to the final passing of the Act, a Special Committee of the House of Commons considered the whole question, and in the session of 1854 reported as follows:—

"'Your committee, however, taking into their consideration the state in which members of unregistered societies find themselves, from being unable to have recourse to a remedy by law if they should be defrauded or unfairly treated by the officers of these societies, are of opinion that this, which has hitherto been treated as a privilege, ought rather to be extended as a matter of justice and policy, for the protection of the members of all societies; and they have therefore introduced a clause, giving a power to members of all friendly societies, unregistered as well as registered, to sue and be sued as amongst themselves and their own officers.'

"Upon this recommendation the Bill of 1855 was prepared and introduced into the House of Commons by Mr. Sotheron Estcourt, and ultimately became law, the 44th section of which was specially framed to carry out the recommendations of the committee.

"Trade societies, since that date, have acted with great confidence upon that section, and under it various prosecutions against defaulters have been successfully enforced.

"We therefore believe that the withdrawal of that protection, which hitherto they have enjoyed under the Friendly Societies' Act is only calculated to impress still deeper the artizan class of this country with a sense of injustice; and we believe that it will also tend to embitter the future struggles between employers and employed, which the largest and best organised trades have for years been unceasing in their efforts to prevent.

"We therefore trust that you will see fit to give your concurrence to the Bill introduced by Mr. Neate, which will temporarily prevent such results, and at the same time that you will undertake the introduction of a permanent measure, such as will ensure the trades' societies the protection which for the last twelve years they thought they possessed, and thus do an act of justice to thousands of industrious and deserving workmen.

"Signed on behalf of the undermentioned societies:—

SECRETARIES.	NAME OF SOCIETY.	No. of Members.
"William Allan	Amalgamated Engineers'	33600
"Daniel Guile	Iron Founders' Society	11150
"R. Applegarth	Amalgamated Carpenters'	8500
"G. Odger	Amalgamated Cordwainers	8000
"Mathew Lawrence	London Operative Tailors'	7000
"E. Coulson	Operative Bricklayers'	6000
"H. Self	Consolidated Society Compositors'	3300
"Beaumont Cole	Curriers' Society	1400
"William Burn	Brickmakers'	1000
"Thomas Joseph Dunning	London Consolidated Bookbinders, 5, Racquet Ct.	658
"William Bockett	Bookbinders, Day-working	420
"Isaac Merryweather	Tin Plate Workers' Society	300
"Thomas Spelling	Vellum Binders	266
		81594

36

At the next Meeting of the Conference it was resolved: That a deputation be appointed to seek an interview with Mr. Gladstone, as that gentleman had given permission to the Trade Societies to deposit their funds with the Government: and it was thought probable that his valuable aid might be enlisted in furtherance of our endeavours to secure the protection we were seeking for our funds. Mr. Gladstone was immediately communicated with, and at once replied that he would have much pleasure in receiving the deputation on the 7th of March. Subjoined is the substance of what took place.

TRADES' SOCIETIES' FUNDS IN POST OFFICE SAVINGS' BANKS.

Interview with Mr. Gladstone.

The deputation included Mr. Allan, Secretary of the Amalgamated Society of Engineers, &c.; Mr. Danter, President of the London Trades' Council; Mr. Odger, Secretary of the same body; and Mr. Applegarth, Secretary of the Amalgamated Carpenters' Society.

Mr. Gladstone having received these representatives with the greatest courtesy,

Mr. Allan said: It will be in your recollection, sir, that in 1864 Mr. Danter and I waited on you, with regard to our society depositing their funds in the post-office savings' banks. You then granted our request, giving us permission to bank in them to any extent. We have since then deposited a very considerable sum of money in those banks. But the late decision of the Court of Queen's Bench has placed us in an awkward position, for defaulting trustees may now rob us with impunity.

Mr. Gladstone: It is, indeed, a very awkward position.

Mr. Allan: We placed our money to the great extent we have done, in consequence of the arrangement made with you, as Chancellor of the Exchequer. We thought we were acting in accordance with law, and certain of legal protection; and more especially so as, in answer to your questions, we stated all the purposes to which those funds were collected and might be applied; that amongst others we paid men when out of work, having refused to accept a lower rate of wages.

Mr. Gladstone having given a general assent to these remarks, asked if the rules were registered.

Mr. Allan: We have deposited a copy of them.

Mr. Gladstone: Yes, deposited a copy of them. I regarded it as a sort of friendly society. Of course it was not for me to determine what was legal or illegal. My impression was that you had complied with the requisitions of the law. I am as much surprised at this decision as anybody can be. I cannot understand it. Do you know on what precise point the judgment turns?

Mr. Danter: Simply that we support men when out on strike.

37

Mr. Allan handed to Mr. Gladstone the opinion of the Lord Chief Justice, when Attorney-General, as to the rules on which the judgment turned.

Mr. Applegarth directed the right hon. gentleman's attention to the rules now declared by the late judgment of the Court of Queen's Bench as operating in restraint of trade.

Mr. Gladstone: You have not the exact terms of the judgment. I have never seen the exact judgment.

Mr. Allan: "Operate in restraint of trade" were the exact terms of the judgment.

Mr. Gladstone said restraint of trade was a very large phrase. He referred to a rule which at one time existed amongst wholesale publishers of books as to the terms of discount upon which alone they would consent to supply the retailer. It had been given up, but at the time it might have been regarded as in restraint of trade. If one of those retailers had refused to pay for a parcel of books, was the wholesale publisher to be defrauded because the books had been delivered under a rule in restraint of trade? A man might devote his land to purposes perfectly legal, but which might be in restraint of trade. He confessed he did not know where that was to stop. The question before him in 1864 was whether a legal society might be permitted to bank to any extent in the post-office savings' banks. He then understood they had complied with the regulation which made it a legal society, namely, by depositing the rules, and that they could not deposit rules which contained anything illegal.

Mr. Allan: Mr. Tidd Pratt says he has nothing to do with seeing whether the rules are legal or not.

Mr. Gladstone: So that an illegal society might deposit its rules?

Mr. Danter: The Act gives him power, and he examines the rules to see whether they are legal or not.

Mr. Gladstone: So I thought.

Mr. Allan: The Society I represent has deposited upwards of £40,000 in those banks.

Mr. Gladstone: The question before me at the time was, whether you might not be allowed to bank as a friendly society. Well, why are you not a friendly society?

Mr. Allan: We have tried it, and ascertained that we must be registered in Scotland and Ireland, as well as England, and then we would be three different societies.

Mr. Gladstone: What is there in point of law which makes you otherwise than a friendly society?

Mr. Allan: We are a corresponding society, going through various countries.

Mr. Gladstone: I want to know whether there is any legal impediment to your becoming a friendly society. Would that rule about supporting men when on strike prevent you?

Mr. Odger explained that the judgment was to that effect.

Mr. Applegarth: Take our societies as a whole, they are for the purpose of assisting our members to take their labour to the best market. Now that embraces the supporting them when on strike or when locked out.

But along with that we have many benevolent objects. We believed as you did, sir, that Mr. Tidd Pratt would not have allowed our Rules to be deposited if our society had not been a legal one.

Mr. Allan said they believed they could not be wrong in acting upon the legal opinion of the Attorney-General of England.

Mr. Gladstone asked the proportion that the amount paid to men on strike bore to the expenditure of the societies for benevolent purposes during the past ten years.

Mr. Allan said not one-tenth, he believed, was paid to men on strike. He then explained that the deputation did not ask Mr. Gladstone to undertake to remedy the grievance complained of, but that he would be kind enough to render his most valuable assistance generally so far as he could to Mr. Neate in carrying the bill he had introduced through Parliament, that temporary protection might be obtained for their funds pending the permanent settlement of the matter.

Mr. Gladstone said the principles of combination were of very great importance. Whatever was permitted by law ought to entail no legal grievance. He was completely puzzled by the matter as it then stood before him, and requested to be furnished with a copy of the judgment ot the Court of Queen's Bench, which the deputation promised to send. It was, he said, a very interesting matter, and he would give it his best attention.

The deputation then thanked Mr. Gladstone for his kindness and withdrew.

THE CONFERENCE AND THE ROYAL COMMISSION.

On Friday, March 15th, the Royal Commission of Inquiry into Trades' Unions met at their office, 2 Victoria Street, Westminster. A deputation from the Conference consisting of representatives of the Amalgamated Engineers, Amalgamated Carpenters, Ironfounders, Bricklayers, and other important trades of the country, had an interview with the Commissioners, and requested that when any particular trade or society was the subject of inquiry, a representative from such trade or society should be allowed to attend.

This was granted, as will be seen by the following copy of a letter sent by the Secretary of the Commission to Mr. Applegarth, who is acting as secretary to the Conference of Trades' Representatives :—

"2 Victoria Street, Westminster,
"March 15th.

"SIR,—With reference to the application made this day by the deputation of the Trades' Delegates Conference, introduced by yourself, I am directed to inform you that the Commissioners have agreed to the following resolutions: 1. 'That in addition to the presence of such persons as the Commissioners may deem expedient at any time, the Commissioners will be ready, as a general rule, to admit some one person connected with or representing any trade or society which seems likely to be affected by the inquiry, to be conducted on a given day.' 2. 'That, when printed, a copy of the short-hand writer's notes of the examination of witnesses each day be sent to every person so attending, and to be supplied, as soon as convenient, to any member of the press who may apply for it.'

"I am, &c.,

"J. H. PATTESON, *Secretary.*

"To Mr. R. Applegarth."

In addition to the above the following communication was received by the Conference :—

"We thank the Conference for their offer, made through their deputation, of assistance in our inquiry, and we will avail ourselves of that offer as far as we think expedient, and we will receive from time to time any application, on behalf of any member of the Conference, for him to be present at any sitting of this Commission."

It will be seen from the above that the Conference has done every thing in their power to secure a fair and impartial examination.

GREAT MEETING OF LONDON TRADES IN EXETER HALL, IN REFERENCE TO THE SHEFFIELD OUTRAGES.

We now come to a very serious part of our narrative, namely, the conduct of the conference upon the appalling disclosures at Sheffield becoming known to the public. Each member of the conference brought the question to the notice of his respective society, and a vote was unanimously adopted at the whole of the large societies to assist the London Trades' Council in calling a meeting at Exeter Hall immediately to express the surprise, indignation, and horror of the trade societies at the crimes perpetrated by Broadhead and his accomplices at Sheffield. The following is a report of the meeting :—

On Tuesday night, July 2nd, at eight o'clock, when the chair was taken by Mr. W. Allan, secretary to the Amalgamated Engineers, the great hall was densely crowded with artisans, and on the platform were seated most of the leading trade unionists of the metropolis. Amongst others who occupied seats on the platform were Professor Beesly. Mr. Beales. and Colonel Dickson.

The Secretary read the following letter from Mr. Ludlow :—

"MY DEAR SIR,—I regret much that I shall not be able to attend the meeting of this evening, but an engagement in my own parish, for a co-operative purpose, hinders me from doing so.

"The Trades of London do well to meet on this occasion, and I trust the present meeting will be followed up by a series of similar ones, both in London and elsewhere, which shall bring the public opinion of the working class to bear upon the men of Sheffield with such weight of earnestness, that not only the recurrence of atrocities like those paid for by Broadhead and his compeers shall become impossible, but that the guilty parties shall feel that there is no room left for them in the midst of honest, God-fearing English artizans.

40

"The present, be assured, is a trial-time for Trade Societies. Oblivious of the fact, that what we see now revealed are not the first, but, as I trust, the last of trade outrages, which were formerly common to many other trades and towns than the six or eight tainted trades of Sheffield,—and that the inquiry which has at last uncloaked these iniquities was asked for by your societies themselves,—foolish men are raising a cry, some for the legislative suppression of trade societies, others, who fancy themselves more moderate, for still withholding from them those legal guarantees for their safe and open working, the absence of which forms at present the only avowed colour of excuse for the outrages in question, and tends necessarily to transform into secret conspiracies institutions otherwise of an essentially friendly, though one-sided character. This temporary crisis in public opinion needs above all to be met by a course of conduct in moral reprobation of outrage which shall overcome prejudice and defy evil, as well as by a firm and temperate indication of the principles on which your societies rest.

"And since it appears to me certain that legislation and fear, hostile legislation—on the subject of trade societies will be attempted in the next session of Parliament, I cannot but earnestly urge working men to be prepared to send members of their own order to the next House of Commons; men of unblemished character, of experience, firmness, and good sense, whether drawn from the actual ranks of trade societies, or even from among such persons as, through various circumstances, may either have left or never joined those ranks, but who entertain (as many such do) a full sense of the need which exists for trade organization, of the benefits which it confers on the working class, and of the guarantees for good character in the workman which trade societies generally supply.

Depend upon it, the time is come when the working man must fight his own battle, in or out of Parliament. You have amongst you good men and true, as I well know. You have the means of sending them to Parliament, and maintaining them while there. If you do not send them, yon know as well as I that it will be your own fault, the fault of those petty jealousies and prejudices, of that narrow-mindedness and exclusiveness, which unfortunately are as prevalent in your class as in almost any other.

Meanwhile, as one who for many years now has paid some attention to trade societies and their proceedings, I am glad to express my perfect conviction, and should have willingly done so, if present, from your platform to-night, that the great bulk of these societies are entirely free from all complicity or sympathy with such practices as have hitherto disgraced the Borough, of which Mr. Roebuck is the long-chosen representative.

"Yours truly,

"(*Signed*) J. M. LUDLOW.

"To Mr. William Allan."

The Chairman, in a brief speech, said that the London Trades' Council had convened this meeting in consequence of the recent disclosures at Sheffield, which had to a great extent involved trade societies that had no conception of the deeds which had been perpetrated. Broadhead appeared to be the leader in these affairs, than which nothing could be worse. These disclosures must to a certain extent damage their societies, as they would cause people to distrust them to some degree. He was exceedingly glad that these doings had been brought to light—(cheers)—and hoped that means would be taken to put a stop to them in the future. He boldly declared that if trades' unions had to depend upon murder the sooner they ceased to exist the better. (Hear, hear, and cheers.) Their presence there showed that they did not countenance these atrocious outrages, but that they repudiated them with sentiments of detestation and horror. Besides the murders, there were the cases of shooting and maiming, which, in his opinion, were almost worse than death itself, as through these means the unfortunate victims were deprived of earning their livelihood.

41

Mr. Guile (ironfounder) moved the first resolution, and in doing so complained of the efforts made to connect all unionists with the Sheffield outrages, and after an effective speech moved—"That this meeting of members of trades' societies not only expresses its utter abhorrence of the crimes perpetrated by Broadhead and his accomplices, but feels bound to say that it cannot understand how the delegates, who composed the Council of Organised Trades of Sheffield, did not discover for so great a length of time, such an unparalleled system of notorious crimes as has existed in connection with some of their societies."

Mr. George Odger, shoemaker, and a member of the Reform League, seconded the motion. Broadhead had not acted for the benefit of trade societies, but for his own purposes; and it was only another of his crimes to pretend that he was trying to benefit the class with which he was connected. He had no purpose to serve except his own. Whether he had been a trades' unionist, or had been associated with the law or the church, or whether he had been a statesman, he would have been just as great a criminal as he was as a trades' unionist. (Cheers.) Was it right that all members of trades societies should be condemned because of the acts of those scoundrels who had confessed their crimes before the commissioners. (No, no.) Had not society been outraged by the crimes of men of all classes? by members of the learned profession? Would it have been just for society to condemn a whole class on account of their individual acts? (Cheers.) On the same ground it was monstrous for newspaper writers and others to charge the whole of their trades' societies as being guilty in this respect. So far from the London Trades' Council having any sympathy with crime, he would remind them that seven months ago the council sent a committee to Sheffield for the purpose of inquiring into an outrage which was then committed—that of blowing up a house. Mr. Danter and himself went down, and discovered that rattening had been encouraged by some of the trades. They made a report which was sent to the London newspapers. One or two took notice of it, but the others did not. In the report they stated:—"We felt it to be our duty to visit the council of the 'Organised Trades' of Sheffield, for the purpose of putting such questions to them as we thought consistent with the responsible position in which we were placed. The result of which is that Mr. Broadhead unequivocally admits that rattening, to the extent of taking away bands and tools, and detaining them until some fines and contributions are paid, has been the custom of some of the societies since they had an existence." The fact that the London Trades' Council made this report fully proved that the system of rattening was unjustifiable, and their desire that it should be put a stop to.

Mr. Howell, secretary of the Reform League, said that a large portion of the press of the country appeared to denounce crimes only when they were committed by a certain class. After strongly condemning the conduct of the trade societies of Sheffield, which did not take sufficient care to inquire into the manner in which their funds were distributed, he remarked that, not to speak of murder and assault, taking even the milder case of rattening, there could be no doubt that such acts were not only great crimes but great blunders. (Cheers.) They must be prepared that evening to denounce all who had connection with those notorious crimes; but if any portion of society thought that on account of the Sheffield outrages they could with impunity assail the trade unions, they committed a great error. (Cheers.) The working men of England were not prepared to allow the principle of combination for trade purposes to be interfered with.

Mr. Harry (carpenter) spoke in the strongest terms of the crime of rattening. When stealing and burglary were once commenced, worse crimes would follow, and murder would finally result. He condemned the language used by the journals which sought to fix on a body the crimes of a few individuals.

Mr. Dunning (bookbinder) said that just now they were before the bar, not of the press, but of public opinion. Public opinion very seldom judged wrongly, but just at present there could be no doubt that there was somewhat of a *furore* against the trade societies. Broadhead was at best a sorry villain, with a keen eye for the main chance, and was at the time that he was plotting great crimes endeavouring to embezzle a few pounds for his own purposes. He did not blame the Government for giving Broadhead and the others certificates of indemnity, but he should express his strong hope that the outcome of the inquiry would not be that the rogues were let off and the honest men punished. (Hear, hear.) The error committed by the men of Sheffield was in not separating themselves from the societies in which the crimes were committed.

The resolution was unanimously adopted.

Mr. Applegarth (carpenter) proposed—"That we, the members of the metropolitan trades' societies here assembled, beg to express our most cordial approbation of the services rendered to the country generally, and to trades' unions especially, by Mr. Overend and the whole of the gentlemen forming the Sheffield Commission, in having brought to public light, and thereby to just execration, a system of organised crime in connection with certain trades' unions in Sheffield, which has been a source of terror to the people of the town and a means of bringing distrust upon all trades' combinations." He would desire to add to that, that all that had ever been written in condemnation of the acts done in Sheffield he fully concurred in, but he did not concur in the censures which had been lavished on the entire population of Sheffield. There was one fact to be remembered, that these horrible trade offences were year by year growing less. Within the last five years there was only one serious outrage in Sheffield; within the previous five years a dozen were committed. He defended the conduct of Mr. Dronfield and others, men whom the press sought to drag down to the level of Broadhead, and who were as honest, as moral, and conscientious as any in England. He then referred to the efforts which the working men in Sheffield had made to discover the causes and perpetrators of outrages, and which had broken down owing to the conduct of the masters. How was it, he asked, that working men, every day at their business, could be expected to discover crimes which skilled police officers had failed to discover? He trusted that the commission would continue the work it had commenced fairly and impartially, and extend their enquiries into the acts of the masters as well as those of the workmen. (Cheers.)

43

The resolution was seconded by Mr. Dodd (zincworker), and supported by Mr. Coulson (bricklayer).

Professor Beesly said he had listened with great satisfaction to the resolutions which had been moved, and to the able speeches which had been made in support of them. He was glad to see this great meeting pledge itself to such excellent sentiments, but he might be allowed to say that when that meeting was over he thought that nearly enough would have been said about this matter. There was no good to be attained by using exaggerated language even about Sheffield outrages. He did not mean to say that any exaggerated language had been used at this meeting. He thought that all the speakers had been animated by a very proper spirit. Murder was a great crime. They were all agreed about that. But after all it must not be forgotten that a trades' union murder was neither better nor worse than any other murder. The wealthy classes of this country, in whose eyes unionism of itself was a crime, naturally considered that any crime connected with unionism was doubly criminal. But clearly that was a doctrine which this meeting could not admit. He was aware that such an observation would excite the anger of those who thought that all the workmen in England ought to be kept sitting in sackcloth and ashes for what had been done in Sheffield; and if his remark should find its way into the newspapers he should very likely see himself denounced as an apologist for murder. Well, that would not be very agreeable. But if we were to shrink from saying the truth because there were base persons always ready and eager to place an odious misconstruction on our words, we had better hold our tongues altogether, and make room for others who might have greater courage and a stronger sense of duty. What he wished to point out was that a man who is perpetually protesting his innocence from crime must inevitably become demoralised. It was impossible that he should retain his self-respect if he had always to assume an apologetic attitude. It was most desirable that this remark should be made by some one, and perhaps it came better from him than from the other speakers. He was no apologist for murder. During the last twelve months he had subscribed his money and given what other assistance he could to bring a great murderer to justice—a murderer whose hands were red with blood not of two or three victims, but of more than four hundred. That murderer had committed his crime in the interest of employers, just as Broadhead had committed his crimes in the interest of workmen. The wealthy class of this country had been called on to express their opinion on the crimes committed by wealthy men in Jamaica just as the poorer classes in London were now called on to express their opinion on the crimes committed by poor men in Sheffield. And what opinion did they express? Did they summon a meeting in Exeter-hall and proclaim aloud that they abhor the crime, and that though they wished to protect property and wealth they repudiated such means of protecting it as Governor Eyre had adopted? (Cries of "Question" from one or two individuals, which was immediately drowned in a roar of applause.) Did they do so? No! but they offered him banquets; they loaded him with honours; they made his deed their own. That murderer was now at large in England, not because he had a certificate of indemnity like Broadhead, but because

44

a bench of magistrates of his own class had refused to send him before a jury. London workmen had acted differently. It was not their class to which the stigma of encouraging murder attached. He did not see that they need take blame or shame to themselves for what had been done at Sheffield. He did not see why they should hold their heads less high or recede one inch from the claims they were making on the legislature. The middle classes would go on talking about the Sheffield outrages for a long time, and they would be glad to keep workmen talking of them in order to divert their attention from the state of the law as affecting unionism. It had already been pointed out to-night by more than one speaker that the law refused protection to union funds. Thus the middle class invited thieves to do what they were afraid to do themselves. They set on thieves to steal the union funds just as Broadhead set on his agents to steal bands. In short, the only difference between rattening as practised by trades' unions and rattening as practised against trades' unions was that the law punished the one and encouraged the other. Again, take the law respecting so-called "intimidation." No one disputed that the man who struck another, or even shook his fist at him, or used language calculated to provoke a breach of the peace, ought to be punished. But such was the state of the law at present that the magistrate at Marlborough-street had been able to commit more outrages in a week than Broadhead had for twelve months. If a tailor told another tailor that he should refuse to work with him, the magistrate would commit him for trial and exact £100 bail from a poor man earning twenty or thirty shillings a week, and then when he came for trial, a jury of employers and a bench of middle-class magistrates would send him to the treadmill. These were monstrous grievances, and the middle-class would try, by raising a great dust about Sheffield, to prevent workmen from getting redress for them. They would make a great mistake if they assumed a more humble tone or confined themselves to a defensive attitude. A defensive attitude was always a weak attitude. They were now likely to have some means of influencing the legislature. They should get bills drawn to remedy the two grievances he had mentioned, and at the next election, under the Reform Bill, every popular constituency should pledge candidates, not to such rotten tests of radicalism as were now proposed, but to these two bills. If such were passed he believed they would hear of no more outrages even at Sheffield, for it was the sense of wrong legally inflicted which made unionists outstep the limits of legality.

Mr. Edmond Beales, who was greeted with the most enthusiastic applause, addressed the meeting in eloquent terms, remarking that he would indeed regret any efforts he had made to obtain the enfranchisement of the people if they were the sympathisers with crime some persons sought to represent them. (Cheers.)

The proceedings closed with the usual vote of thanks.

45

DELEGATE MEETING OF THE LONDON TRADES' COUNCIL.

The delegates at the Annual Meeting of the London Trades' Council considering that the speech of Professor Beesly at the Exeter Hall meeting, on July 2nd, had been much misrepresented deemed it their duty to express their opinion on the matter.

The following resolutions were unanimously adopted, and the opinions therein expressed are based on the result of years of intimate acquaintance between Professor Beesly and the Trades' Council:—

First Resolution—

"That this meeting of delegates and members of the trades' unions of London desires to express its deep sense of respect and gratitude to Professor E. S. Beesly who for the past seven years has been the patient, zealous and judicious friend and adviser of the trades' unions of the united Kingdom, and who has invariably dealt with the many difficult, exciting, and irritating questions that have arisen between workmen and their employers in the spirit of truth, kindness and moderation, and has thus done much to calm and assuage bitter feelings on the part of the unionists, and to check the spread of that class antagonism which he has always recognised to be one of the most deplorable results of the present unsatisfactory relations between labour and capital."

Second Resolution—

"That this meeting also desires to express its sorrow and indignation at the mode in which certain influential portions of the public press have attempted to destroy the character, position, and influence of Professor Beesly, by misrepresenting and taking advantage of a few isolated passages in his speech at Exeter Hall, on the 2nd of July last, which passages, if they afford the slightest excuse for the denunciations for which they have furnished a text, do so only by the incompleteness with which they expressed his views, and which ought by all fair men to be interpreted in accordance with the honourable, just and useful character of his public life; and this meeting is assured that no impression was conveyed to those who heard his speech at Exeter Hall, that he had offered, as is alleged, the slightest extenuation of the Sheffield crimes."

Third Resolution—

"That copies of these resolutions be forwarded to the institution with which Professor Beesly is connected."

In giving them insertion here we desire to express our hearty approval of the opinions contained in these resolutions.

CONCLUSION.

We have thus given to the members we represent, and to trade societies generally, a true and faithful account of our transactions in connection with these important questions; and we feel ourselves justified in saying that the effect of our proceedings has been, throughout the whole of this most critical period, to secure the general approval of the press and the public. In conclusion, we take this opportunity of stating that we shall follow up the work which we have been so much engaged in until the Royal Commission has completed its labours, and make known the result to the trades from time to time, or call a conference of trades generally, should it be found necessary.

We remain,

On behalf of the Conference of Amalgamated Trades,

Delegates.	Societies.
MR. W. ALLAN,	Amalgamated Society of Engineers.
" R. DANTER,	
" D. HENDERSON.	
" W. ROBSON,	
" D. GUILE,	Ironfounders' Society.
" J. HEMLEY,	
" T. POWELL,	
" R. APPLEGARTH,	Amalgamated Society of Carpenters and Joiners.
" T. W. HUGHES,	
" J. D. PRIOR,	
" E. COULSON,	Operative Bricklayers' Society.
" T. SPELLING,	Vellum Binders' Society.
" G. ODGER	London Trades' Council.

W. ALLAN, *Chairman.*

R. APPLEGARTH, *Secretary.*

REPORT

OF

ST. MARTIN'S HALL

United Kingdom Trades' Conference Committee

ON THE

TRADES' UNION INQUIRY COMMISSION

TILL 7TH AUGUST, 1867.

FELLOW-WORKMEN,—The Royal Commission on Trades' Unions at Westminster, having at their last Meeting adjourned for some months; the Committee appointed by your representatives in Conference have thought this a convenient time to bring before you in a report a summary of the proceedings which you appointed them to watch on your behalf. The Committee desire, also, to draw your attention to the following Minutes of their proceedings.

TRADES' CONFERENCE COMMITTEE.

OFFICIAL MINUTES OF PROCEEDINGS.

LONDON, *May 17th, 1867.*

The United Kingdom Trades' Conference Committee met this morning, according to circular, at the BEEHIVE Office, Bolt Court, Fleet Street. All the Members were present, Mr. CONNOLLY presiding. The Minutes of the previous Meetings were read and confirmed.

The Chairman made a statement as to the proceedings of the Trades' Union Commission of Inquiry, which he had attended as representative of this Committee. The financial position was thereupon taken into consideration, and after a full statement, it was found that, although not quite satisfactory, we were in a much better position than was anticipated.

The Secretary reported that the amount due from the Glasgow Trades' Council had been sent up by him, and he would pay the same over to the Treasurer. Edinburgh had not as yet sent anything, but he was of opinion that there was no doubt but Edinburgh would

2

maintain her honour in this matter; at least he (the Secretary) strongly hoped so.

Committee took into consideration the course to be pursued towards this inquiry, when a full report of the statements of the various witnesses was much desiderated. With a view to giving our constituents the fullest opportunity of defending themselves, the following resolutions were adopted, viz:—

1. "That this Committee respectfully urge upon the Commission, the desirability of examining the employers of labour in the first instance, and that when such evidence is published, that a copy be furnished to the next operative witness of that trade to be examined, in order that he may have an opportunity of giving explanations, and replying to statements of the other side; or in the event of this mode of procedure interfering with the arrangements of the Commission, that the representatives of operative Trades' Unions have an opportunity of replying to statements made against their respective Associations; this privilege to be granted to those witnesses who have been examined already.

2. "That with the view of having this investigation as fair and exhaustive as possible, and that the case of the various Trades' Societies be fully put before the Commission and the public, that the selection of witnesses to give evidence on their behalf be left in the hands of the trades themselves.

3. "That with the view that the trades of the United Kingdom may continue to be as efficiently represented as possible, this Committee shall endeavour to make such arrangements as shall enable any trade to comply with the preceding resolutions; and shall also send, from off this Committee, to watch the proceedings on behalf of the workmen, the Member best qualified for the interest of the trade under examination.

After a long conversation as to the re-arranging the officers of the Committee, the matter was delayed, and the Committee adjourned at six o'clock till nine o'clock next morning.

May 18.

Committee assembled this morning. Resumed consideration of change of officers; but after a long discussion, it was finally agreed, with one exception, to continue the Committee as at present constituted. Mr. Holmes, of Leeds, our respected Treasurer, insisted on retiring, on the score of ill-health, and the urgent representations of his medical adviser that he should do so. It was with great reluctance the Committee were forced to acquiesce in this decision, but ultimately, Mr. Holmes agreeing to continue a Member of Committee, his resignation as Treasurer was accepted, and Mr. M'Donald appointed in his place.

With the view of obtaining the pecuniary and moral aid of the various trades in conducting our operations, it was resolved, that each Member of Committee shall use his influence to get such trades as have not hitherto joined us to do so; and urge those trades represented at the conference, but who have not contributed to the levy, to do so at once. And that in order that the trades may be fully informed of all the evidence given before the Com-

3

mission, the BEEHIVE newspaper be requested to publish a full report, and we would respectfully call on all Trades' Unionists to give this organ of labour their utmost support.

It was next resolved that, in order to strengthen the hands of Mr. HARRISON, our representative on the Commission, that we render him every information in our power to enable him thoroughly to accomplish our work, and we hereby instruct our Secretary, as far as practicable, to furnish Mr. HARRISON from time to time with such information.

There having as yet been no audit of the Treasurer's books, it was agreed Mr. Holmes consenting, that he shall continue to receive all monies, and continue to discharge the duties of Treasurer, till next Meeting, when audit may be made, and the books handed over.

After a vote of thanks to the Chairman, the Committee adjourned *sine die.*

6th August, 1867.

Met, according to circular, the United Kingdom Trades' Conference Committee. Present—Messrs HOLMES, of Leeds; ALLAN, of Liverpool; WOOD, of Manchester; and LEIGH, of Hyde—Provincial representatives; and Messrs POTTER, LEICESTER, and CONNOLLY, of London, The other provincial Members, viz.,—Mr. M'DONALD, of the National Miners, was absent on business; Mr. KANE, of the Iron-workers, from not having got the intimation in time; and Mr. PROUDFOOT, of Glasgow, the Secretary, was at the Commission, and under examination; Mr. CONNOLLY presided. The first business was the consideration of the position and duty of this Committee towards the Commission of Inquiry, in consequence of the exclusion by the Commission of Mr. CONNOLLY, (who has from time to time represented this Committee there) from their Meetings in consequence of some criticisms passed by Mr. CONNOLLY, at a Public Meeting, on the political conduct and consistency of a Member of the Commission. The representatives of the London Trades, at a Meeting, had appointed a deputation to wait on the Commission, with a view to procure the rescinding of their resolution. Mr. LEE, of the London Compositor's Society, on the part of the deputation, was present, to report to the Committee.

Previous to receiving this report, Mr. CONNOLLY made a statement as to the proceedings of the Commission, the evidence of the various witnesses up to the date of his exclusion on the 5th July, and reported the facts relative to his exclusion.

Mr. R. LEE reported from the deputation that they had an interview with the Commission, who declined to rescind their resolution respecting Mr. CONNOLLY, but were quite willing to receive another representative from the Committee in his place.

After receiving this report, the Committee at six o'clock p.m., adjourned till to-morrow morning at ten a.m.

August 7th, 1867.

The Committee re-assembled this morning, and in addition to those present yesterday, Mr. PROUDFOOT, the Secretary, was present. Mr. LEIGH, said he thought their first business was to consider what future connection this Committee should hold with the Trades' Union

4

Inquiry Commission, or in the event of resolving to hold no further communication, to dissolve this Committee. Mr. PROUDFOOT made a statement as to his experience of the Commission, and thought it would be suicidal policy on our part to dissolve the Committee, until it had fulfilled, or found it impossible to fulfil, the objects of its appointment. In his opinion, nothing whatever has occurred, to alter the position of this Commiteee in regard to the important inquiry before the Commission, and if the Commissioners had (as in his opinion they had), somewhat compromised their dignity, by becoming parties to a dispute which was purely personal, between one of their number and Mr. CONNOLLY, or constituted themselves judges *ex parte* in the cause, that was no good reason why we should forget our dignity, or the important interests committed to our charge. Moreover, he, Mr. PROUDFOOT, could not see how we could, in honour to those who appointed us, abdicate our functions without calling them together and consulting them, and giving them the opportunity of bidding us go on, or removing us, or electing others in our room. This Committee had various duties to perform to their constituents, besides sending a Member to sit at the Commission. It was, he Mr. PROUDFOOT, considered a part of that duty to keep the trades fully informed of every movement, and to advise and instruct such as had explanations to make, and information to give. Also, after the Commissioners had given in their report, it would be the duty of the Committee to make a report to the trades, as to the action to be taken by Trades' Unionists on the report of the Commission; in fact all the more important work of the Committee was yet in the future, he was therefore decidedly opposed to the idea of dissolving the Committee.

Mr. ALLAN, Mr. LEICESTER, Mr. WOOD, and Mr. HOLMES all concurred in the views enunciated by Mr. PROUDFOOT. Mr. POTTER also concurred, and stated further, that, as the Commission had adjourned for some months, this Committee should issue a report of the progress of the Inquiry up to this date, and moved a resolution to the following effect:—"That as the Trades' Union Commission had adjourned its sittings till November next, this Committee adjourn for the present the consideration of what future connection they will hold with the Commission; and in the meantime, this Committee shall prepare and issue to the trades an opinion as to the line of policy they would recommend to be adopted in future. And further, that to enable the Committee to carry on its work, a levy of One Farthing per member (in accordance with resolution of Conference), be now called for." Mr. PROUDFOOT seconded the resolution, which was adopted unanimously, after some discussion. Mr. LEIGH, of Hyde, was appointed Treasurer in the room of Mr. HOLMES, who retired at last meeting in consequence of the state of his health—Mr. M'DONALD, who was elected to the office at last meeting, and agreeing, now declined to fill the office. Mr. HOLMES submitted an account of income and expenditure from opening of Conference till 18th May last, when he retired, which was audited by the Secretary and Mr. Wood, and found correct, the books and balances since that date being handed over to Mr. LEIGH.

The Secretary was instructed to prepare a report to the trades on

5

the Trades' Union Commission, in conformity with the resolution of Committee, as early as possible, and the proceedings of this Committee (as far as not published in the Conference report), relative thereto, and also a statement of income and expenditure.

After the transaction of some routine business, the Committee adjourned *sine die.*

(Signed) JOHN C. PROUDFOOT, *Secretary.*

TO THE TRADES REPRESENTED AT ST. MARTIN'S HALL CONFERENCE, AND TO THE TRADES OF THE UNITED KINGDOM GENERALLY.

BRETHREN,—From the repeal of the Combination Laws in 1824-25, whereby the toiling millions of this country were allowed to exercise in freedom the right to agree together as to the conditions on which, the proportion of time, and the price at which we would sell our labour,—till now, there has only been one crisis of any importance, in regard to legislative action, respecting those associations which we have established for the protection of our industrial interests, and for maintaining our right to sell our labour, which is our capital, upon the conditions which we may mutually think most conducive to our moral, intellectual, and physical welfare. That crisis was, I need hardly remind you, in 1838, when a Select Committee of the House of Commons was appointed to inquire into the subject of Trades' Combinations, under the chairmanship of the late DANIEL O'CONNELL. The appointment of this was consequent upon the great Strike of the Cotton Spinners of Glasgow, the year before. Trades' disputes were then, and had been for many years previous, characterized by violence and outrage, than which those recently disclosed at Sheffield were not more atrocious, horrible, and inhuman. The Committee was appointed to make inquiry into the operation of the Act 6, Geo. IV., chap. 129, and generally into the constitution, proceedings, and extent of any Trades' Unions, or combinations of workmen, or employers of workmen in the United Kingdom, and report their observations to the House. Another crisis of still greater importance has again occurred after the expiry of two decades of years. We say of greater importance, because the ramifications of Trades' Unions are vastly more extended than at any former period of their history; and their business is carried on in a manner, and with a completeness that would do credit to some Imperial Governments; and, as a general rule, with a care and a deference for the rights of individuals which, whatever our enemies may say to the contrary, is almost unimpeachable. We have once more an inquiry into the subject of Trades' Unions this time by Royal Commission, who, by Royal warrant, and in pursuance of Act of Parliament, are to inquire into, and report on, the Organization and Rules of Trades' Unions and other Associations, whether of workmen or employers; and to inquire into and report on the effect produced by such Trades' Unions and associations on the workmen

6

and employers respectively and on the Relations between Workmen and Employers, and on the Trade and Industry of the Country; with power to investigate any recent acts of intimidation, outrage, or wrong, alleged to have been promoted, encouraged, or connived at by such Trades' Unions, or other Associations; and also to suggest any improvements to be made in the law with respect to the matters aforesaid, or with respect to the relations between workmen and their employers for the mutual benefit of both parties. Happily, in these latter days, with some very trifling exceptions (and any connection with these exceptions the great body of Trades' Unionists most emphatically repudiate) we have no cases of outrage or wrong promoted, encouraged, or connived at by us for the Commission to investigate. The inquiry, however, is not on that account of the less importance to us as Trades' Unionists; nor is the result of the inquiry likely to be of such a nature as that we can safely leave the Commission to pursue it without any attention being paid on our part to its progress, the witnesses brought up to give information, and the nature of the evidence tendered by them to the Commissioners.

As we took occasion to state to you in our former report, this inquiry having been entered upon, it is of the utmost importance that it should be fair, free, and exhaustive; and that if the opponents of those institutions of ours which are deemed essential to the due protection of the interests of labour against the overwhelming influence of capital, are to be allowed and invited to rake up from the ends of the earth all sorts of possible, impossible, and ridiculous charges against them, you, who are the creators and natural conservators of these institutions, should be watchful, and have equal opportunity, and be equally ready by your representatives to defend to the utmost whatever you may think unfairly attacked; and as your enemies will by no means be over-scrupulous whence they draw their information, very nice as to its absolute correctness, or generous in its application, it must be seen to by you that, if such a course is likely to be, or is being pursued, that you will not allow it to go to the world without such contradiction and explanation as shall convince all men that there is nothing connected with our Associations which we desire to conceal, and nothing which we deem of importance that we are not prepared to defend. If there be anything in the constitution of our Societies, their operation, or effects of those operations, which is proved to be against public policy, and injurious to the great body of the public, we are sure no one will rejoice to have the matter corrected more heartily than the great operative class, who have established these institutions for the protection of their labour.

The inquiry has already, as all of you are aware, been going on for five months, so that we are no longer left to guess at the nature of the evidence (and the spirit shown in it) which is to be brought against us. The campaign was opened by the examination of some of your own representatives. Mr. ROBERT APPLEGARTH, of the Amalgamated Joiners and Carpenters, was the first witness, and without agreeing in all the doctrines enunciated by Mr. APPLEGARTH, we may say that his positions were generally well taken, and ably argued. He was examined as to the origin, constitution, and mode of management

7

of that Association, as to the payments by, and benefits guaranteed to members, and as to the relations which the operations of his Association produced between employers and employed; but as we have already published all his evidence given at that time in the *Beehive* newspaper, we need not at present condescend on particulars further than to say that a great portion of the examination being directed to elucidating particulars respecting the friendly portion of their Association was not, in our opinion, strictly germain to the purpose for which the Commission was appointed.

Mr. GEORGE POTTER, of London, a member of our Committee, was the next witness, and was examined at great length on the constitution and practice of his local Society, as well as on his experience of Trade Societies generally, and his acquaintance with other Associations of an analogous kind; also as to piece-work, overtime, and the practice of fixing a minimum, or average rate of wages, which Mr. Potter justified on the principle that while one man might be more highly skilled in one department of labour than another, another might be more valuable in some other department of labour in which strength was of more avail than skill, and to which work the more highly skilled artizan would not apply himself, and at which all his skill would render him no advantage over the less skillful workman, and that men were usually put to that class of employment which produced the highest profit to the employer. Mr POTTER gave very lucid evidence on all points, and specially that it is quite untrue, as alleged by the enemies of Trades' Unions, that their effect is to prevent the more highly-skilled and prudent artizan to better his condition, but that every man was at full liberty to appraise his own labour, and act accordingly, although, as a general rule, we should never forget that our motto should be "each for all and all for each."

The main, or at least one motive cause for the appointment of this Commission, was the loud outcry made by certain interested though rather obtuse individuals, that the effect of the extensive combinations of the operative classes was sending a great portion of our most important industries to be executed abroad, and which would eventually tend to drag us from our position as the great workshop of the world. This "cuckoo" cry was repeated *ad nauseam* by a portion of the press whose delight it is, in the interests of the moneyocracy, to yelp at the heels of that class by whose toil all wealth was originally created, and by whose patient industry it still progresses with such giant strides, and has now attained to such magnitude that it looks with disdain upon its humble parents, and "and knows them not," except as created for its profit. To expiscate some information relative to this, among other matters, Mr. WM. ALLAN was put under examination as to the Society of Amalgamated Engineers, and explained to the Commission that practically he had no fear of foreign competition, as although the engines about which there had been so much talk had been put together abroad, the material for them was really manufactured in this country; and that although the foreign workman nominally received a lower rate of wages, it was not practically so, as he did at least one third less work than the English workman, and the franc of France and Belgium, in point of fact, procured there more of the necessaries and conveniences of life than an Englishman's shilling at home. Mr. ALLAN was examined at very great length as to the constitution of his Association at home and abroad, its effects and influence, and proved that it had all along been part of their settled policy to prevent strikes and promote good feeling and confidence between employer and employed; and that instead of their vast ramifications and large funds instigating them to be quarrelsome, these facts had quite an opposite effect. We beg to refer to the evidence as published from week to week in the columns of our organ. After Mr. ALLAN, followed Mr. HARNOTT, Secretary of the Operative Society of Stonemasons of England, who was examined in regard, *inter alia*, to certain rules of their Society respecting a practice called "chasing," and the publication of black lists. After Mr. HARNOTT came Mr. THOMAS CONNOLLY, of the same Association—a member of our Committee, who gave evidence as to the question of "chasing" and overtime. Mr. COULSON, of the Bricklayers' Association, was the next witness, and Mr. GEO. HOWELL, of the same Society. Mr. CHAS. WILLIAMS, of Liverpool, followed, and gave explanations as to certain lockouts and disputes at Bradford. Mr. Howroyd's sons, about whom the dispute originated, while reaping all the advantages derived by the members of the Union, refused to become members, and thereby contribute from their increased earnings towards the expenditure in virtue of which their earnings were so increased, and in consequence the workmen withdrew. In looking at this matter from a purely commercial point of view, there is much to be said for the Bradford Plasterers' Association, and however obtuse Mr. ROEBUCK may have been in apprehending Mr. WILLIAMS's illustrations, Mr. HOWROYD's sons being in enjoyment of the increased remuneration consequent upon the action and expenditure of the Trades' Union, were most undoubtedly "reaping where they had not sown." Operatives are entitled to sell, or to refuse to sell their labour, at any time, for any reason, or no reason at all. The only good ground of complaint the master plasterers of Bradford or Mr. HOWROYD could have had, was if they could have proved that by intimidation or coercion their men were withdrawn, and by the same means were prevented from getting others to supply their place. But the mere fact that one man or one hundred men refuse to sell their labour for what seems good and sufficient reason to themselves, can form no legitimate ground of complaint on the part of the person or persons to whom they refuse so to sell. Mr. HOUSLEY, of the Bricklayers' Society, Sheffield, was the next witness, then Mr. M'DONALD, of the Manchester and Salford Alliance of House Painters, and Mr. ROBERT LAST, of Manchester, Secretary of the General Union of Joiners and Carpenters. This list of witnesses completed the operative portion heard first after the opening of the Commission, and the Commissioners proceeded to hear representatives of the employing class. Mr. THOMAS PIPER, Secretary to the London Master Builders' Society, was examined, but nothing of any importance in respect to the inquiry was elicited, neither from this nor the two following, viz., Messrs. GEO. SMITH and G. F. TROLLOPE, excepting that the last two expressed an opinion that the action of Trades' Unions had led to estrangement between employers and employed. Mr. ALFRED MAULT was the

9

next witness, and he appeared fully charged to devastate the ranks of Trades' Unionism, and to give to our protective Societies their final *coup de grace* and consign them to the limbo of "things that were." To attempt to give a summary of this gentleman's evidence would be adding one more to the labours of Hercules. Indeed, that old hero, notwithstanding all his wondrous feats, might well have shrunk appalled from the task; and it would be "cruelty to animals" to impeach his personal courage for so doing. Mr. MAULT's evidence, if it has no weight in any other respect (and we are far from saying that it has not) has great weight from the mere quantity of it. The evidence is mostly all second-hand, having been supplied by employers from the northern and midland counties for Mr. MAULT, to put into shape and lay before the Commission; and whatever may be said of the value of some of it, Mr. MAULT deserves every credit for the labour he has bestowed upon the work. There is, no doubt, much valuable information in Mr. MAULT's evidence, but it is wrapped up in such a cloud of dust as to be very difficult to get at. We have no desire to defend every action of every Trades' Union in the country, nor indeed every action of any one, conscious that there are some things in most, and much in many which could be well spared, but we do not see that the principle of Trades' Unionism is to bear the blame of all the unjustifiable actions of men who are Trades' Unionists, even when such actions are done by them either as individuals or associations; and no class in this country will have greater satisfaction, if by exposing such reprehensible practices to the light of public opinion, through the evidence before the Commission, or otherwise, Trades' Unions may be relieved from the grievous burden of being afflicted by such practices, and having a stigma attached to what is legitimate and just in itself, in consequence of the conduct of certain individuals or associations, who, over anxious about conserving and extending their own rights, are forgetful of the rights of others.

In speaking thus we are not for a moment to be supposed to admit as proved all the statements of Mr. MAULT, framed as they are, and intended to damage the principles of Trades' Unionism, nor do we mean to say that very much of what he, in his anxiety to damage our cause has charged against us, may not be justified on the principle that every man, or body of men, have the right to exercise their freedom in the manner most conducive to their own interests, so long as by their exercise of that right they do not infringe on the freedom of others. And we most sincerely hope that all our Trades' Unions will take the earliest opportunity of rebutting such charges as are laid against them, and affording such explanations to the uninformed in these matters, as shall silence all calumniators, and in respect to other practices which, in their own nature, are unjustifiable, they will take such measures, so that as in all matters we should be "without fear," so we should also be "without reproach."

But let us ask the denouncers of our Unions, while so very anxious to take the mote out of our eye, they have seen to it that there was no beam in their own. May not very much of what they complain of be the result of the spirit of retaliation. If a greater moderation and conciliation had been adopted towards the operative class, by employers; if they had shown a disposition when the claims of the operatives to an increased remuneration were obvious and undeniable, to meet such claims with a good grace, instead almost always grumblingly and grudgingly, a better spirit might, in return, have been exhibited by the operatives; but this has never, or very rarely, been the case.

We take it, that even now, when such an outcry is making about the tyranny of Trades' Unions, few men of sense will be inclined to argue that the increase of the wages of the operative class has kept pace with the increase of the gross wealth of the nation, or the increase of income (chargeable with Income-Tax) of the monied classes for the last twenty years. Without being prepared to say what exact proportion of the profits of industry properly belongs to the labourer, we have no hesitation in saying that notwithstanding the great improvement which has taken place of late years, he is still far short of receiving that share in the profits of his own labour, to which its value in the production of wealth ought fairly to entitle him. We may be met by the query, what would labour be without capital? Very true, we reply. And just put the converse, what would capital be without labour? To use an illustration of Mr. MILL, on another branch of the same broad subject, the production of wealth—Capital and labour are as necessary to each other as the "two legs of a pair of scissors are in the act of cutting," and it is "just as rational to dispute whether five or six does most in making the product of thirty." Yet, the employing class do not hesitate to argue that the productive labourer ought to have no word at all in the matter of selling his labour, and whenever, along with his fellows, he ventures to state the conditions upon which he is willing to be employed, the cry is raised that he is interfering with the mode in which a master chooses to conduct his business, whereas all the while it is purely in reference to his own business, viz., the sale of his labour about which he engaged. We most sincerely hope that many of the statements made by Mr MAULT are fictions, or, at all events, so highly coloured as to approach the fabulous. And as with the view of enabling the trades to know all that has been advanced of a nature damaging to the interests of the various Unions implicated, we have from time to time published the evidence, we hope those trades will be thoroughly prepared to meet such statements, either with a direct denial, or afford such explanation as shall set their position right before the Commission and the public, or at once cry *peccavi*, and prepare in the interests of justice and good policy to correct all abuses in connection with themselves, being well assured that that which cannot be justified ought to be at once excised. Such abuses, if retained, only serve to bring upon all the objects of our Associations that condemnation which should only apply to the abuses themselves, but as long as such things are not only retained, but cherished, we cannot expect the public to distinguish between the good and the evil, and all combinations of the operative class, therefore, are indiscriminately denounced. As the Committee representing the interests of labour, appointed at the great Conference at St. Martin's Hall, in March last, we have left nothing undone which we deemed it essential to your interests to do. At our meeting in May last, we came to certain resolu-

11

tions as to the manner in which it would be desirable that evidence in regard to Trades' Unions should be taken; first, by hearing the charges, the opponents, and denouncers of Unionism had to bring, and thereafter the defence of the operatives. On speaking of the matter to a very esteemed member of the Commission, we were informed that such a course was the course, which the Commission as a general rule, intended to pursue, and that every facility would be afforded to those Trades' Unions against whose rules, conduct, and constitution, charges have been made, to meet such charges by rebutting evidence. In order that the Trades might be fully aware of all such charges, it was resolved to publish from week to week in the "Beehive" Newspaper, the minutes of evidence as printed for the Commission; and also, that the Commission should be requested to leave the selection of witnesses on the part of any Trade, to the members of the Trade, which course has been followed by the Commission It only remains therefore with the Trades' Unions of the United Kingdom to give us such aid, moral and material. as shall enable us to make such a defence for them as the magnitude of their interests demands

In the middle of Mr. MAULT's evidence, an American gentleman, an Iron Manufacturer, gave some very useful information as to the condition of labour in the United States, and the relations between Employees and Employer, and stated in answer to question 3777, that the Employer and workmen in America agree in this. "In England they have degraded their labour to very low wages, too low for families to maintain themselves properly and comfortably. In America we are determined never to see that state of things introduced as long as we can possibly help it." And further, that English competition and the race after cheapness is all very good and sound, except that it leaves out of question, the "condition of the labourer." Mr. HEWITT also stated that he accounted for the absence of strikes and disputes in the establishments with which he is connected, by the fact that "our personal relations with the men are very good, and we do not hesitate frankly to talk over the condition of trade with them." How different the state of things is in Britain we all know. It would be deemed *infra dig* in an Employer of labour in this free country to talk over the condition of trade with a body of his operatives, and as to the question of the "condition of the labourer" entering into his calculations, such a thing could not for a moment be thought of. Mr HEWITT also exposed the fact, that in the race after cheapness, and the scurry of competition, and in the country whose proudest vaunt is that she leads the van of civilisation, and in this later half of the Nineteenth Century, women and boys' labour is largly employed to do the work of men, and at the miserable pittance of 10d and 1s 3d a day. An attempt is made to excuse this state of things by saying, "it is the women themselves who desire the work" This may be true, but what is the cause, plainly that the renumeration given to their fathers, husbands, and brothers is so very low that they cannot keep their females without forcing them into competition with themselves, and thereby affording a pretext for a still further reduction of their already starvation wages. We strongly commend this gentleman's evidence to the attention of the operative class. Mr. JOHN M'DONALD of Glasgow, Chairman of the Glasgow Master Brick-builders' Association came next. This gentleman delivered judgment *ex cathedra* against all Trades' Unions whatsoever. His "points" as he called them, were mostly directed against what he termed a monopoly of labour, as sought to be established by the Union of Operative Bricklayers. He endeavoured to prove this by stating that the Union of Operatives limited the number of apprentices, and by making it incumbent on an apprentice, after being six months at the trade to become bound to the Employer for a term of years, and to pay to the Union of the workmen, an apprentice fee of Two Pounds. He also stated that the effect of Unions was to prevent an energetic, skilful workman from improving his condition, although how, Mr. M'DONALD did not or perhaps could not very clearly show. One thing Mr. M'DONALD certainly did show, viz., that the Society of which he is the honoured head do that which he charges against the Union of the operatives, by levying a fine of Five Pounds against any Master who pays any one man in his employment, or allows him to be paid either directly or indirectly in any form, one penny more than the regular agreed on rate of pay. Mr. M'DONALD goes on to say that in his opinion, and from his experience, wages would have been just as high or higher if there had been no union on the part of the men, and yet he admits that every increase of wages was conceded in consequence of an appeal from the workmen. He also says that supply and demand will, and ought to regulate wages. We thoroughly concur with Mr. M'DONALD in this doctrine, but before it will do so that NATURAL LAW must be put in operation by the workmen. Mr. M'DONALD most effectually proves this by stating that, during a very busy season, when in the ordinary and natural course of things, the competition between the different masters for men ought to have been very strong, not one Employer, although all were most urgently needful, would offer one penny more wages to induce more men to come to them, thus most clearly proving that Employers are very ready to put the law of supply and demand into operation in a falling market, but can never see their way spontaneously to do so in the interest of the operative. Mr. M'DONALD seems to have very confused notions regarding certain "unalienable rights" the operatives have according to the showing of the Secretary of his own Association, to combine together for "mutual council and support." Mr. M'DONALD does not think that such "unalienable right" at all extends to agreeing among themselves, as to what price they will sell their labour at, or how much of their own time they will sell, as he thinks such conduct interfering with the sovereign rights of the Employers, although how a man or any number of men have not the right to say what price they will consent to take, for what is most emphatically their own property, Mr. M'DONALD does not seem to know. How men can be possessed of unalienable rights, and yet have no right to exercise those rights we leave Mr. M'DONALD to decide. He also says that in his opinion the workmen would in most cases have got what they wanted without a strike. This seems a very odd statement for any man of common sense to make, when as everybody who has any knowledge of the matter at all knows, that

13

a strike only intervenes when the operatives fail to obtain what they think suitable terms. The Employers must be gifted with a greater amount of stupidity than we gave them credit for, if they submitted themselves to the loss and inconvenience caused by a strike rather than give something, which all the while they were willing to concede for the simple asking,—such are specimens of Mr. M'DONALD's logic. This gentleman also entered a complaint against the Operative Bricklayers' Union, for joining with the Master Bricklayers, to have all bricks made the same size. Now this is a very odd combination. Most of the Master Bricklayers in Glasgow make their own bricks, at least all the more important Employers, and the workmen are accused of combining with the Masters as Brickmakers against the Masters as Brickbuilders. This surely is confusion worse confounded. The truth seems to be that it was a Master Brickbuilder, and one of the largest contractors and Employers of labour in the United Kingdom, who first started the question of the sizes of bricks; but it comports very well with all Mr. M'DONALD's reasoning to blame the operatives for what was actually the work of an Employer.

Mr. E. ASHWORTH, of Manchester, was examined after Mr. MAULT and gave evidence as to certain strikes at Manchester and the neighbourhood. This gentleman, while deprecating many of the actions of Trades' Unions, frankly admitted that they had been successful in raising wages, and maintaining the increase in a great many instances. Mr. JAMES WILSON, Chairman of the Leeds Branch of the General Builders' Association, was the next witness, and gave evidence as to the doings of certain Trades' Unionists in Leeds and Yorkshire generally. This witness also unreservedly added that the effect of Trades' Unions had been to increase the renumeration of the workmen. Mr. WILSON was examined at very great length as to the influence the operations of Trades' Unions had on the relations of Employers and employed, and their general effect upon trade; and he was reluctantly forced to make many admissions as to the value of Trades' Unions, to the Operative Class, but as it would be impossible to give a summary without doing so at great length, we must refer Trades' Unionists to the evidence itself. The entire evidence up to the 6th August, is published in a Blue Book, and will form a very important document for Trades' Societies. After Mr. WILSON came Mr. RICHMOND, an Inspector of County Police in Lancashire, who gave evidence as to certain Trade outrages committed by Brickmakers, and as to a rule among the Brickmakers of the Manchester and Stockport districts that does not permit Bricks made in one district from being carried into the other, or used there. Mr. JOHN BRISTOW was the next witness who detailed certain differences and disputes between himself and the Unionists of Manchester, both Joiners and Bricklayers, he also gave an acconnt of personal injuries done to a friend of his, a Mr. BARLOW, a Brickmaker, and certainly, if the facts which Mr. BRISTOW relates be true, it betrays a state of things very much to be deplored, and approaches more nearly to the outrages of Sheffield than we had conceived possible anywhere else in the country. Such conduct not only brings disgrace on Trades' Unionists (that is, if done, or en-

14

couraged, or connived at by them), but on humanity itself, and it is as much the interest of Trades' Unionists to put down such practices as it is the interest of every other good citizen. Mr. G. HARROP, another Inspector of Police, was examined in corroboration of the two previous witnesses in regard to Trade outrages, for information as to which we must refer to the Blue Book. Mr J. KETTLE, the foreman Bricklayer at the Manchester County Prison, during the strike there, was the next person examined, and gave a history of that transaction Mr JAMES WEST, Carter, of Manchester, was examined as to his not being allowed to cart bricks in the Manchester district, and corroborated Mr. BRISTOW's statements. Mr. JOSEPH BARLOW was next examined as to the Trade outrages mentioned by Mr. BRISTOW, and confirmed that gentleman's statements. Mr. MICHAEL JOSEPH O'NEIL, general secretary of the Bricklayers' Union in Manchester, was examined in explanation of certain statements of Mr. Robert KETTLE, regarding the dispute or strike at the Manchester Assize Courts, and said that the dispute originated between Mr KETTLE and the labourers on the job, because Mr. KETTLE had appointed some labourer to be the foreman of the labourer's "gang" over some other senior, who, according to custom, ought to have got the position. The Bricklayers being thrown out of employment, in consequence of the refusal of the labourers to work under Mr. KETTLE's nominee, they demanded, previous to re-commencing their work at the New Prisons, that the wages they had lost in consequence should be paid by the contractor. With all due deference to Mr O'NEIL, and the Manchester bricklayers, we are of opinion that they put the saddle on the wrong horse, as by their own showing the cause of the loss lay not with the contractor but with the labourers, who assumed to dictate to Mr KETTLE, and through him to the employer, who should be their foreman. The supreme right of choosing who shall do any work most clearly belongs to the person who has to pay for the work being done, and no custom, nor any amount of customs, neither can or ought to override this right. Upon the whole, we are of opinion that Mr. O'NEIL's statements, considered in the light of common sense, justice, and candour, do not improve the case. Mr. JAMES CLARKE, Secretary to the Stockport Bricklayer's Union, was the next witness. He denied that there had been, since he became secretary to the Union, any serious disputes, nor anything at all approaching outrage or intimidation. Headmitted that there was an understanding between the bricklayers and makers, that the former would lay no bricks made by non-unionists, and the makers, on their part, would not supply any master bricklayer, with their bricks, when there was a dispute between such master and his operatives. The witness also admitted that it was the practice to put pickets on to watch work in cases of dispute, or strike. This of course for the purpose of peaceably persuading operatives not to seek employment during the currency of the dispute. The witness following Mr. CLARKE was a Mr. TUCKER, actuary and accountant in the city, who made a statement as to the payments of the members of the Amalgamated Carpenters and Joiners, and the inadequacy of these payments to meet the promised benefits; and that in consequence of such inadequacy, that Society, and others similarly situated must, sooner or later, become insolvent. Mr. TUCKER, however, had not had the needful time to go so minutely into the accounts as he would like, and requested further time to consider from the constitution of such Associations, how soon that state of insolvency would begin to show itself. This course of examination was entered on in consequence of some statements of Mr. MAULT, that men were induced to join these Associations by the promise of benefits, which it was quite impossible for the Societies to keep, and that for the legislature to permit such a state of things was to lend themselves to a fraud Mr. ROBERT APPLEGARTH, who was present, was examined as to Mr TUCKER's statements and very clearly gave it as his opinion, that from certain circumstances peculiar to Trades' Societies, and between which and ordinary assurances there was no strict analogy—his association would be quite able to meet all its obligations. He was also examined as to the picketing system, and very strongly defended the right of picketing as quite within the "peaceable persuasion" permitted by the Amended Combination Act, 1859, provided always that no intimidation nor coercion of any kind whatever was resorted to * Mr. WILLIAM ALLAN, of the Amalgamated Engineers, was further examined as to the payments and benefits of his Society with reference to the statements of Mr. TUCKER. Mr FINLAYSON, the Government actuary, was called in on this question, but owing to that gentleman not having had the necessary papers long enough, he was not prepared to go into the question. The Commission desired Mr. FINLAYSON to look carefully into the matter and report. Mr JAMES MATSON, of Glasgow, general Secretary to the Associated Carpenters and Joiners of Scotland, was the next witness He put in copies of the Rules of the Association, also copies of the Monthly and Annual Reports issued to the members since the formation of the Society. Mr MATSON explained very clearly the financial operations of the Association, and the general operations and influence of the Society. Mr. JOHN C. PROUDFOOT, of Glasgow, a member of the same Association, Secretary to the Trades' Council of Glasgow, and also Secretary to our Committee, was the next and last witness examined. Mr PROUDFOOT stated that he had had very considerable experience of Trades' Unions, having been Secretary of the Association (on which the Society for which Mr. MATSON is now Secretary was founded) in 1854, and had for many years made a study of such questions. Mr. PROUDFOOT stated that he could not presume to defend every action of any one Trades Union, but as a rule he was most decidedly of opinion, from long observation and experience, that judiciously conducted Trades' Unions had been of great benefit to the operative class; in fact he attributed the comparatively improved position of the entire operative class very greatly to the influence of the Unions taking advantage, of course, of the progress of Society, and the increase of the national wealth. On the question of limiting the number of apprentices by some Unions of operatives he stated that the Society with which he is connected does not restrict the number, but he entered on an abstract argument on the subject, and strongly defended the right of the operatives to do so, where, as is almost universally the practice the operative is the person called on to teach

* See note at the end

such apprentices their trade or profession, and that certainly therefore the operative had an undoubted right to say how many persons he would be instrumental in bringing into the labour market to compete against and undersell himself. He most severely criticised the statements made by Mr. M'DONALD of Glasgow, as crude and illogical, and defended the operative class from the charge of that gentleman, that the Unions legislated for the employers as third parties. Mr. PROUDFOOT claimed as the "unalienable right" (mentioned by the Secretary of Mr M'DONALD'S Society) of the entire operative class whether one or five hundred, the right to say on what conditions, how much of, and at what price they would sell their toil. This they did as first parties legislating for themselves, and then they come to the purchaser or second party, the employer, and made offer of their labour on these terms, and prepared to make a bargain about it. If the terms are suitable they are accepted, if the contrary, rejected. The master keeps his wages and the workman his labour. It was farther shown that all increases of wages had within living memory been the result, more or less, of combinations of the operatives, whether these operatives came together in a temporary manner for some one particular object, without having previously contributed to any common fund, or in the more permanent form of an ordinary recognised Trades' Union for general purposes in connection with any trade. The first is as virtually a Trades' Union as the second. In regard to picketing, Mr PROUDFOOT thoroughly concurred with Mr APPLEGARTH, as to its legitimacy in the light of the Combination Act, 1859; or otherwise the Act 6 Geo. IV. c. 129, as to conspiracy had been amended for "naught and in vain." If men are to have the liberty "peaceably to persuade," it naturally follows that they have the liberty to discover who the parties are who need such persuasion, and this can only be done by picketing. It is, no doubt, a very difficult thing in the case of a Lockout or Strike, to keep all persons engaged in that Strike strictly within the bounds of law in regard to strong language, a natural consequence of irritated temper, but it seems very like "Justices' justice" to visit the Office-Bearers of any Union, with condemnation and punishment for the faults of temper exhibited by members of the Union doing picket duty; and for whose conduct in performing that duty, the said Office-Bearers cannot in fairness be held responsible. Indeed picketing exists not in consequence of any determination to that effect on the part of Office-Bearers, but in consequence of a resolution of the whole number of persons engaged in any Strike, or dispute, and instead of Office-Bearers encouraging or conniving at intimidation, or coercion, all their influence is, and ought to be, devoted to restraining the members within the due bounds of law. In the recent case of the Office-Bearers of the London Tailors' Protective Association, we solemnly believe this to have been the case, and that these officers of the Union were no more responsible for, nor cognisant of, any coercion or intimidation which pickets may have been guilty of, than Mr. BARON BRAMWELL himself, or the jury who returned a verdict of guilty against them. This verdict too was delivered on the most shadowy distinction, (or rather the line was so very fine that it had no shadow,) as to what was to be construed intimidation, and what

17

"peaceable persuasion." Assuredly they are doubly blessed who are in possession of such powerful legal microscopes. The chemical feats of Burns' Doctor Hornbook, were nothing to this. His

> "Mite-horn shavings, filins, scrapins,
> Distilled *per se*,
> Sal-alkali, o' midge tail clippins,
> An mony mae,"

were nowhere in comparison. This hairsplitting forcibly recalls to our memory Swift's sarcastic lines—

> "Strange that such difference should be
> 'Twixt tweedle-dum and tweedle-dee."

We sincerely hope the learned BARON'S bow may "long abide in its strength," and that the sauce which has been found so suitable for the *goose*, may also be served for the *goose's* master. Perhaps the magistrates of Great Marlborough Street will seriously lay this to heart, and may be discover that conspiracy and intimidation are not absolutely confined to combinations of operatives.

Mr. PROUDFOOT was further examined as to the effects of the operations of Trades' Unions on the relations between Employers and Employees, and stated that, as far as he knew, there might, just at the time of a dispute about the rate of remuneration or any other condition of labour which the operatives deemed it their interest to insist upon, be a little irritation of feeling. Still, so far as his experience went, such a feeling was, by no means permanent. He had taken for some years previous to 1857 a rather prominent part in trades' questions, and yet was always on the best terms with his Employer, who, he thought, as a general rule, did not think worse of any man for standing out for what he thought was right, so long as it was done in a reasonable manner, and with a due respect for the rights of others.

We would again advise Trades' Associations to provide themselves with a copy of all the evidence tendered up till the adjournment of the commission, for the information of officers and Members. In a report of this nature, we can only give the merest outline, indeed, we do so mostly as a foundation for criticism, explanation, and reply.

Having thus, brethren, given you a short and necessarily imperfect outline of the doings of the Commission, we are desirous in a few last words, to urge upon your attention the absolute necessity of close attention being paid to the progress of this enquiry, and that no effort nor sacrifice should be spared on your part, in laying your case in the best possible manner before the Commission and the public. We believe we are justified in saying for you, that you have no desire to extenuate any of the least of those outrages which have been perpetrated in the name of Trades' Unionism; and while desiring perfect freedom for yourselves, you can have no desire to fetter the freedom of others. We believe, also, we are justified in saying for you, that you sympathise deeply and sincerely with those trades of Sheffield who, neither by their office-bearers nor otherwise, had any connection with any of the atrocities recently disclosed. And in particular, would we desire to sympathise with those office-bearers of the United Kingdom Alliance of organised trades, who

18

were associated in the government of it with the notorious Broadhead, of whose character and doings, we believe, they were as profoundly ignorant as those writers in a portion of the press, who most unfairly and dishonestly desire to include the Alliance in his guilt. We also earnestly hope that the exposure of such doings, and the culpable neglect of Members of Committees which made such doings possible, will have the effect of not only eradicating such things from the Unions of Sheffield, but from every other Union of operatives in this country, for we much regret to say, that there are grave suspicions, that there exist practices either connected with Trades' Unions, or at least countenanced by professed Unionists, which, if at all different in degree (?) are not so in kind.* If we are to ask such a recognition in law as shall make us incorporated bodies, enabling us to sue and be sued by our office-bearers, we must make our institutions such as the law can legitimately recognise.

From the tone adopted by a portion of the public press of this country, which has always been notorious for its virulence in opposition to every attempt of the labouring population to obtain a somewhat fairer share of the profits of their own toil, you may very clearly perceive the spirit which animates the patrons of that press in the part taken by them in the Commission inquiry. It is, perhaps, rather late in the day to attempt a revival of the laws against combinations of operatives for purposes which are in themselves legitimate; but that such is the drift of the opponents of working class trades' unions, there can be no manner of doubt. If, by the remotest chance, any matter damaging to the character of a trade's union or unionist can be raked up, twisted, and construed to suit a foregone conclusion, with what gusto and smacking of the lips it is done, as if, forsooth, any association could be answerable, or could in reason be expected to be answerable, for all the actions of all its various members; and, upon the whole, much of what is objectionable in the practises of unionists, or combinations of operatives, is due to the contemptuous and supercilious manner in which their just and reasonable claims have been met. If, occasionally, in some of their actings they have outraged the laws, let it not be forgotten that they have been ostracised and placed beyond the pale of the law. It is, therefore, scarcely to be wondered at (although deeply to be deplored), that they should have adopted such means in default of the law, as most readily appeared to their minds suitable or likely in the circumstances to answer their purpose, and be a law unto themselves, judge, jury, etc. We desire to draw the attention of the entire operative class to the paragraph in the Royal Speech respecting trades' unions, at the prorogation of Parliament; and to the likelihood of some legislation being attempted on the subject next session. We have no doubt the Royal Commission will very early in the session present, if not a final, at least a report of their conclusions on the evidence taken by them up to the commencement of the session. As the committee watching over your interests in this matter, we must call on you to give us your utmost and most cordial support We are of opinion that any important legislation on this question ought, in fairness, to be delayed

* This was written previous to the Manchester disclosures.

till after the general election consequent on the Reform Bill. We shall then, by our own representatives, be in a position to take part in deliberating on those laws which we will be called on to obey; and on which depend the "safety of our property, our liberty, and our life."

It would be highly imprudent, as well as impolitic, in our opinion, for the present moribund Parliament to proceed to legislate on a class of questions from the consideration of which the class of persons most interested is entirely excluded. We also hope that at the next general election of members to serve in the Commons House of Parliament, the operative class will make such use of their enfranchisement as shall secure that it is really a representation of the Commons of the United Kingdom. If landocracy, railwayocracy, moneyocracy, and all other ocracies be represented, why should the wealth-producingocracy be awanting? Send men to represent you, brethren, who, by experience, know your needs, your wishes, and your ideas. Then you will have pleasure in obeying loyally those laws which, by your representatives, you have assisted to frame. This is a most important matter. The longer experience we have of the Royal Commission of Inquiry, the more are we struck with the mistake made in not putting, at least, one or two intelligent men belonging to the class to be put on their trial, among the representatives of other classes of the community. A great deal, in that case, seems confused and misty, would have been made clear; and the inquiry been more satisfactory to the class who are most deeply concerned. We have no fault to find with any member of the Commission. We willingly admit their anxiety to bring out the whole truth, and to deal towards both sides with the most scrupulous fairness; but we cannot shut our eyes to the fact that their practical knowledge does not enable them fully to accomplish that which they so anxiously desire. Let us see to it, fellow-workmen, that this remedy is provided in the future legislative councils of the nation; and that henceforth the laws affecting the welfare of the great labour-selling class of this country, shall not be left exclusively to the making of the labour purchasers.

As the work of our Committee cannot be carried on without some pecuniary resources, we desire to direct the attention of those trades represented at the St. Martin's Hall conference, to the resolution of Committee calling upon them (in terms of resolution of conference), for a levy of a second farthing per Member. The business has hitherto been managed on the most economical principle (too economical perhaps for thorough efficiency), and the various trades may be assured that, consistent with efficiency, the same system of economy will be pursued for the future. We desire, also, to say to those trades or societies who were not at the St. Martin's Hall meeting, that our cause is also theirs. Brethren, there should be no standing aloof from one another at this crisis; we should join hand in hand and put shoulder to shoulder, and work earnestly together for the accomplishment of our object, alike important to us all. The pecuniary pressure is not heavy, and more labourers make the work lighter. Who would not sacrifice a mere fraction of his earnings to defend his right—to take the best means apparent to him for enhancing the value of the labour by which he lives, or in the words of ADAM SMITH, "the right of the men who feed, clothe, and lodge the whole body of the people; to reserve to themselves so much of the profits of their own labour, as that they themselves should be suitably fed, clothed, and lodged?" We hope you will respond to this appeal as early as possible, and so enable us to discharge fully and efficiently the duty to which you did us the honour to appoint us. We also desire that you will, as soon as possible, by letter addressed to the Secretary, 1, Holmhead Street, Glasgow, indicate any opinion which your various associations may have formed, as to what should be our future course of action in respect to the Commission of Inquiry.

Signed for Committee.

JOHN C. PROUDFOOT,
Secretary.

INCOME,

TRADES' COUNCILS &c.

Edinburgh Trades' Council, - - - - - - -	£0	0	0
*Glasgow United Trades' Council, - - - - - -	5	4	2
*Association of Organised Trades of Sheffield, - - -	6	5	0
*Liverpool United Trades' Protection Society, - - -	2	17	1
*Manchester and Salford Trades' Council, - - - -	9	2	8
Wolverhampton Trades' Council, - - - - - - -	0	14	7
*Preston Trades' Council, - - - - - - - -	1	19	10
Halifax Trades' Council, - - - - - - - -	0	0	0
Nottingham Building Trades' Council, - - - - - -	0	10	0
*Association of Organised Trades of Nottingham, - - -	1	8	0
*Sheffield Trades' Defence Committee, - - - - -	2	1	8

TRADE SOCIETIES.—PROVINCIAL.

*Boiler Makers' and Iron Shipbuilders' Executive Council, -	9	7	6
*Associated Ironworkers of Great Britain, - - - - -	2	5	8
*Operative Stonemasons (Executive Council), - - -	18	10	6
Glass Bottle Makers' Protection Society (Yorkshire), - -	0	14	0
*Northern Counties' Association of Cotton Spinners (Manchester)	6	5	0
*Millmen's Association, West Bromwich (Executive Committee),	1	13	3
Copper Plate Printers of the Staffordshire Potteries, - -	0	5	0
*Flint Glass Makers' Society of Great Britain and Ireland, -	1	14	9
*National Association of Plasterers (Executive Council), -	5	14	2
Painters' Society (Woolwich), - - - - - -	0	0	0
Lace Makers' Society (Nottingham), - - - - - -	0	18	9
Canterbury Joiners' Society, - - - - - - -	0	0	0
Leicester Joiners' Society, - - - - - - - -	0	4	2
Crate Makers' Society, Hanley, Staffordshire, - - -	0	6	3
*West Yorkshire Miners' Association, - - - - - -	2	1	8
Amalgamated French Polishers' Association (Manchester), -	0	3	4
Cotton Spinners' Association, Bolton, - - - - - -	0	16	8
*Miners' National Association, - - - - - - -	37	10	0
Power Loom Carpet Weavers, Kidderminster, - - -	0	10	6

21

*Amalgamated Tailors of England,	£12	11	2
Birmingham Joiners' General Union of Carpenters,	0	8	4
*Amalgamated Association of Malleable Iron Workers,	3	2	6
Power Loom Weavers' Association of Preston,	0	0	0
*Manchester Alliance of House Painters,	4	3	0
LONDON.			
*London Society of Compositors,	3	8	9
United Branches of the Silver Trade,	0	0	0
*Amalgamated Union of Bakers,	3	9	0
United Brushmakers' Society,	0	0	0
*Shipwrights of the Port of London,	1	15	6
Horse Shoe Fund Society,	0	0	0
Amalgamated Sadlers' Society,	0	14	0
Steam Engine Makers,	0	0	0
Plumbers' West End Society,	0	0	0
Co-operative Tin Plate Workers' Society,	0	0	0
Leather Shavers' Society,	0	2	1
Friends of Freedom Carpenters' Society,	0	0	0
Silver Cup Carpenters' Society,	0	4	8
Running Horse Carpenters' Society,	0	2	4
Progressive Carpenters' Society,	0	7	1
Old Lambeth Carpenters' Society,	0	5	0
Hand of Fellowship Carpenters' Society,	0	5	0
Adam and Eve Carpenters' Society,	0	6	0
Perseverance Carpenters' Society,	0	1	1
Independent Carpenters' Society,	0	5	0
Moore Arms Carpenters' Society,	0	0	0
West End French Polishers,	0	2	1
Braziers Hand in Hand,	0	5	0
King's Arms Gilders,	0	0	0
Three Doves Gilders,	0	0	0
Union Gilders,	0	0	0
United Order of Smiths,	0	0	0
Wood and Tin Packing Case Makers,	0	0	0
Printers and Warehousemen's Association,	0	0	0
Paper Hangers' Society,	0	2	3
Alliance Cabinetmakers,	0	16	8
Amalgamated Society of Smiths,	0	0	0
Shipriggers' Association,	0	9	6
Fancy Cabinetmakers,	0	1	0½
St. Martin's Painters' Society,	0	3	0
Westbourne and Bayswater Society,	0	0	0
The Castle Society,	0	0	9
Western Provident Society,	0	5	0
Enterprise Society,	0	0	0
Hope Society,	0	0	0
Grosvenor Basin Society,	0	5	0
Trimmers and Harness Makers,	0	0	0
Gas Meter Makers,	0	2	11
London Farriers,	0	0	0
Hatters' Society,	0	17	8½
Bookbinders' Consolidated Lodge,	0	12	6
Garment Dyers' Society,	0	0	0
*General Council of Painters,	3	2	6
Portmanteau and Trunk Makers,	0	0	0
National Association of Rope and Twine Spinners,	0	12	6
Ropemakers' Society,	0	2	1

22

East End Cabinetmakers, - - - - - - - - -	£0	10	0
Operative Brickmakers,- - - - - - - - - -	0	0	0
Carmens' General Protection Society, - - - - - -	0	1	1
French Polishers' Trade Association, - - - - - -	0	4	0
Hammermens' Society, - - - - - - - - -	0	0	0
Upholsterers' Society (Leeds), - - - - - - -	0	10	0
United Composition and Ornamental Trade Society, - -	0	10	0
Brassfinishers' Protective Society, - - - - - - -	0	0	0
Corkcutters' Society, - - - - - - - - - -	0	8	2
Coach Wheelers' Society, - - - - - - - - -	0	4	6
London Tailors' Protective Association (on Strike since Conference)	0	0	0
Tallow Chandlers,- - - - - - - - - - -	0	0	0
Beehive Lodge General Union of Carpenters,- - - - -	0	2	1
United Curriers' Society, - - - - - - - - -	0	8	4
Seamans' Mutual Benefit Society, - - - - - - - -	0	7	4
Lightermen and Watermen of River Thames,- - - -	0	0	0
Tin Plate Workers, Bell Inn, Old Bailey, - - - - -	0	4	0
Tin Plate Workers (Leeds), - - - - - - - - -	0	10	0
Working Men's Association, - - - - - - - - -	0	12	6
Shipwrights (Bishopwearmouth,) - - - - - - - -	2	0	0
Cloth Dressers (Leeds), - - - - - - - - - -	0	2	0
Packers' Friendly Society (Burslem), - - - - - -	0	5	0
Lithographers (Leeds), - - - - - - - - - -	0	1	0
Amalgamated Joiners (Leeds), - - - - - - - -	0	1	4
General Union Joiners (Leeds), - - - - - - - -	0	1	11
French Polishers' Trade Protection Association, - - - -	0	2	1
	£165	11	4

* NOTE.—The representatives of those Societies paid, in addition to the sum of 5s. each at the opening of Conference, the farther sum of 2s. per 1000 of their Constituents.

ABSTRACT OF INCOME AND EXPENDITURE

TILL 8TH AUGUST, 1867.

INCOME.

To Amount paid by Delegates at Conference, as per last Balance Sheet in Report, - - -	£26	15	0
,, 2s per 1000 additional levy for Conference, paid by Societies of over 1000, - - - -	9	0	0
,, Amount of first farthing per head levy, including some payments towards Conference expenses not previously paid, - - - -	165	11	4
	£201	6	4

EXPENDITURE.

By Conference Charges as per Conference Balance,	£45	9	5½
,, Accounts come in since close of Conference, -	0	14	0
,, Printing Circulars, Reports, &c., - - -	27	13	6
,, Stationery, Postages, Post Office Orders, Carriage of Parcels, and Rent of Rooms for Committee on 15th and 16th March, - - - -	8	13	0
,, Expenses of Committee, including Railway and other fares, - - - - - - -	99	9	7
,, Cash in hands of Treasurer to meet Charges since 18th May, - - - - - - -	19	16	9½
	£201	6	4

(Signed) J. C. PROUDFOOT, W. H. WOOD, } *Auditors.*

24

The Committee most cordially thank the Trades for the ready manner in which the last call was responded to, and would most respectfully urge upon all the desirability of an equally ready response being made by them to the present, that the Committee may thereby be enabled to proceed with full freedom and energy in defence of your cause. All the more important bodies represented at Conference have paid the voted levy, with one or two exceptions, which will be seen on reference to the detail of income, and it is earnestly to be hoped that all who have not paid their proportion will, for their own honour, yet do so without delay, as it is no less than dishonest to impose upon others the expense of doing work in which all have an equal interest, and to the expense of doing which all the Trades represented at the Conference are equally obligated to contribute.

Post-Office Orders, &c., made payable to Mr. WILLIAM LEIGH, George's Street, Hyde, near Manchester.

(Signed for Committee), JOHN C. PROUDFOOT, SECY.

NOTE.—We are asked to explain that, at the time of the Conference, the Edinburgh Trades' Council was in a state of disorganisation, and had, in point of fact, no representative there. Mr. Troup was sent up to the Conference by the office-bearers of a number of the Trades there, who become personally responsible for his expenses. The Trades' Council is now reorganised, and will, it is hoped, join us in the defence of our Trades Associations, as what is important to the Trades of the United Kingdom cannot be indifferent to our Edinburgh brethren.

After Mr. APPLEGARTH'S last examination, Mr. JOHN MOONEY, of Glasgow, President of the Operative Bricklayers' Association, was examined. He contradicted and explained some of the statements of Mr. M'DONALD, as to the operations of his society, and placed matters in a somewhat different aspect than was done by Mr. M'DONALD—*see Blue Book.*

TO THE
TRADES' UNIONISTS'
OF
GREAT BRITAIN AND IRELAND.

FELLOW WORKINGMEN,—

Nearly four years have elapsed since a small number of workingmen, belonging to different countries, convened a public meeting at St. Martin's Hall, at which the International Workingmen's Association was established. They were doing the right thing at the right time. During that space of time the Association has acquired a position that no other organization has ever attained in Europe. It is neither a rival of, nor in conflict with any workingmen's organizations; on the contrary, it aims at utilising and extending the influence of existing organizations in all countries, by endeavouring to bring about a common understanding and promoting common action between them.

As to its character, though it advocates complete political freedom, it is not a political association in the old acceptation of the term. While aiding all progressive movements it avoids the contending factions, well knowing how futile it would be to expect any real amelioration in the condition of the labouring poor by trusting to the capitalists as a class. Slanderers assert that the Association has provoked strikes, it has not done any such thing; but it has enabled workingmen to resist lock-outs, and successfully terminate strikes rendered inevitable by the encroachments, bad faith, and wanton conduct of employers.

The fundamental principle of the Association is, that the produce of labour ought to be the property of the producer; that the *brotherhood of labour* should be the basis of society; and that the workingmen of all countries should throw aside their petty jealousies and national antipathies, and make common cause with each other in their struggle with capital. Labour is of no country! Working-

2

men have the same evils to contend with everywhere. Capital is but accumulated labour. Why should the labourer be the slave to that which he has himself produced? Too long have the capitalists profited by the national isolation of the sons of toil. Foreign competition has always furnished a plea for the reduction of wages. For a long time the Trades' Unions of this Kingdom sufficed to keep wages up. Free trade has worked a change. The Continental workmen work longer hours, for less money, than the British do. If this country is yet producing cheaper than others it is due to a higher development of her machinery. The distance in the race between the British and the Continental manufacturer for the prices in the markets of the world, is rapidly diminishing; the British is ahead, but only just ahead.

These facts ought to convince the British workmen of the importance of the International Association. It has everywhere on the Continent fostered the *formation of Trades' Unions,* and served as an engine for their common and fraternal action. In France its action in the lock-out of the Paris bronze-workers was characteristic, 1,500 men were locked out and 4,000 more threatened to be locked out, unless they abandoned their newly-formed trade society —one of the first that was formed after the British model. By the aid of the International Association that combination of employers was defeated. Since then trades' unions have become naturalized in France, and the government, alarmed at the progress of the International Association, has again and again tried to suppress it by fining and imprisoning the executive committee of Paris. In Switzerland the strike in the building trades of Geneva resulted in a reduction of the hours of labour and an increase of wages per day, but the masters only gave in after they had ascertained to what extent the men received supplies from abroad. Instead of crushing the Association, the action of the master builders has tended to increase its influence.

In Belgium the International Association has played an equally prominent part. In consequence of the general crisis in the metal trades, the directors of the mines resolved upon working only four days a week. To make sure of the shareholders' dividends they gave peremptory notice of a ten per cent. reduction of wages. The miners refused to continue working on such conditions, and the over-anxious Government tried the persuasive influence of powder and lead upon them; many were killed, many more were wounded, and

3

imprisoned. At that stage the Brussels committee steppéd in. They procured medical aid for the wounded, pecuniary support for the bereft, and counsel for the imprisoned. Since then they have succeeded in establishing a miner's union in the coal basin of Charlerois. The untutored miners have thus been brought within the bonds of labour's brotherhood which will be a safeguard against their oppressors riding rough-shod over them on a future day.

In Germany, on the occasion of Count Bismarck proposing a reform of the tariff, the chamber of commerce of Barmen and Elberfeld (the Prussian Manchester) objected on the ground that the Prussian manufacturers could not compete with the English without a reduction of wages—a course that could not be pursued without danger in the face of the rapidly spreading influence and prompt action of the International Association.

In countries where the development of modern industry has not yet led to open war, its members content themselves with quietly propagating the principles held by their fellow-workmen in the more advanced countries. The London Council is also in correspondence with the officers of the great Labour Reform Movement in the United States.

To British trades' societies it has rendered signal service, by furnishing special information from abroad when required, and by circulating correct accounts of their disputes all over the Continent, and thus prevented the employers obtaining foreign labour to supplant that of their own men. In the accusation against the Paris committee the imperial prosecutor stated, one of the chief reasons for demanding the condemnation of its members was that they had not only brought an excessive influence to bear upon all strikes in France, but had efficiently supported those in foreign countries, and as instances he stated that during the strikes of the English zinc-workers, tailors, and railway *employeés,* the Paris committee had prevented French workmen from proceeding to England.

The ever-ready cry of the British capitalist that wages must be reduced because the workmen on the Continent work longer hours for less money than the British, can only be effectually met by endeavouring to approximate the hours of labour and the rate of wages throughout Europe. This is one of the missions of the International Workingmen's Association, and its annual congressses one of the most efficient means to accomplish it. At those gatherings the spokesmen of the working classes of different countries meet each

4

other face to face. The exchange of ideas which is brought about in the private conversations outside the regular meetings exercises as great, if not a greater, influence than the regular debates. It is there where everybody says what he has to say, and makes enquiries as to what people think of kindred topics elsewhere. At the London Conference of 1865 the French and Swiss delegates expressed it as their conviction that trades' unionism would never take root on the Continent. At the Congress of 1867 there were upwards of 40 delegates representing Continental trades' societies formed on the British model. The seed that had been sown in London had borne fruit.

The next congress will assemble on the first Monday of September next, at Brussels, a few hours' journey from England. Brussels was selected at the last Congress with the view of enabling the British workmen to send a greater number of delegates than they had been able to send to Switzerland. To make the British delegation a really respectable one the Council urges the affiliated societies to send as many delegates as possible.

Societies joining before the end of August will be entitled to send delegates of their own. Affiliated societies who do not consider it advisable to send delegates of their own, and trades societies desirous of rendering assistance are solicited to contribute towards defraying the expense of delegates appointed by the Council.

Amongst the questions that will be submitted for deliberation are; 1. Reduction of the hours of labour. 2. The influence of machinery in the hands of capitalists. 3. Property in land. 4. The education of the working class. 5. The establishment of credit institutions to promote and facilitate the social emancipation of the workingclass. 6. The best means to establish co-operative production.

By order of the general Council of the International Workingmens Association.

H. IUNG, CHAIRMAN.
R. SHAW, TREASURER.
J. GEORGE ECCARIUS, GENERAL SECRETARY.

256, High Holborn, London, W.C.

Money orders to be made payable to the Secretary at the Charing Cross Office, W.C.

SIR W. ERLE ON TRADE UNIONS.

From the "Spectator," February 20, 1869.

The title of Sir W. Erle's book is the *Law Relating to Trade Unions.* He says in his introduction that he "intended, in writing this memorandum, to state the law as it is, and only rarely to express an opinion as to what the law ought to be." But he had said a little before that, "in stating the common law," he had "aimed to assign some of the reasons for it, both with a view to show that the law is as stated, and also in order to foster loyalty." Sir W. Erle does not seem to perceive that when a writer only assigns "some of the reasons" for a law, and not those against it, he, in fact, impresses upon others, if he does not himself express, the opinion that the law *is* what it *ought* to be. Hence, his book consists really of two elements—law, *plus* advocacy of the law, and there is great danger that what belongs to the second element will be misconstrued, or knowingly misapplied, as belonging to the first; that the ex-Chief Justice's mere theories will be trumpeted forth as autoritative law. And it is surely a serious question how far the head of a Commission, before the report is issued, or probably even agreed on, is justified in putting forth to the public, not a merely impartial statement of a portion of the law which forms, amongst other things, the subject of that Commission's inquiries, but one garnished with arguments in its favour. Such an act can but serve to bias public opinion beforehand, and although Sir W. Erle may offer the memorandum on his "own responsibility," the mere fact of its publication, being what it is, commits him as chairman of the Commission, and the Commission *pro tanto* with him.

That these are no merely speculative objections every page almost of Sir W. Erle's book is sufficient to show. Space, however, fails us for dwelling on the many misstatements into which his advocacy of the existing law has led him,—*e.g.*, his extraordinary attempt to read modern political economy into our old common law, and to represent the latter as upholding the "right to a free course for trade" during a half-millennium of restrictive legislation, and long before the very idea of labour as a commodity, and no longer as a mere personal service, has ever been conceived. But there is one such misstatement which repuires to be adverted to, as being the most specious of all, and in fact pervading the whole book.

Great allowance must, no doubt, be made for the fondness of a judge full of years and honour towards a law which he has long administered. See how, in the midst of his dry *Fourth Institute*, old Sir E. Coke breaks out into something very like poetry, whe he speaks of "the laws of England, the *golden netwand* whereby all men's causes are justly and evenly measured"! But when Sir W. Erle in his introduction informs us that, "in law, labour and capital are not contradistinguished, neither the rights nor the duties of a person are affected by being placed in the class of either labourer or capitalist;" or, again, at p. 12,—this time in all the solemnity of italics,—that "every person has a right under the law, as between him and his fellow subjects, to full freedom in disposing of his own labour or his own capital according to his own will," the only answer is: 'It is not true. The law does contradistinguish labour and capital, does withhold from the labourer in many ways that freedom in disposing of his labour which it generally grants to the capitalist in disposing of his capital; nay, in some cases withholds such freedom even from the latter.'

The power of disposing of a thing according to one's own will includes surely that of placing it at the disposal of another. A cannot be said to have "full freedom in disposing of" a sovereign if he cannot hand it over to B to give away. But we are expressly told by Sir W. Erle, some twelve pages further on, that this does not apply to the two articles, labour or capital. "A person," he says (pp. 24–5), "can neither alienate for a time his freedom to dispose of his own labour or his own capital according to his own will,"—referring to "Hilton *v.* Eckersley,"—"nor alienate such freedom generally and make himself a slave,"—referring to Sommersett's case,—"it follows that he cannot transfer it to the governing body of a union."

The case of "Hilton *v.* Eckersley" affords, indeed, one of the most obvious arguments in favour of the impartiality of the law between labour and capital. It was that of employers entering into mutual bonds to carry on their business in accordance with the resolutions of a majority, and the bonds were eventually held incapable of enforcement; first, as being "in restraint of trade;" secondly, because the Court saw "no way of avoiding the conclusion that if a bond of this sort is capable of being enforced at law, an agreement to the same effect amongst workmen must be equally legal and enforceable," and so they would "be giving a legal effect to combinations of workmen for the purpose of raising wages, and make their strikes capable of being enforced at law." The latter ground of decision (to the former we shall revert hereafter) amounts to this,—that a particular form of association is to be withheld from the capitalist, lest a certain analogous one should become open to the labourer. Now, it so happens that the particular form of capitalist's contract invalidated by 'Hilton *v.* Eckersley," that of a bond with a penalty, has been since—as Sir W. Erle well knows from the evidence taken before him—replaced to great advantage for employers' associations by a deposit of promissory notes,—an expedient perfectly easy for the capitalist, quite nugatory, as a rule, for the labourer. But what if the form of association which the law in "Hilton *v.* Eckersly" refuses to ratify be only one among many which are open to the capitalist,—if the similar one for the labourer be the only one generally available to him? Would the law then be impartial, which ignores these two specific forms and enforces the others? And if "restraint of trade" under our present social circumstances may be caused as well by one capitalist as by many,—if it can be caused by many capitalists in many ways without legal hindrance, is the law impartial which uses it in one case against the associated capitalists, precisely in order not to sanction the association of labourers?

It is not true that "a person cannot alienate for a time his freedom to dispose of his own capital." With the single exception of the case of combination among independent employers, the law has given this power, within ever expanding limits, to the capitalist. Take the ordinary contract of partnership. Two or more men allow to each other equal rights in disposing of their respective shares of capital. Or they vest special powers of control over the capital generally in one or more managing partners; or one of them is a sleeping partner, and leaves his share at the entire disposal of the rest. In all this the law sees nothing amiss; approves, helps, enforces. Take the Joint Stock Company. Hundreds, it may be thousands, of capitalists throw their contributions together, with or without limited liability, placing the disposal of the capital, with the reservation to the great bulk of them of generally very slender powers of check or control, at the disposal of a few as a Board of Directors, or in many cases of one or two only as Managing Directors. Here the law not only helps and enforces, but expressly regulates and superadds privileges, often exorbitant ones.

Now, what is virtually the working-man's Trade Union but a form of partnership,—the only one generally available to him,—of which labour is the subject-matter, the regulation of its terms of sale the business, and the object,—instead of the money profit of the capitalist,—the advantage expected to be derived by the members through such regulations in and from the sale of their labour? But the mere forming of such a partnership was long treated as a crime; it has been made as yet dispunishable for two purposes only; and the aid of the law is wholly withheld for enforcing even such purposes. Nor is this all. The law directly supports the capitalists' associations under the various forms above indicated, against the associated labourers; gives the former full legal power to fix the terms for the purchase of the article labour; denies to the latter all legal power to fix the terms for its sale. A provision, out of which legal action may arise, that no apprentice, clerk, or servant shall be engaged or employed in or about the business by one partner without the consent of the other, is inserted as a matter of course in a partnership deed, unless by special instruction the power to buy labour—for this is what the clause really refers to—is vested in a particular individual. In a joint-stock company's deed of settlement, the power to appoint, employ, dismiss clerks, servants, &c., is vested in the Board of Directors, or in a Managing Director. Wherever the business is large, whether carried on by a company, by a partnership,—nay, by a private individual,—the regulation of the terms on which labour is to be bought is even vested in persons who may be mere strangers, whether managers or foremen. But if labourers, even when associated together for purposes for which the law itself supplies a specific form of recognition,—*e.g.*, as a Friendly Society,—attempt to make the regulation of the sale of their labour one of the incidents of their association, the law withholds its sanction altogether. And thus, whilst the associated capitalists, be they few or many,—the firm, the railway, or other joint-stock company,—may, without fear of the law of conspiracy, either by themselves in common, or through a stranger, fix any terms, not in themselves illegal, on which alone they will purchase labour—may, if those terms are not accepted, and whatever may be the result of their action as respects the restraint of trade to others, perhaps even though it should arrest the whole traffic of one of those monopolized highways called a line of railroad, throw thousands of labourers out of work; if two or more of the labourers who sell their labour to such associated capitalists seek, by any binding form of union, to fix the terms on which they will sell their labour, they fall at once under the shadow of that law of conspiracy which—like some arithmetician who should maintain that noughts linked together become units—makes crimes out of no crimes by mere association of individuals, and which may be called into action against them by the very associated capitalists themselves. And if they refuse in common to sell their labour on any terms but their own, the law, by the mouth of its luminaries, talks big about "restraint of trade," the "right to a free course for trade," quotes the case of "the negro Sommersett" to prove that any such combination for selling labour on specified terms is a form of slavery; and, in short, dwells from every point of view, and in every form of words, on the inestimable privilege which is conferred upon every individual stick by forbidding its being bound or binding itself in a bundle with others,—except, indeed, by that wisp of straw, itself only recognized by the law in certain cases, an agreement which every person agreeing can break!

But now let us look for an instant at that imposing plea of "restraint of trade" on which Sir W. Erle dwells at such length, which is so solemnly invoked against two or more individual capitalists agreeing to carry on not one, but several businesses according to mutual agreement, in order that labourers may be debarred from seeking to dispose in like manner of their labour. What does it amount to in presence of existing social circumstances? Surely it needs only to open one's eyes in order to see that at the present day such a plea is practically one *for* the capitalist, *against* the labourer. As between capital and labour, is not the accumulation of capital, in the hands of one individual or of several, exactly equivalent to the association of labourers? A B, a single capitalist, employs a thousand men. The mere fact of his doing so shows that, as against those thousand individual labourers, he has the united power of a thousand individual employers. Supposing that he enjoys (and instances of this would be easy to find) the practical monopoly of a certain trade, at all events within a given district, and that because all or some of those thousand ask more or refuse to take less for the sale of their labour than he choses to give, he shuts the door of his establishment. In what is the "restraint of trade" less than if those thousand men by common consent had struck work? Or, to steer in the very wake of "Hilton *v.* Eckersley," in what is it less than if, instead of one capitalist employing 1,000 men, there were two employing 500 each? Yet where is the legal sage who would talk of restraint of trade against the one man, even though, for fear of the thousand combining, he might talk of it against the two?

It has been with great pain that the writer of this paper has been compelled to speak thus unfavourably of a work by a judge of high character, and one whose judgment in "Hilton *v.* Eckersley" (though he seems now to consider that he was mistaken) gave indications of far greater breadth of thought than that of the majority by which he was overruled. But we dare not forget his position as head of the Trades' Unions' Commission, nor can we repress the fear that this entirely one-sided Memorandum of his is an augury of the Commissioners' coming Report. Should this be the case, it will be a real misfortune to the country. It is time that the labour of England should have as full and free power of association as its capital. Whatever Sir W. Erle may say, it has not that power yet. J. M. L.

[I am informed that the writer of the foregoing article is our eminent and esteemed friend, J. M. Ludlow.—E. C.]

MISCELLANEOUS.

The *Builders' Trade Circular*, of February 11th, 1869, gives the following as the Manchester Trade Union Rules and Regulations:—*Firstly as to Brickmakers.* No person is allowed to make bricks by machinery. *Secondly, as to Bricklayers.* 1.—No bricklayer is allowed to set machine-made bricks. 2.—No master bricklayer to have more than three apprentices. 3.—No bricks to be used in Manchester that are made beyond an arbitrarily fixed boundary line, averaging about three and a half miles from the Exchange. 4.—No bricks to be wheeled in a barrow. 5.—Labourers not to go up one ladder and come down another. 6.—Masters must employ men resident in Manchester at all their works within fifteen miles of the city. 7.—Every bricklayer to have one labourer to attend upon him whether there is work for the latter to do or not. 8.—All foremen to be members of the Union. [We believe that it is for the interest of the Manchester Bricklayers' Society to appoint delegates—six or twelve—to meet the same number of builders in conference, with full powers to discuss and agree to a code of trade rules, which will be a more creditable, business, and satisfactory mode than resorting to an exciting and expensive strike.—E. C.]

The builders of the Potteries reported that the Operative Bricklayers had sent notice for alteration in the first and second rules, viz., to leave off work at half-past five instead of six, and to receive an advance of wages of one half-penny per hour. The employers replied agreeing to the first alteration but declining the second. The matter will go to arbitration, the men not accepting the terms offered by the masters.

It is thought that the report of the Commission on Trades' Unions will not be laid before Parliament until after the Easter vacation. There are, it is stated, two if not three reports, one signed by a majority of the Commissioners, the others by the minority of two or three. The London Trades' Council and the Conference of Amalgamated Trades have completed their arrangements for convening a special delegate meeting of the London trades immediately upon the appearance of the report to take it into consideration. The various Trades' Councils throughout the provinces have adopted the same course.

TRADES' UNIONS

AND

THE COST OF LABOUR.

SPEECH

DELIVERED BY

THOMAS BRASSEY, Jun., M.P.

IN THE

HOUSE OF COMMONS

7th July 1869.

WITH ADDITIONAL STATISTICAL DETAILS.

'The liberal reward of labour, as it is the necessary effect, so it is the natural symptom of increasing wealth.'
Adam Smith.

LONDON:
LONGMANS, GREEN, AND CO.
1870.

INTRODUCTION.

YIELDING to the request of several Members of Parliament, and to a wish expressed by many other persons outside the House who take an interest in the great Labour question, I am induced to publish my speech of last Session on the Second Reading of the Trades' Unions Bill.

If the speech be worthy of publication in the form of a pamphlet, of which I am by no means confident, it is solely as a compendious statement of facts not very generally known that it deserves to be preserved. From this point of view, I think it possible that the results of my father's vast and varied experience in the employment of labour, and the statistical details which, by diligent research, I have been enabled to bring together from other sources of information, are not devoid of public interest. I have availed myself of the opportunity afforded me through the republication of the speech, to give in more ample detail tables of wages and statements of figures, which may be thought valuable for the elucidation of the subject.

I have also thought it my duty to insert a few additional observations on the danger to the working-classes of the restrictions, imposed on the free supply of labour by the dictation of ill-informed agitators. Few persons are more indebted to the labourers and artisans of this country than I am: and I hope I may ever be found ready to sympathise with their difficulties, and to vindicate their rights. In the speech, which I now publish, I have anxiously laboured to prove the injustice of condemning British labour as more costly in its results, simply because the daily rates of pay are higher in England than on the Continent. But I know that I should be doing but half my duty, if I failed to point out and strongly condemn the errors, into which our operatives are in risk of being betrayed by the ill advice of leaders, who have no just claim to the influence which they unhappily possess.

In using this language I am anxious not to be misunderstood, as intending to draw an indictment against the great body of the working-classes of this country. The cases to which I refer have been exceptional blots on our industrial history: and, though a cloud now seems to rest, at least in some less favoured districts, on British enterprise, I attribute the temporary collapse of trade not so much to Trades' Unions as to other circumstances, for which the operatives are not responsible.

I know that there are many who fear that the day of England's commercial glory is departed. I can see no reason for sharing in these alarms.

O passi graviora! dabit deus his quoque finem.

I place a firm reliance on the industry and common sense of our working-people ; and I believe that happier and brighter days are yet in store.

SPEECH.

I AM ANXIOUS, as the son of an employer on a large scale, to offer a few remarks on the important subject which is now under the consideration of the House. While I feel myself impelled by many and potent influences to take an employer's view of this question; on the other hand, I cannot forget that the working-classes, of whom vast numbers have for many years rendered honest and faithful services to my father, have great claims on my sympathy and my regard.

Before entering upon the discussion of the details of the subject, I would venture to remind the House that the tendency to combination, for the purpose of promoting their mutual interests, is no new thing among the industrial classes. The Guilds of the middle ages were but the forerunners of the Trades' Unions of to-day; and the strikes of modern times have had their counterpart in the Jacquerie riots of the fourteenth century. And when we take into view the great changes which have been brought about in the industrial organisation of this country during the present century, the substitution of steam for manual power, and of machinery for hand labour;

and remember that the resources of machinery can be most fully developed only when applied on a large scale, the reasons why workmen have gathered together in recent times, in numbers so vast, round our great industrial centres are not far to seek.

Now, when operatives have thus been assembled together in great numbers under the same roof, tending the same machine, working at the same table ; is it not natural, nay, reasonable, that they should confer and take action together on all questions of mutual interest?

In this most legitimate tendency, Trades' Unions have had their origin. It is further to be observed that no law, however severe, has been successful in destroying such combinations. The penalties with which combinations in restraint of trade were punishable under former laws, now happily repealed, simply had the effect of causing proceedings to be conducted in secrecy; for which it is most desirable, in the interests of society at large, to secure the greatest possible publicity.

It is now, however, generally believed that the legislature would not be justified in refusing to the operative classes the right to combine voluntarily—for the purpose of regulating, as far as they can, the terms on which they shall sell their labour.

At the same time, it is universally admitted that every possible protection which the law can afford against coercion and intimidation on the part of trade combinations, ought to be given to the operative, who wishes to preserve his independence.

I am not, I may add, insensible to the great errors and follies which have marked the policy and conduct of some of the Trades' Unions. In regard to trade interests generally, as distinguished from benefit objects, their influence has too often been essentially illiberal, anti-social, and calculated to establish among the industrial classes of this country that subdivision of caste which has been the great curse of India.

There is a general tendency amongst the Trades' Unions to ignore the interest of the master, as if his success were not essential to their own prosperity. For example, in his evidence before the Trades' Union Commissioners, Mr. Connolly, of the Masons' Society, made the frank confession that their rules were for the men not the masters. 'They want,' he said, 'the greatest profit, we the highest wages.'

On the other hand, I am not afraid of the Trades' Unions. On the contrary, I believe that their power both for good and for evil has been greatly exaggerated. When the demand for labour is increasing, and employers begin to compete against each other for the supply of labour, wages will rise from natural causes, without the intervention of Trades' Unions.

Their organised and united action may secure an advance of wages at a somewhat earlier date; but eventually, in the great majority of cases, the competition among employers would bring about equally beneficial results to the working-people. The advantage to the working-classes of obtaining the advance at an earlier date is not, in my honest opinion, sufficient to compensate for the expense of perpetually maintaining, by heavy subscriptions, a Trades' Union organisation; still less to compensate for the great loss which is caused by unsuccessful strikes.

I have admitted that Trades' Unions may have the effect of obtaining an advance of wages at an earlier date; but the most protracted strikes in which the working-men have been engaged have generally taken place, not for the purpose of securing an advance, but for the purpose of resisting a fall, in wages. Resistance to a proposed reduction of wages was the cause of the Engineers' strike in 1852; of the strike at Preston in 1853; of the strike in the Iron Trade in 1865; and of the strike of the Colliers at Wigan in 1868. In each of these cases the masters had found it necessary, in consequence of the depressed state of trade, to reduce the wages of the men; but the men, ignoring the circumstances of the trade, and looking only to what they believed to be a degradation of their position as workmen, refused to accept the reduced rate of wages. They, therefore, went out on strike; but, after a protracted struggle, were compelled to accept the reduction originally offered by their employers. In point of fact, employers as a general rule, from motives of kindness and consideration towards their workmen, are anxious to avoid, if they possibly can, reductions of wages. They rarely ask their workpeople to accept a lower rate of wages, until the condition of their trade has become so unfavourable as to make the reduction absolutely necessary. And here I may observe that the power of combination has been proved, by experience of its results, to be at least as much for the advantage of the masters as the workmen.

The defeat of the shipwrights on the Thames in 1852, and, more recently, the failure of the ironworkers' strike in Staffordshire, are conspicuous examples of the power which the masters acquire by combination among themselves.

The great evil however of Trades' Unions, in their action in regard to the wage question, arises from the fact that, while the leaders of these societies exercise great authority over the members of the Trades' Unions, they have no corresponding power of obtaining for their clients what they tell them they ought to have.

No virtue is so freely displayed as the virtue of generosity, when exercised at the expense of others. Trades' Union agitators have too often sought to win the admiration of their auditory by thoughtless declamation against the alleged rapacity of employers, and by loud professions of sympathy with the wrongs of their industrial brethren. Their credulous hearers have been apt to forget that, when the trade in which they are employed is yielding no profit, or is perhaps being carried on at a loss, no uncommon case, when workmen have been fighting against a reduction of wages; it is better for the employer to abandon for the time a business, in which he has hitherto persevered, only in the hope of an ultimate revival of trade, rather than consent to give rates of pay, which must inevitably involve him in disaster.

The leaders in several protracted strikes have exhibited a melancholy ignorance of the state of their own trade, and even of the market value of the goods, in the production of which they are engaged. How much suffering might have been spared to the working-classes if they had but known, before they engaged in a hopeless struggle, the true merits of their case! I was once present at a meeting of employers, during a large strike in the coal trade. I had the means of knowing that the wages which had been offered were the highest which the employers could afford to pay, and that the markets were so overstocked that it was a positive advantage to stop for a time the working of the pits. But the facts, which I had the means of knowing, were apparently unknown to the miners; and it was pitiful indeed to see the hard-earned accumulations of many years exhausted in an obstinate resistance to a reduction of wage, which had not been proposed by the employers until it had been forced upon them by the unfavourable condition of their trade.

Though the co-operative principle and the payment to the workmen of a percentage of profits, which has been adopted by Messrs Briggs, give some reason for hope that in future the workmen may possess more information in regard to these matters, I fear that the day is still far distant, when the leaders of the Trades' Unions will have such opportunities of knowing the real merits of their case; as could alone justify their claim to exercise authority over the actions and opinions of their fellow-labourers. The propriety of asking for an advance, and of striking if it is refused, depends entirely upon the state of trade, the amount of business in prospect, and the profits which the employers are enabled to realise—circumstances of which the Trades' Union agitators are too often wholly ignorant. But, while I have thought it my duty to condemn the unreasonable proceedings of the Unionist agitators, on the other hand I rejoice in the conviction that some of the most trusted leaders of the Trades' Unions have profited by past experience, and are strongly averse to strikes. Mr. Allen, of the Amalgamated Engineers, stated to the Trades' Union Commissioners that their Executive Council was always opposed to strikes. He added—and this is surely one of the best reasons for giving that protection to their funds which is proposed in the Bill introduced by my honourable friend, the member for Frome—that their large accumulations, amounting to 149,000*l.*, only made the members of his society so much the more anxious not to waste their money in injudicious conflicts with their employers. Doubtless, some of the leaders of the Trades' Unions are wiser men than their followers, and would be only too glad to restrain their ardour, if they dared.

For these reasons I venture to indulge the hope that, when their status has been improved, through the recognition of their Societies by the State, their increased influence will enable them to exercise a more effectual restraint upon the unreasonable demands of the workmen in the Unions.

The truth cannot be too forcibly impressed on the members of the Trade Societies, that it is to the operation of the laws of supply and demand, rather than to the action of Trades' Unions, that workmen must look for improvement in their wages. An increase of wages can only take place when trade is prosperous, and when the supply is not sufficient to meet the increasing demand for labour. In support of that view, I may point to the fact that, whenever Trades' Unions succeed in imposing arbitrary regulations, which, though by indirect means, practically raise the price of labour, their restrictive rules are only accepted by the masters, in consequence of the scarcity of labour, and the necessity of obtaining the services of every available man; and in the natural course of events, without the intervention of Trades' Unions, a rapid rise in wages would inevitably result from an increased demand for labour.

For example, the building trades in Manchester have succeeded in enforcing a most arbitrary code of rules; but these would never have been accepted by the masters, had not the building trade in that city been in a condition of unprecedented activity during the last five years.

As an example of the dictatorial attitude assumed by workmen under these circumstances, I may quote, from the 'Builders' Circular' of February last, the rules to which the master-builders of Manchester have temporarily submitted, under the circumstances already explained:—

Rules of Trades' Unions, Manchester.

Firstly, as to Brickmakers.

No person is allowed to make bricks by machinery.

Secondly, as to Bricklayers.

1. No bricklayer is allowed to set machine-made bricks.
2. No master-bricklayer to have more than three apprentices.
3. No bricks to be used in Manchester that are made beyond an arbitrarily-fixed boundary line, averaging about three and a half miles from the Exchange.
4. No bricks to be wheeled in a barrow.
5. Labourers not to go up one ladder and come down another.
6. Masters must employ men resident in Manchester at all their works within fifteen miles of the City.
7. Every bricklayer to have one labourer to attend upon him, whether there is work for the latter to do or not.
8. All foremen bricklayers to be members of the Union.

Thirdly, as to Masons.

1. No quarry-worked stone allowed to come into Manchester.
2. No ashlar (walling stone) to be worked by machinery.
3. All masons, bricklayers, and plasterers (whatever they are worth) are to have the same rate of wages according to their respective trades.
4. No piece-work allowed.
5. Masons not to work overtime on any consideration.

Though the state of business may have induced employers to submit for a time to this arbitrary code of rules, it is equally clear that the workmen will ultimately be the greatest sufferers by their own folly.

By the system which they have adopted, the cost of building must be needlessly increased, and the demand for houses will be proportionately diminished. Sooner or later labour in the building trades will, in consequence, find but scanty employment; and, under the pressure of necessity, workmen will be compelled to accept more reasonable terms of remuneration. In recent years there has been in London, as in Manchester, a considerable increase in the rate of wages in the building trades.

The amount of this increase is shown in a statement, which has been kindly prepared for me by Messrs. Lucas, Brothers, the well-known builders.

'We find,' they write, 'for some years previous to September 1853, that the rate of wages was as follows:—

	For Mechanics, Masons, Bricklayers, Carpenters, and Plasterers	Labourers
Previous to 1853 .	5*s.* per day of 10 hours.	3*s.* per day of 10 hours.
From September 1853 to March 1861	5*s.* 6*d.* . . .	3*s.* 4*d.*
March 1861 to September 1865	7*d.* per hour, or 5*s.* 10*d.* per day.	4¼*d.* per hour, or 3*s.* 6½*d.* per day.
September 1865 to May 1866 .	7½*d.* per hour, or 6*s.* 3*d.* per day.	4½*d.* per hour, or 3*s.* 9*d.* per day.
May 1866 to present time	8*d.* per hour, or 6*s.* 8*d.* per day.	4¾*d.* per hour, or 3*s.* 11½*d.* per day.

and we consider that the price of building is twenty-five to thirty per cent. more now than it was in 1853.'

It should, however, be observed that, contemporaneously with this advance in the rate of wages, payable in the building trades in London, there has been a considerable increase in the rent of the dwellings of the labouring-classes. I am informed by Sir Sydney Waterlow, that in London working-men, earning from fifteen shillings to forty shillings per week, pay on the average one day's wages in rent; that, below twenty-five shillings a week, this proportion is rather more than less; and that the increase, in the latter case, during the last twenty years, has ranged from twenty to thirty per cent., chiefly owing to the augmented local taxation, which, in the case of the labouring-classes, is not less, on the average, than one-fifth of their rent.

It is too often asserted that the organised dictation of the Trades' Unions is the cause of the advance in wages in the building trades in London; though in truth that advance, as I have already said, has been but the necessary result of the altered relation between the supply of workmen and the demand for labour.

The metropolitan railways, the growth of Kensington, Bayswater, and other suburbs, are visible evidences of the pressure of the demand of the master-builders of London upon the supply of labour in the trades which they employ.

Abundant evidence in support of this explanation of the cause of the rise of wages in all branches of trade is supplied by employers, by no means friendly to the Trades' Unions.

Mr. Trollope, for example, made the following admission in his evidence before the Commissioners:—'I am bound to say that hitherto there has been

B

such an enormous pressure for work, that almost every man who can handle a tool has been taken on at an unreasonable rate.'

Again, speaking of the advance in wages in the building trades in the provinces, Mr. J. Mackay, an experienced agent in my father's employ, says in a report he has made to me on the subject, 'Wages have risen during the last twenty years from twenty to twenty-five per cent; but, by the force of circumstances, they would have risen as much or more if Trades' Unions had never existed.' To the same effect, Mr. Robinson, the Managing Director of the Atlas Works, Manchester, in his evidence before the Commissioners, says, 'I do not think the Unions have altered the rate of wages; the changes are rather due to the demand for labour in particular branches.' Between 1851 and 1861 no advance took place in the wages of the Engineers, though theirs is the most powerful of the Trades' Societies; but, in the case of the Boiler-makers, wages rose from twenty-six to thirty-two shillings and sixpence, in consequence of the extension of iron shipbuilding, and the great amount of iron bridge-work.'

Many interesting examples of the effect of unusual pressure upon the labour resources of a sparse and scanty population have presented themselves in the course of my father's extensive experience. The advance of wages which has occurred in such cases, from the natural operation of the laws of supply and demand, surpasses the most golden dreams of the Trades' Unions.

When the Grand Trunk Railway was being constructed in Canada, my father sent out, at his own expense, a great number of operatives from this country. Men were engaged in Lancashire and Cheshire; and, on landing in Canada, received forty per cent. more for doing the same work than they had been earning in England; the reason being that the supply of their labour in England was abundant, while in Canada skilled artisans were comparatively scarce.

In Spain, a few years ago, in the construction of the railway from Bilbao to Tudela, the wages earned by labourers, which at the commencement of the contract were one shilling a day, rose before the works were completed to three shillings a day. On the same contract, the wages of the masons increased in the corresponding period from one shilling and fourpence to five shillings a day. Take again the evidence of Mr. Mault, the Secretary to the Builders' Association of Birmingham. Mr. Mault stated to the Trades' Union Commissioners that of the 900,000 men employed in the building trades, not more than 90,500 were members of the Trades' Unions; and, that while the Trades' Unions professed to aim at securing uniformity of wage throughout the country; yet the wages of masons varied in different parts from fourpence-halfpenny to sevenpence three-eighths per hour, the wages of bricklayers from fourpence-halfpenny to eightpence, and those of carpenters from fourpence five-eighths to eightpence an hour. These figures conclusively prove the fallacy of the idea that Trades'

Unions can secure for their clients an uniform rate of wages, irrespective of the local circumstances of the trade in which they are engaged. Again, it is said by a numerous section of alarmist employers that Trades' Unions have succeeded in raising wages in this country to a point far beyond anything which has been attained in the corresponding trades on the Continent. But I maintain, unhesitatingly, that daily wages are no criterion of the actual cost of executing works or carrying out manufacturing operations. On the contrary, it is the result of my father's experience that there is a most remarkable tendency to equality in the actual cost of work throughout the world. In the industries which compete against the manufacturers of the Continent for the supply of the neutral markets of the world, it is clear that Trades' Unions cannot raise the cost of production in this country beyond the cost of producing an equivalent quantity of work abroad, without diminishing the relative rate of profit of the British manufacturer ; and if the rate of profit which could be obtained by an investment of capital in this country were to be reduced below the profit accruing from a corresponding investment abroad, the immediate result would be the withdrawal of capital from this country. In point of fact, the amount of daily wages affords no real measure of the actual cost of work ; and it is quite possible that work may be more cheaply executed by the same workmen, notwithstanding that their wages have largely increased. In illustration of this view, I will state to the House what occurred in the execution of a railway in an agricultural district in England, remarkable for the low rate of wages there prevailing among the agricultural labourers.

At the commencement of the construction of the North Devon Railway, the wages of the labourers were two shillings a day. During the progress of the work their wages were raised to half-a-crown and three shillings a day. Nevertheless it was found that the work was executed more cheaply when the men were earning the higher rate of wage than when they were paid at the lower rate. Again, in London, in carrying out a part of the metropolitan drainage works in Oxford Street, the wages of the bricklayers were gradually raised from six shillings to ten shillings a day; yet it was found that the brickwork was constructed at a cheaper rate per cubic yard, after the wages of the workmen had been raised to ten shillings, than when they were paid six shillings a day.

And now, crossing the Channel, I will give to the House an example, derived from my father's experience in the construction of the Paris and Rouen Railway, in 1842.

The Paris and Rouen Railway was the first large railway work executed on the Continent. About 10,000 men were employed in its construction, of whom upwards of 4,000 were Englishmen. Perhaps, so remarkable an exodus of English labour to Continental Europe never before occurred, and it is improbable that it will ever be repeated. A special effort was made to secure the services of English workmen on this particular contract, because it was a question whether native workmen could be obtained in sufficient numbers, and it was still more doubtful whether they would possess the necessary skill and experience for carrying out railway works, which, at the period to which I refer, were a novelty even to English engineers, and entirely unknown on the Continent.

Under these exceptional circumstances, a large body of Englishmen were sent over to Normandy. It is scarcely necessary to observe that the employment of English manual labourers abroad must always be a costly and somewhat doubtful policy; but, in this particular case, it was not found to be disadvantageous in a pecuniary point of view. Though these English navvies earned five shillings a day, while the Frenchmen employed received only half-a-crown a day; yet it was found, on comparing the cost of two adjacent cuttings in precisely similar circumstances, that the excavation was made at a lower cost per cubic yard by the English navvies than by the French labourers. Extending the investigation from France to other European countries, to Italy, Austria, Switzerland, Spain, Germany, Belgium and Holland, it has been found that in the cost of railway work, executed by unskilled labour, there is hardly any perceptible difference, notwithstanding the striking differences in the rates of daily wages prevailing in those countries.

The same remarkable tendency to equality of cost

exhibits itself even in India. On the Delhi and Umritsur Railway, it has been found, as I am informed by Mr. Henfrey, my father's resident partner in India, that, mile for mile, the cost of railway work is about the same in India as it is in England, although the wages, if estimated by the amount of daily pay, are marvellously low. The coolies, or common labourers, for example, earn about sixpence a day; their wages, before the railways caused an increased demand for labour, having ranged from fourpence to fourpence-halfpenny a day.

Earthwork, it is true, is executed by the coolies at a cheaper rate than in England; but native-skilled labour is more expensive; while the necessity of employing a numerous body of English foremen greatly enhances the cost of the supervision, which in India is found to amount on an average to twenty per cent. on the entire outlay. In Italy, as in India, it has been found that a numerous but unskilled population, in a climate where the necessaries of life are inexpensive, can undertake the mere manual labour at a cheaper rate than in England; but this is only true when works are not pushed on so rapidly as to require the importation of labour from a distance. When the local labourers are alone employed, the Italian villagers, men, women, and children, carrying earth to and fro in baskets on their heads, and as ignorant as the coolies themselves of the resources and appliances of mechanical science, can execute earthwork about as cheaply as in India.

On the other hand, masonry, and other work

requiring skilled labour, is rather dearer in Italy than in England. In the Mauritius, the result of the experience acquired in the construction of a railway in that island by my father's partner, Mr. Longridge, established the same result as in the cases already quoted. Though the nominal wages are low, yet, when you take into account the extra supervision, the cost of earth work, rock-cutting and masonry is quite as great; and skilled work, as for example, carpentry, is from twenty to twenty-five per cent. more costly in the Mauritius than in England.

The facts which I have quoted are the result of the large experience of many practical men, who have been engaged during the last quarter of a century in executing railway works of the greatest magnitude in every quarter of the globe.

But theirs is no new discovery. It was written a hundred years ago, by the great practical philosopher Montesquieu, in the 'Esprit des Lois,' 'Il y a dans l'Europe une espèce de balancement entre les nations du Midi et celles du Nord. Les premières ont toute sorte de commodités pour la vie et pour les besoins; les secondes ont beaucoup de besoins et peu de commodités pour la vie. L'équilibre se maintient par la paresse qu'elle a donnée aux nations du Midi, et par l'industrie et l'activité qu'elle a données à celles du Nord.'

In the recent interesting publications of Mr. Lothian Bell, in a report to Congress of Mr. Commissioner Wells, the Special Commissioner of

Revenue in the United States in 1868, and in a report of Mr. Redgrave, one of the inspectors of factories, I have found many other equally remarkable cases in various trades, which tend to prove that the cost of labour cannot be conclusively determined merely by reference to the rate of daily wages paid in the respective industries.

Mr. Lothian Bell, in a recent address read at a meeting of ironmasters in the north of England, gave the result of his investigations as to the cost of smelting pig-iron in France, which he said distinctly established the fact that more men were required to do an equivalent quantity of work in France than in England. He stated that from a very careful inquiry at a large establishment in France, he had ascertained that forty-two men were there employed to carry out the same amount of work which twenty-five men were able to do at the Clarence Factories on the Tees. In spite of the actual labour on a ton of pig-iron for smelting being twenty per cent. cheaper in France than in England, the entire smelting charges were sensibly greater in France than in the general run of work at Middlesbro'. And, taking into account the saving in respect of fuel, the cost of producing pig-iron in France was twenty shillings, in some cases even thirty shillings, more than that exhibited by the cost-sheets of the manufacturers at Cleveland.

So, too, Mr. Hewitt, an American ironmaster, stated that the price of iron was one pound sterling per ton higher at Creuzot than in England. And

Monsieur Michel Chevallier, in his introduction to the Reports of the Jurors of the French Exhibition, in 1867, says that rails are from twenty-five to thirty francs dearer per ton in France than in England. To the same effect, Mr. Lothian Bell says that whereas labour in Westphalia costs from twenty to twenty-five per cent. less than with us, the labour-saving arrangements are much neglected; and a ton of iron smelted in the Ruhrort district cannot be produced for less than fifteen shillings a ton above the cost upon the Tees. A similar difference of price is shown in the price of the rails recently purchased for the Mont Cenis Railway, the price of which at the works in France was from seven pounds twelve shillings to eight pounds per ton, while the price in England was seven pounds per ton. In proof of the conscious inability of the French ironmasters to compete with our manufacturers in an open market, I may mention that the import duty in France on rails is two pounds eight shillings per ton.

Mr. Commissioner Wells, in an able report to the American Congress, has discussed in minute detail this most important question. Taking the puddling of iron as the representative process of the trade, he says that he found that the average price of labour per day for puddlers was from seven shillings and sixpence to seven shillings and tenpence in Staffordshire; six shillings and fourpence in France; and from four shillings and ninepence to five shillings in Belgium; yet the average price of merchant bar-iron was six pounds ten shillings in England, seven pounds in Belgium, and eight pounds in France. In a recent report on the condition of the textile industries in England, Mr. Redgrave, one of Her Majesty's Inspectors of Factories, says that while the foreigner is under the same conditions as to the raw material as the English manufacturer, and his fuel is more expensive, his workpeople do not work with the same vigour and steadiness as Englishmen. Consequently the same number of operatives, employed upon the same machinery, do not produce the same quantity of yarn as in this country. 'All the evidence that has come before me,' he says, 'has gone to prove that there is a great preponderance in favour of this country. Comparing the work of a British with a foreign spinner, the average number of persons employed to spindles is—in France, one person to fourteen spindles; in Russia, one to twenty-eight spindles; in Prussia, one to thirty-seven; in Great Britain, one to seventy-four. But I could find many cotton-spinning factories in my district, in which mules containing 2,200 spindles are managed by one minder and two assistants.' 'I have recently been told,' he continues, 'by one who had been an English manager in a factory at Oldenburgh, that though the hours of work were from 5.30 A.M. to 8 P.M. every day, only about the same weight of work was turned off under English overlookers as would be produced in a working day from 6 A.M. to 6 P.M. in this country. Under German overlookers, the produce was much less. The wages were fifty per cent. less in many cases than in England; but the number of hands, in proportion to machinery, was much larger. In some departments it was in the proportion of five to three. In Russia the inefficiency of the labour of the foreign, as compared with the labour of the English, operatives, is even more strikingly manifested; for on a comparison of the wages, supposing the Russian operatives to work only sixty hours a week, as they do in England, instead of seventy-five, as they do in Russia, their wages would not be one-fourth the amount earned in England. But the wage must be taken into account with the power of the operative as a producer; and herein will be found an advantage of the English operative over the foreign competitor, sufficient, with some qualification, to counterbalance the mere cheapness of wage.'

Mr. Wells, in the report to which I have already referred, confirms the view expressed by Mr. Redgrave. He says that, 'whereas female labour in the cotton manufacture is paid at from twelve shillings to fifteen shillings a week in Great Britain; at from seven shillings and threepence to nine shillings and sevenpence in France, Belgium, and Germany; at from two shillings and fourpence to two shillings and elevenpence in Russia; the one thing which is most dreaded by the Continental manufacturers everywhere is British competition. The demand for protection is loudest in France, Austria, and Russia, where the average wages reach their minimum.'

The operation of the protective duties imposed by foreign countries upon British goods, is strongly condemned by Mr. Murray, one of the British Commissioners for the French Exibition in 1867. In his report on the manufacture of cotton goods in this country, he says that, 'In 1865 the value of the English imports into France was twenty-five million five hundred thousand pounds, the value of the raw materials being eighteen million pounds, and the value of British manufactures not three million pounds, and those manufactures paid an average duty of ten per cent., ranging from twenty-seven and a half per cent. on pottery, to seven and a half per cent. on metal work. On the other hand, France exports to us twenty-one million pounds of her manufactures. She takes an unmanly advantage of our adoption of a large and noble principle.' I have said that the mere rate of daily wages does not give any indication of the cost of executing work. It is equally true that the hours of labour are no criterion of the amount of work done. In 1842 Messrs. Hornby, at Blackburn, made a calculation that, even if their operatives were paid the same for working sixty hours as for sixty-nine hours per week, the increased cost would be so small as not to be weighed in the balance against the advantage to the operatives themselves of a larger amount of leisure. And, more recently, Monsieur Dolfus, of Mülhausen, reduced the working hours in his establishment from twelve hours to eleven hours per day; and promised the men that no reduction should be made in their wages if the amount of work they performed was equal to what it had been before. After a month, it was found that the men did in eleven hours, not only as much work, but five per cent. more than they previously did in the day of twelve hours.

The examples which I have already quoted will sufficiently show that higher wages and shorter hours of work may not be found incompatible with a diminished cost of production ; and that low wages and long hours may sometimes prove less advantageous to the employer than shorter hours of labour and a higher rate of wages. This apparent anomaly is partly explained by the necessity of giving to the labourer, who has to undertake severe manual exertion, the means of procuring a generous diet. In Belgium the workmen are not so expensive in their habits as the English artificer. They consume less meat ; their bread is seldom purely wheaten ; and they work for lower wages ; but, on the other hand, it cannot be expected that, under these conditions, they can have the same physical vigour as the English labourers, who are better fed.

Mr. Hewitt, to whose evidence I have already referred, speaking on this subject, remarked that at Sireuil the rate of wages of the common labourer will only admit of his having meat once a week ; and yet the manufacturers were not making money. He also stated that there was a deplorable look of hopelessness among the lower class of workmen at Creuzot ; though this was not discernible among the better-paid men.

It must, I think, be admitted that these facts furnish a conclusive answer to the often-repeated argument, that, unless the rate of wages in this country be reduced, British industry must be beaten in competition with foreign manufacturers for the supply of goods to neutral markets. And I think it is equally clear that the higher rates of wages which are paid in England, so far from being a concession to the pretended power of Trades' Union organisation, a power so persistently asserted by the agitators, and so much too readily acknowledged by some employers, are, on the contrary, well and justly earned by the superior vigour and ingenuity of British labour. If English employers pay high wages, it is because in the long run they receive an adequate return in the quantity and in the excellence of the work which is executed for the money.

I will next proceed to examine the statements, widely circulated, and largely accepted by the public, to the effect that there has been a greater advance in the wages of operatives in recent years in England than in the corresponding period abroad ; this increase being, it is alleged, entirely attributable to the powerful organisation of the Trades' Unions. If it were true that the workmen in Union possessed a great power which they could not exercise without the aid of the Unions, the development of these societies would become a question of the utmost gravity to the commercial interests of the country. I have, therefore, thought it most important to ascertain how far these statements are justified by the facts. For the purpose of elucidating this portion of the subject under discussion, I have obtained tabulative statements, extending over the last sixteen years, of the comparative rate of wages in several important manufacturing establishments in this country, in which many of the workmen employed belonged to the much dreaded Trades' Unions. Full and accurate information on the actual rate and progressive increase of wages abroad, where Trades' Unions until very recently did not exist, is contained in the valuable reports on these subjects, which were specially made to Lord Stanley, by the Secretaries of Embassy and Legation abroad. I will begin by stating how the case stands as regards the engineering trade in England. In this trade it is an undoubted fact that since 1852 there has been practically no augmentation whatever in the wages earned by the operatives. The reason why this is so is obvious. The rate of wages in England is limited, as I have already shown, by the necessity of competition with the foreign manufacturers. Employers in England, as elsewhere, only employ labour on the assumption that they can realise a profit by their business ; and, in the engineering trade, in consequence of the impossibility of increasing the cost of production to any great amount without losing the trade in the neutral markets, there has been no possibility of making an advance of wages. On the other hand, the active competition between the numerous body of manufacturers in the country has reduced profits to a rate so moderate that, if it were to be further reduced, the trade would no longer offer any inducement for the investment of capital. I give in a tabular form a statement of the rates of wages, which have been paid by my father to the skilled workmen in his employ at the Canada Works at Birkenhead, since the formation of that establishment.

The average number of hands employed is six hundred ; a sufficient number to afford a fair opportunity of testing the average wages in the mechanical trades throughout the country. It will be seen, on examination of the table, that my statement to the effect that there has been no appreciable improvement in the rates of pay in recent years is fully borne out by the figures it contains.

Again, the experience of the same establishment is equally conclusive, in proof of the opinion, already so frequently urged, that the rate of wages is regulated of necessity by the ratio between the supply of labour and the demand for labour; and that, when the supply exceeds the demand, wages must fall, if the depression in trade is long protracted. Since the contraction of trade, consequent upon the late financial crisis, the price of piece-work has been reduced at the Canada Works sufficiently to allow of the construction of locomotives and bridge work at a cheaper rate now than at any time during the last fifteen years; thus fully establishing my view that Trades' Unions can never succeed in advancing wages, except when the prospects of trade are favourable.

On the other hand, the advances which have occurred in the rate of wages abroad, prove, that without the intervention of Trades' Unions, and solely in consequence of a greater demand for skilled labour, through the development of manufacturing industry on the Continent, the wages of the working classes have risen more rapidly than in any industry in this country. While I contend that the rise of wages

Average Rates of Wages Paid to Skilled Workmen at the Canada Works, Birkenhead.

	1854	1855	1856	1857	1858	1859	1860	1861	1862	1863	1864	1865	1866	1867	1868	1869
	s. d.	*s. d.*	*s. d.*	*s. d.*	*s. d.*	*s. d.*	*s. d.*	*s. d.*	*s. d.*	*s. d.*	*s. d.*	*s. d.*	*s. d.*	*s. d.*	*s. d.*	*s. d.*
Fitters . . .	29 0	28 3	29 0	30 6	28 10	27 6	27 6	27 0	27 10	28 0	28 0	28 1	31 0	32 6	31 0	30 0
Turners . .	29 4	30 3	31 3	33 0	31 6	31 0	32 0	31 6	32 0	31 6	31 6	31 5	31 6	31 0	30 0	29 4
Coppersmiths and Braziers	31 6	30 10	28 10	29 0	28 0	30 0	31 0	29 6	28 6	28 1	31 6	31 7	32 6	32 0	32 0	30 9
Grinders . .	27 0	27 0	27 0	24 0	24 0	22 0	26 0	25 6	27 0	27 6	27 6	32 0	28 6	32 0	26 6	23 0
Smiths . . .	31 0	31 5	32 0	31 0	30 0	29 6	30 3	30 0	29 6	31 0	30 6	30 3	31 9	32 9	31 6	30 0
Boiler Smiths .	34 0	34 0	35 0	34 0	32 6	33 0	33 8	33 0	32 6	33 0	33 0	34 6	36 0	37 0	36 0	36 0
Bricklayers . .	34 0	34 0	34 0	34 0	34 0	34 0	34 0	34 0	34 0	34 0	34 0	34 0	34 0	34 0	34 0	34 0
Saddlers and Belt Makers	26 0	27 0	26 0	26 0	27 0	26 0	27 0	27 0	27 0	27 0	27 0	25 6	24 0	24 0	25 0	26 0
Forgemen . .	36 6	37 0	36 0	33 6	—	—	33 0	36 0	35 6	35 0	34 6	33 0	32 9	33 0	32 6	32 6
Painters . .	24 0	23 0	24 0	26 0	26 6	25 0	27 0	26 0	25 6	25 6	25 8	26 6	27 6	24 6	24 0	23 0
Moulders . .	32 0	31 6	33 0	33 0	32 0	31 6	31 6	32 6	32 0	32 6	33 0	33 0	32 9	34 6	34 0	31 6
Joiners and Pattern Makers	28 0	28 6	29 0	28 2	27 6	29 0	29 6	30 0	29 6	29 6	29 0	30 0	30 6	31 4	30 9	30 0
Boiler Makers .	31 6	31 0	30 6	32 6	32 0	30 6	31 0	31 6	31 0	31 6	31 3	31 9	34 2	33 0	32 0	32 0

abroad is mainly caused by increased demand for labour, I admit that it is partly due to the augmented cost of living, which increase has been much more rapid abroad than in England in the last twenty-five years.

In France, Belgium, and Prussia, the three great competing countries with England, prices are from twenty to thirty per cent. dearer than twenty years ago; and this increase in the cost of living tells immediately upon the price of all labour, especially of common or unskilled labour. Twenty years ago labourers were content to work for eighteen pence a day. At the present time, from two shillings and two pence to two shillings and four pence is the ordinary rate of pay. Mr. Fane says, in his report to Lord Stanley, that 'the general rate of money wages in France has increased about forty per cent. in the last fifteen years in those industries which compete with foreigners in the neutral markets. This rise in the money wages has been accompanied by a considerable rise in the price of food and clothing; still the relative proportions in which money wages and the price of commodities have risen leave a margin in favour of the former.' In reference to the same subject Mr. Wells says that already in France and Germany the drain of labour from the rural districts to engage in manufacturing industry in towns has caused much embarrassment to agriculture, and that a further supply of labour from the same source can only be obtained by the payment of higher wages.

I will now quote several examples of the increase in the cost of skilled labour on the Continent. The first case to which I shall refer is the manufacture of machine tools in France. These tools are made to the value of two millions of francs annually; but, though the raw materials used are much cheaper since the negotiation of the treaty of commerce with this country, yet the selling price continues the same, owing to the increasing dearness of labour. I shall next refer to the case of the tailoring trades. At the military clothing establishment of Monsieur Dusantoy, in Paris, there are 3,300 persons employed, 800 of them being men, 2,000 women, and 500 children. The amount of wages paid in 1866 amounted to one hundred thousand pounds in the year. The daily wages for men ranged from three shillings and four pence to eight shillings and four pence; while in London the rate is stated by Mr. Levi, in his essay on the wages and earnings of the working classes, to be from four shillings to seven shillings. At Monsieur Dusantoy's, women earn from two and a half to four francs a day. In London the wages of girls are one shilling, and of women employed as seamstresses from two shillings to half-a-crown a day. The children at Monsieur Dusantoy's earn from one to two francs a day. In London their wages would be about one shilling. Again, comparing the present and former rates of wages paid at the famous engine building establishment at Creuzot, Mr. Fane states, that the number of persons employed at the date of his report was

9,950; while the annual expenditure in wages was four hundred thousand pounds a year. The mean rate had risen between 1850 and 1866 from two shillings to two shillings and eleven pence per head, per day, or thirty-eight per cent.; and some men earned from six shillings and eight pence to eight shillings and four pence per day. In addition to their money wages, great facilities are given to the workpeople, at the expense of the proprietors, for feeding, clothing, and educating themselves and their families: 700 families of the operatives are lodged by the company at fifty per cent. below the normal rate of house rent, and 700 gardens are let at the nominal rent of two francs per annum. Compare what has occurred in this country with what has taken place at Mr. Schneider's, at Creuzot, and it will be clearly proved, as I think, how small is the power of a Trades' Union in comparison with the natural effects of an increasing trade and increasing competition among masters for the supply of labour. At Monsieur Schneider's, without the assistance of a Trades' Union, the working people have obtained, during the last seventeen years, an augmentation of wage of thirty-eight per cent. In England, in the corresponding period, the most powerful of all the Trade Societies, with an accumulated fund of one hundred and forty-nine thousand pounds, has found it impossible to secure any increase in the earnings of its members.

On the other hand, it should not be forgotten, that if we were unduly to diminish the rate of wages earned by our operatives, either by a combination among the masters, or by any other artificial means, we should see the result immediately in a more active emigration to the United States. The increasing flow of emigration to the United States has of late been marvellous. Between July 1, 1865, and December, 1868, a million emigrants entered the United States; and during the last five years Mr. Wells declares that there has been a greater development of the industry of that country than at any former period. As a necessary consequence of this, there must be an increased demand for labour.

Take, for example, the railways of the United States. Since 1865, 8,000 miles have been constructed, and the present state of increase is double what it was before 1860. That increase is now 1,156 miles a-year, while the growth in the goods traffic is sixteen times greater than the growth of the population. The production of pig iron has been increased from 913,000 tons in 1860, to 1,550,000 tons in 1868.

The import duties have tended greatly to raise prices, and the profits of the manufacturers have been very large, and these circumstances have artificially stimulated production. The effect of these influences upon the wages of the artisans employed is shown in a statement made by Mr. Hewitt to the Trades' Union Commissioners. He told them that the wage for puddling in Pittsburgh was from twenty-one to twenty-seven shillings per ton, as compared with eight shillings and sixpence in England; there being, notwithstanding the great increase in the cost of provisions in the United States, no corresponding difference in the cost of living. Mr. Wells gives a comparative statement of the wages in the United States, as compared with the rates prevailing in England; showing that, in the iron rolling mills in 1868, wages were forty per cent. higher; in the foundries in 1867, fifty-eight per cent. higher; in the ship yards forty-eight per cent.; in the cotton mills twenty-nine per cent.; and in the woollen mills twenty-five per cent. higher than the corresponding rates in England. We must further bear in mind that there is a constant emigration of operatives from the manufacturing districts of the United States to the unsettled territories in the Far West.

The emigrants are men who have accumulated sufficient means to embark in agricultural enterprise, and are wearied with the toils of an industrial life. To these well-to-do operatives, the boundless tracts of fertile land, still uncultivated and unoccupied, are offered for sale at a dollar and a quarter per acre.

Who, indeed, can wonder that the seductive charms of a pastoral life should be found so irresistible to the toil-worn labourer at the anvil or the loom?

Nor is the effect of emigration to the United States on the rate of wages and the supply of labour in the manufacturing industries of this country prospective merely. It is now and has long been felt. It is remarkable that our artisans, who earn such ample wages in this country, should be disposed to emigrate in such large numbers, and to give up employment in establishments in which there is a certainty of constant work for every well-conducted operative.

Messrs Herries and Creed, in their pamphlet, 'Handicraftsmen and Capitalists,' express an opinion exactly coinciding with that which I have formed from information which has reached me from other sources. Admitting the impossibility of retaining skilled artisans in this country, if wages were to undergo any sensible reduction, they say that, 'in South Wales the value of labour is, as compared with other districts, cheaper in the extreme, which cheapness it is to be feared will not be maintained, owing to the growing feeling for emigration manifested and manifesting itself among the Welsh iron-workers.'

Once more, comparing the rate of wages at home and abroad, we ought to have regard to the fact that abroad employment is steadier than at home. The very spirit of enterprise, which has made England what she is, tends to produce great fluctuations in the labour-market. When trade is good, our iron-founders and cotton spinners are too ready to increase the productive resources of their establishments. This leads to over-production, and a consequent cessation of demand from abroad. It cannot be doubted that this spasmodic and fluctuating character of our trade produces an unhappy effect upon the operatives who are subject to its influence. They suffer a constant fluctuation in wages: the increase and the reduction of their pay following the varying course of trade in natural sequence it is true, but still with very inconvenient results to the internal economy of their homes. In proof of the preference which is felt among the working classes for a more moderate rate of wage with constant employment, it is interesting to compare the rate of wages in the dockyards with the wages paid in the private shipbuilding yards, on the banks of the Thames. The following table, kindly compiled for me by Admiral King Hall, C.B., gives the rate of wages in Sheerness yard, in the years 1849, 1859, and 1869; and I am indebted for the table showing the current rates of wages at the corresponding period in the private yards on the Thames, to Mr. John Hughes, the manager of the Millwall Works:—

Sheerness—Rates of Wages.

		1849	1859	1869
s. d.		*s. d.*	*s. d.*	*s. d.*
	Shipwrights	4 0	4 6	4 6
	Caulkers	4 0	4 6	4 6
	Joiners	3 6	3 10	3 10
8 0	Forgemen	7 0		
5 9	Furnacemen	5 0		
4 8	Assistant Furnacemen	4 0		
5 2	Steam Hammermen	4 6		
6 4	Anchor Firemen, 1st class . . .	5 6		
5 6	" 2nd "	4 9	same	same
5 6	Double Firemen	4 9		
4 10	Single Firemen, Vicemen, and Fitters .	4 3		
4 4	Hammermen, 1st class	3 9		
3 9	" 2nd "	3 3		

The figures in the left hand margin show the pay of smiths employed 10 hours a day.

Average Rates of Wages paid at Millwall Iron Works.

	Rates of Wages during Years 1851 to 1869							
	1851		1861 to 1865		1865 to 1869		1869	
	per week		per week		per week		per week	
	s.	s.	s.	s.	s.	s.	s.	s
Fitters . . .	33 to	38	33 to	38	35	40	33	38
Planers . . .	30	33	30	33	32	34	30	33
Drillers . . .	22	27	22	27	23	28	22	27
Smiths . . .	30	42	30	42	30	42	30	42
,, Helpers. .	22	24	22	24	22	24	22	24
Moulders. . .	36	38	36	38	36	40	36	40
Pattern Makers .	36	39	36	39	39	42	36	39
Joiners . . .	36	39	36	39	36	42	36	42
Shipwrights . .	42	48	42	48	39	42	36	39
Platers . . .	36	42	36	42	36	42	36	42
,, Helpers. .	21	24	21	24	21	24	21	24
Rivetters. . .	30	32	30	32	30	32	30	32
,, Helpers .	20	24	20	24	20	24	20	24
Caulkers . . .	30	33	30	33	30	33	30	33
Chippers . . .	28	30	28	30	28	30	28	30
Angle Iron Smiths .	38	40	38	40	38	40	38	40
Boiler Makers . .	36	42	36	42	36	42	36	42
,, Helpers	21	24	21	24	21	24	21	24
Painters . . .	21	30	21	30	21	30	21	30

Hours of work, 58½ hours per week.

Average rent of men's houses in 1851 was about 16*l.* per year; and in 1865 about 20*l.* per year. These are six-roomed houses, and in most cases more than one family occupied them.

During the corresponding periods, shipwrights in London have been earning from six shillings and sixpence to seven shillings a day, and the shipwrights in Sheerness yard, men at least as skilled as those employed by the private ship-builders, could at any time have put their tools into their baskets, and, at the end of an hour and a half's journey by rail, have obtained employment from the private ship-builders at the higher rate of wages.

They preferred, and with good reason, a more moderate rate of wage, with a certainty of employment, to a higher rate of wage without the certainty of permanent occupation. The same preference for regular employment at a moderate rate of wage over a less certain employment with a higher wage manifests itself abroad as much as in England. I will take the case of the Sotteville engine works, established originally to supply the locomotives for the Paris and Rouen Railway. The employés of a railway company may look upon their employment, at least during good behaviour, as being almost as certain as that under a government. Thus it has happened that at Rouen there has been no advance of importance in wages for the last twenty-five years in the class of labour employed by locomotive builders; although, as I have shown already, there has been a very great increase in the wages paid by Messrs. Schneider and other private employers. The rates of wages at Sotteville are quoted in the subjoined table.

It may be interesting to compare them with the wages earned by the same trades in England:—

Sotteville Works.

	s.	d.	
Erecters, Fitters, and Turners	24	0	per week.
Smiths	27	0	"
Strikers	18	0	"
Joiners	22	0	"
Modellers	23	0	"
Moulders in the Foundry	22	6	"

The wages paid at a similar establishment on the

largest scale in England are shown in the following table:—

Average Rates of Wages paid to Skilled Workmen, Locomotive Works, England.

	1859		1869	
	s	d.	s.	d.
Fitters	28	3·15	28	7·69
Turners	28	4·57	29	3·76
Braziers	28	6·85	28	7·06
Grinders	27	6	28	10·50
Smiths	28	5	26	10·35
Boiler Smiths	31	8	30	4·50
Bricklayers	24	5·10	30	·57
Saddlers	19	8	20	3
Forgemen	34	3	34	4·05
Painters	22	10	23	1 60
Moulders	29	4·50	28	5·58
Joiners, Pattern Makers, and Sawyers	24	6·18	24	4·95
Brickmakers	27	8·44	27	5·28
Total average	27	11·23	28	1·28

I have already alluded to the statements so repeatedly made that the competition with foreigners is producing a disastrous effect on British industry, and I do not deny that the foreign manufacturer is approaching us day by day. Indeed, the race is now so close that, if the Trades' Unions should succeed in forcing upon employers an advance of wages, without any corresponding equivalent in the efficiency and productiveness of labour, it would be impossible to maintain the competition with foreigners in neutral markets in many important branches of trade; and capital, no longer obtaining a fair return in this country, would be invested in foreign undertakings.

In that event I should like to know what the leaders of the Trades' Unions, with all their boasted power, could do for the unhappy operatives whom they had deceived. A heavy responsibility indeed rests on those who possess an influence over the ignorant masses of the population. It were well that the leaders of the Trades' Unions were more fully alive to their responsibilities, and that they would realise to themselves the fearful consequences which would ensue if any important branches of our trade were driven by Trades' Union agitation to take refuge in other lands.

As yet, however, I rejoice in the belief that we have no reason to complain of the measure of success which has attended British enterprise in the competition with foreign industry.

Taking, first, the figures from the Board of Trade returns, as a general indication of the commercial activity of the country, it will be seen that the export of British produce has increased from ninety-seven million pounds in 1854 to one hundred and eighty-eight millions in 1866. It is true that, in consequence of the financial troubles of the years 1867 and 1868, a considerable falling off occurred in those years, to the extent, in round numbers, of eleven million five hundred thousand pounds; but the same causes produced an even greater depression in the export trade of France. Mr. West states in his last report, as Secretary of Legation in France, that, whereas the total value of the exports of France in the nine months ending September 30, 1866, was ninety-five million pounds, it had fallen in the corresponding period of 1868 to eighty-three millions, showing a reduction of eleven million seven hundred and eighty-one thousand pounds. These figures show conclusively that the recent dulness of trade was not the exceptional misfortune of this country; but that the same causes produced even more discouraging effects in those countries which are our most formidable competitors in commerce. On the other hand, the Board of Trade tables, issued for this year, show that, whereas in the five months ending May 31, 1868, the total value of the exports of British and Irish produce was seventy million six hundred and sixty-eight thousand pounds, the value has been increased in the corresponding period of the present year to seventy-four million seven hundred and forty-four thousand pounds. It is not a little remarkable that a great part of this increase has taken place in the iron trade, in which, as has been alleged, the pressure of foreign competition has been most particularly felt. The total value of the exports of iron and unwrought steel for the five months ending May 31, 1868, was five million five hundred and forty-two thousand pounds; but that value had increased in the corresponding period of the present year to seven million one hundred and sixty-eight thousand pounds. It has been said that the Belgians are now able to roll rails at prices so low as to prevent our iron-masters from competing with them; but it appears that the value of railroad iron exported from England has increased from a total of one million eight hundred and nine thousand pounds during the five months ending May 31, 1868, to a total in the corresponding period of 1869 of two million four hundred and eighty thousand pounds. Of the total quantity exported, the proportion sent to the United States in 1868 was valued in the Board of Trade tables at seven hundred and ninety-three thousand pounds; and this amount had been increased in 1869 to a total value of one million and fifty-four thousand pounds. Taking, again, steam-engines and machinery, much has been said of the falling off in that branch of trade between this country and Russia; but it appears that while the value of the steam-engines exported to Russia in the first five months of 1868 was only sixty-one thousand four hundred and eighty-nine pounds, the value of our exports in the same class had increased in 1869 to a total sum of one hundred and three thousand three hundred and ninety pounds. While the more recent revival of our trade must be admitted to be highly satisfactory, yet when we look back upon the commercial progress of England through an extended cycle of years, the growth of our trade becomes still more remarkable. Monsieur Michel Chevallier, in his introduction to the reports of the jurors for the Exhibition at Paris in 1867, quotes an interesting calculation, made by an eminent French statistical authority, Monsieur Chemin Dupontes, for the purpose of showing how England has obtained of the increasing trade of the world, a proportion immensely larger than has fallen to the share of any other manufacturing country. Monsieur

Chemin Dupontes states that in 1800 the imports and exports of the western nations, including the United States, to and from the East, amounted in value to sixteen million pounds; that the total value of the trade between the West and East in 1860 amounted to one hundred and twelve million pounds; and that this large sum was again increased in 1866 to one hundred and sixty-one million pounds—an increase, in round numbers, of forty-nine million pounds in an interval of only six years. The whole of the recent gain has been secured by Great Britain. Owing to the disturbance of their commercial relations, consequent upon the war with the Southern States, the commerce of the United States with the East has made no advance in the last eight years; and if a comparison be made with France it appears that, while in the last sixty years the trade of England with the East has increased ten-fold, that of France has not even doubled. Again, Mr. Wells says that England makes two-thirds of the entire product of textile fabrics in Europe, and that her cotton goods are as thirty-seven to one hundred of the total value. Do not these figures prove that during the last fifty years there must have been such a demand for labour, to meet the requirements of our increasing trade, as would fully explain the rise in wages, without the intervention of the Trades' Unions; and that the demand for labour, notwithstanding the more general use of labour-saving appliances, must have increased more rapidly than the supply which our increasing population affords?

For the Trades' Unions to pretend that it is to their powerful organisation alone that the improvement in the condition of the labouring classes is attributable, is a most vain and shallow pretence. No, it is not the Trades' Unions, but the enterprise and resources of the British capitalists which have secured for our labourers and artisans a wider field and a higher remuneration for their industry and their skill. As Adam Smith truly observes, 'it is not the actual greatness, but the continual increase of trade which occasions a rise in the wages of labour.'

On a general review of the subject the profits of trade in England in the last quarter of a century cannot but be regarded as satisfactory.

If the returns had been larger, employers would have encountered more severe competition; and, though wages may be a little higher in England than abroad, our superior machinery and greater command of capital as yet compensate for the difference.

I will now give to the House some details upon the particular cases, which have been quoted in the newspapers and elsewhere, in proof of the success with which foreigners have engaged in competition with our manufacturers, in branches of trade in which we formerly enjoyed a monopoly.

The experience of the Consulting Engineers of our Indian railways does not by any means go to prove that foreign iron-masters or engine builders can successfully compete with the iron-masters and engine builders in this country. Their experience, I may add, is the more valuable because the Indian railways afford the most perfect example of a purely neutral market. There is no personal influence acting on the minds of Indian railway engineers and directors prejudicially to our interests; and no Customs' duties, which are protective to our manufacturers, are imposed upon the importation of our manufactures into India. The plant and machinery for the Indian railways are purchased in the cheapest market; and it is certain that the foreigner would be preferred, regardless of national sympathies, if he could compete with the iron trade at home, either in quality or price. Let us then examine into the actual state of the facts as regards the supply of rails and locomotives to the Indian railways.

I shall first appeal to the experience of Mr. A. M. Rendel, who has kindly furnished me with some figures, which are most interesting, and perfectly conclusive as to the pre-eminence hitherto maintained by British industry. In November and December, 1865, tenders were invited by advertisement for a large number of locomotives for the East Indian Railway. Eminent foreign as well as English makers were free to compete, and twenty-two tenders were sent in. The result was that eighty engines, varying in cost from three thousand one hundred and sixty-five pounds to two thousand four hundred and fifty pounds, were ordered from English makers, at an average price of two thousand six hundred pounds each; twenty from Kiessler, of Esslingen, near Stuttgart, at two thousand five hundred and fifty pounds each; and twenty from an English maker at two thousand four hundred and forty pounds; so that the foreign maker received a price intended to be intermediate between those of the English makers. I ought to mention that at the date when the order was given, English houses were very full of work. Not long afterwards, in consequence of the rapid development of traffic on the East Indian Railway, it became a matter of urgent importance to send out additional locomotives as early as possible. Accordingly ten more engines were ordered from an English firm at the price agreed upon in the first tender—namely, two thousand four hundred and fifty pounds; and ten more were ordered from Esscher, Weiss, and Co., of Zurich, who undertook to make them for two thousand five hundred and fifty pounds each,—the same price as had been previously accepted by the other foreign makers. At the termination however of their contract, Messrs. Esscher, Weiss, and Co. made a representation to Mr. Rendel that they had sustained a loss; and asked to be allowed, by way of compensation, to make ten more engines of the same kind, but at the enhanced price of two thousand eight hundred pounds. It is, therefore, very evident that in the results of their competition with the English makers, who were under no pressure in regard to price, all the shops being so full of work that early delivery was an impossibility, Messrs. Esscher, Weiss, and Co. had little cause for satisfaction. Indeed, they admitted a substantial loss. But, even if this contract had been more satisfactory to Messrs. Esscher, Weiss, and Co. than it actually proved, their success would have been largely due to British industry; for the boiler plates, the copper fire-boxes, the wheels, the pig-iron for the cylinders, the tubes, and the frame plates, in short, two-thirds of the materials used in the construction of their engines, came from England in a manufactured state. It was the same with the engines supplied by Messrs. Kiessler. That firm assured Mr. Rendel that they could not think of asking him to accept Prussian iron or copper; and that by far the greater portion of their material came from England. Of course, to a certain extent this was done under the requirements of the specification; but no pressure was needed on the part of the engineers.

The axles and wheel tires were specified to be of Prussian steel; but for this they, too, would have been of English make. But the experience of Mr. Rendel is, I need scarcely say, by no means limited to the purchase of locomotives. Rails, and iron bridge work upon the largest scale, have been supplied in England for the Indian railways for which he has acted; and the tenders have been obtained on all occasions, when a large order has been given, by open advertisement, and all Continental makers have been as free to tender, and would be accepted on the same guarantees as English makers. Yet out of the total expenditure during the last ten years, of from seven to eight million pounds, on materials and plant for the East Indian railways constructed under Mr.

Rendel's supervision, with the exceptions I have made, the whole of these contracts have been obtained by English manufacturers.

Much has been said from time to time as to the importation of iron girders from Belgium into this country, but Dr. Percy, in his evidence before the Committee on Scientific Education, stated that the iron girders recently imported from Belgium would be made here if there were a larger demand. A manufacturer would not alter his mills for a special kind of girder unless there was considerable demand; and he urged, as a sufficient reason why there should be no apprehension on this subject, the remarkable success which has been achieved in England in recent years in the production of armour plates. Another interesting and conclusive proof of the success with which our engine builders can compete for the supply of locomotives, is furnished by the following schedule, kindly prepared for me by Mr. W. P. Andrew, of the tenders received by the Punjaub Railway Company, in answer to a public advertisement in January, 1866, for a supply of ninety-four locomotives and tenders:—

Tenders for Supply of Engines for the Punjaub Railway.

Country from which tender received	Prices per engine and tender
1. Germany	£ 3,156
2. England	2,990
3. England	2,960
4. England	2,950
5. England	2,850
6. England	2,835
7. England	2,810

Country from which tender received	Prices per engine and tender
8. England	2,790
9. England	2,750
10. Germany	2,750
11. England	2,685
12. Germany	2,680
13. England	2,680
14. Switzerland	2,650
15. England	2,650
16. England	2,600
17. France	2,595
18. England	2,575
19. England	2,500
20. Scotland	2,424
21. Scotland	2,395

It is not necessary that I should offer any comment on this most important illustration of the relative powers of British and Continental industry. It is quite evident that, though we are not yet beaten in the race, foreign competition treads closely on our heels. It is for our manufacturers to apply to the best advantage all the resources which a ready supply of the raw material and a large command of capital afford; and our operatives, instead of lending themselves to the delusions and false teaching of Trades' Unionism, must strenuously exert themselves to increase the productive value of their labour, or they cannot hope to retain the proud pre-eminence which they have hitherto enjoyed.

Serious alarm was felt when, in 1865, fifteen engines were ordered for the Great Eastern Railway from Messrs. Schneider. These misgivings would probably have been allayed had it been generally known that at the same time when the fifteen engines were ordered from Creuzot, forty other engines were ordered from English firms, and that when Messrs. Schneider were subsequently asked to undertake the construction of twenty-five more engines at the same price as they had agreed to accept for the fifteen engines originally ordered, the offer was declined.

The eminent English engineer, at whose instance the original order was entrusted to Messrs. Schneider, possesses, from long residence in France, a special knowledge of French workmen, and it is his opinion that the price of that kind of labour in France is not generally cheaper for a given quantity of work than it is in England, while the material, of course, costs at least as much.

It has been also said that Belgian rails are being largely imported into England, and I believe that some six hundred tons for the East Gloucestershire Railway were supplied by a Belgian firm in 1865. The price of these rails was six pounds ten shillings, or thereabouts, delivered at Gloucester. But a solitary instance like this proves nothing as to the general comparative prices of English and Belgian rails. It was because our iron-masters were more fully employed than the iron-masters in Belgium, and because the prices of rails had in consequence fallen more rapidly in Belgium than in England, that the order in question was executed abroad.

Since the year 1865 rails have been made in England at a cheaper rate than that paid for the Belgian rails supplied to the East Gloucestershire Railway. I may also mention that in the present year my father invited tenders for forty engines for a railway in Hungary. The contracts were open to all the world, but of the forty engines thirty-five were given to English, and only five to foreign firms. The prices quoted in the tenders from the German makers were from two thousand five hundred pounds to two thousand seven hundred pounds each, while the prices of the English makers were considerably less.

These statements do not, I would venture to assert, bear out the inferences drawn by the Edinburgh Reviewer in his interesting and able article on the comparative industry of nations.

With reference to the French Exhibition in 1867 the reviewer writes as follows: 'Mons. Schneider et Cie., of the Creuzot Ironworks, exhibited a remarkably well-finished express engine for our Great Eastern Railway; and it was the sixteenth out of an order for forty engines.' Elsewhere in the same article it is stated that, 'another locomotive, also built from English drawings, was exhibited by Messrs. Kiessler, of Esslingen, which was an instalment of an order given by the East Indian Railway Company for twenty engines. These two engines afforded incontestable proofs that English designs can be executed abroad as well as at home and at a cheaper rate.' The explanation which I have already given of the circumstances under which the orders for locomotives were given to Messrs. Schneider and Messrs. Kiessler has been, I hope, completely satisfactory. I need not therefore stay to point out how little cause for discouragement as regards the competitive power of British and foreign industry, is to be found in either of those cases. The reviewer further alludes to the evidence given by Mr. John Robinson of the Atlas Works, Manchester, who had stated to the Trades' Unions Commissioners that the amount of engines supplied by his own firm since 1865 to foreign countries had been constantly diminishing, in consequence of the high rate of wages which he had found it necessary to pay. I have shown, however, that there has been no increase whatever in recent years in the rate of wages paid in England in the trades concerned in the building of locomotive engines ; while on the other hand at Mons. Schneider's establishment there has been an increase during the last fifteen years of thirty-eight per cent. We must not, therefore, look at the increase of wages for an explanation of the reason why we are no longer monopolists of the engine-building trade. The real explanation is to be found in the circumstance that as the railway system was first established in this country, so we were the first in the field as locomotive engine builders.

When, for example, a supply of rolling stock was required for the service of the Paris and Rouen Railway, the first railway of much importance constructed on the Continent, it was thought necessary to create the special engine building works already mentioned at Sotteville, near Rouen, for the purpose of building the locomotives and carriages required for the line. A great number of the mechanics employed at the works were Englishmen; and the direction and supervision were exclusively English. And why was it that recourse was had to English experience in this case? Solely because the science of building locomotives was an occult science at that time on the Continent. Then, as now, labour of all descriptions was not only as cheap, but, as I have shown, much cheaper abroad than in England. It is true that the necessary experience and mechanical skill were not as yet to be found among Continental mechanics; but surely no reasonable man could expect that we were to remain for ever monopolists of a trade, in which the foreigner only required additional experience in order to enable him to compete with our countrymen. I say, therefore, that it is because we were first in the field, and not because at a former time labour was relatively cheaper, that we, for many years after the commencement of the railway system, supplied engines to Continental countries, which now supply themselves. And be it remembered that even now our Continental neighbours would still draw large supplies from England, if they did not protect their own manufacturers by heavy import duties.

I have said much of the harm done by the Trades' Unions in vain attempts to force up the rate of wages by regulations which tend to destroy the free liberty of the labourer; but though I consider that in this respect the Trades' Unions can do but little good to the workmen, and may do some injury to the masters, I believe that in other respects, and especially as benefit societies, the Trades' Unions have effected, and are capable of doing great good to the working people. They encourage a noble spirit of self-help, and in point of fact, devote by far the greater portion of their funds to benefit objects. For example, the Engineers' Society, out of a total income of forty-nine thousand pounds a-year, spends but seven hundred and twenty-four pounds a-year in contributing to the support of the members of the Unions who are out on strike.

It should be the care of a wise statesman diligently to encourage, by a timely legal recognition of these trades' societies, the noble efforts which our working people are making, through their instrumentality, to insure themselves against the risks of scanty employment and the degradation of pauperism.

Surely we must honour and admire the sentiments of fraternal sympathy, which prompt men to promote each other's advancement in life by that mutual aid and support, which these societies are intended to afford; though I admit that these aspirations, however legitimate in themselves, are sometimes manifested in forms not altogether satisfactory. Although it may be sometimes apprehended that the existence of an organisation, framed for the purpose of carrying on a strike, may, like the maintenance of large standing armies, be a provocative of war; on the other hand, I have pointed out that many of the more experienced leaders of the Trades' Unions are strongly opposed to strikes; and, even if a strike should unfortunately occur, it is likely enough that, under the control of a Trades' Union, the conduct of the workmen would be more reasonable and temperate than if no organisation existed to check a lawless disposition.

In this respect, the conduct of the Trades' Unionists, while out on strike, will probably be as much superior to that of the rioters in the manufacturing districts in the early part of the present century, as the discipline of a standing army is superior to that of a guerilla band.

As regards the Bill before the House I intend to vote for the second reading. I coincide fully with the view of the Superior Council of Commerce and Industry in Belgium, as set forth in a memorial which they addressed to their Government on this question. From this document, I extract the following observations :—

The right of property gives to every citizen the right to regulate the price of his labour. The exercise of this right, in concert with others, cannot make that unlawful which it is lawful for every man to do individually. To prohibit simple combination, in the absence of any attempt to coerce the will of others who have taken no part in the deliberation, is an injustice which the representatives of Belgian labour would desire to see removed from their codes. As the general state of the law on this subject, which is very difficult, will be far more ably elucidated by other honourable members, who are more especially qualified to address the House on legal questions, I shall not enter into any details on the subject.

While supporting the Bill, however, I think it my

duty to recommend that, contemporaneously with the repeal of all laws specially directed against the Trades' Unions, the provisions of the law as regards the offence of intimidation shall be carefully considered.

The committee appointed to inquire into the combination laws in 1824 state in their report that it is, in their opinion, 'absolutely necessary, when repealing the combination laws, to enact such a law as may efficiently, by summary process, punish workmen or masters, who, by threats or violence should interfere with that perfect freedom, which ought to be allowed to each party, of employing his labour or capital in the manner he may deem most advantageous.' Lord Althorpe also expressed an opinion in debate to the same effect, and I trust that when we are about to admit and give recognition to the principle of combination, and to protect the funds of those who combine for objects which do not involve a breach of the law we shall at the same time take ample security for the protection of those workmen who wish to preserve their liberty, and to exercise an independent judgment upon their own affairs, unbiassed and unfettered by the dictatorial decrees of Trades' Union Committees.

It has been judiciously suggested in the columns of the 'Pall Mall Gazette,' that we have an article in the penal code, recently promulgated for India, which presents an excellent model for legislation on this subject. The Indian law is as follows:—'Whoever threatens another with any injury to his person, reputation, or property, with intent to cause alarm to that person, or to cause that person to do any act which he was not legally bound to do, or to omit to do any act which that person was legally entitled to do, as the means of avoiding the execution of such threat, commits a criminal offence.'

Again, the practice of ticketing, which is so common during a strike, and so difficult to define or punish under any of our present laws, seems very similar to the offence known to the Roman law as 'convicium.' In Justitian, lib. iv. tit. iv., this offence is defined as being any proceeding which publicly annoys or insults another, such as gathering a crowd around a man's house. Under the Roman law, therefore, it would appear that the petty persecution, to which workmen are too often exposed on the part of Trades' Unions, was a punishable offence under the ancient civil law, and it is worthy of consideration whether similar provisions should not be introduced into English law. This, however, is a question rather for a lawyer than a layman, and I do not think it expedient that I should trouble the House further with my views upon this subject.

It has, I think, been conclusively proved that Parliament can no longer refuse to recognise the right of workmen to combine for the purpose of regulating the terms on which they will jointly agree to work or not to work; and it follows, surely, as a necessary consequence of this concession, that the funds of their trade societies should be protected from embezzlement so long as the societies conform to the law. I think, in addition to this protection of their funds, that certain special privileges might be given to those Trades' Unions which are willing to observe such conditions as the wisdom of Parliament may prescribe. The Royal Commissioners prescribe four conditions, upon which full protection and special privileges might be granted; but I am not sure that all these can be reasonably enforced.

It would, however, scarcely be consistent with public policy to do more than to give a bare protection of its funds to a society which refuses to permit the use of machinery. On the other hand, I think that more wisely-constituted trade societies, which have no such rules in restraint of trade, might enjoy the privileges conferred by the 'Friendly Societies Act' on societies duly registered and certified under that Act, and be permitted to buy an acre of land, and to build a hall for the purposes of their meetings.

As far as the rate of wages is concerned, I would repeat once more the indisputable truth, too often forgotten by both masters and workmen, that no legislation, however severe, no Trades' Unions regulations, however arbitrary, can overcome the necessary results of the operation of the law of supply and demand. For the workmen who are willing to accept lower wages than the Unionists I would secure ample legal protection against the tyranny of Trades' Unions. But while I am anxious to grant to the oppressed the ample protection of the laws, I look rather to the development among the working classes of a more generous tone of feeling towards each other and a more universal sentiment of loyalty and good will towards their employers, as affording a far more real security for their mutual happiness and prosperity than any emendation of the law can give, however skilfully devised.

Quid leges sine moribus
Vanæ proficiunt?

Surely it is not chimerical to hope that the spread of education during the last quarter of a century, and the noble efforts, made in Parliament and elsewhere, to enlighten and elevate the people are destined hereafter to produce their fruits in the general recognition by workmen themselves of the free rights of labour.

LONDON: PRINTED BY
SPOTTISWOODE AND CO., NEW-STREET SQUARE
AND PARLIAMENT STREET

TRADES UNIONS:

AN INQUIRY INTO

THEIR RULES AND WORKING,

BASED ON THE EVIDENCE BEFORE THE ROYAL COMMISSION;

SHOWING

THE FOLLY OF ALL ATTEMPTS TO RAISE WAGES BY

VIOLENCE AND STRIKES,

AND THE BEAUTY AND EXCELLENCE OF THE DIVINE LAWS GOVERNING

WORKMEN AND EMPLOYERS.

BY ROBERT JOBSON.

(*Reprinted from the* "FREE LABOUR JOURNAL.")

LONDON:
NEWS-AGENTS' PUBLISHING COMPANY (LIMITED)'
147, FLEET STREET, E.C.
AND MAY BE HAD AT
MANCHESTER: JOHN HEYWOOD. GLASGOW: WILLIAM LOVE.
SHEFFIELD: E. WESTON.

PRICE TWOPENCE.

TRADES UNIONS.

A Prize Essay.

"The (Divine) law requires us only to take another's place, and to consider what our principles would require us to do under the other's circumstances If one is a seller, he is to consider what—if he were a buyer—he would consider it right that a seller should do. If he is a master, what—if a servant—he would expect a master to do. By thus placing ourselves in the position of those with whom we have to do, we learn to be more just and merciful—to demand less, and to give more, in a word, to be more equitable What a different world it would be, if this great law were in any considerable measure the rule of conduct! . . . . . . The law of equity is the practical form of the law of love to the neighbour; practically, to love our neighbour as ourselves is to do to him as we would that he should do to us. This is the law of Heaven."—Bruce.

In this paper we purpose placing before you a clear view of that important question, Capital and Labour, now engaging public attention. In endeavouring to do this, we hope also to show the influence Trades-Unions are exercising on the trade of this country. We shall also take into consideration the character of their rules in relation to fixing the price of labour, piece-work, apprentices, machinery, "chasing," the relationship of master and workman, and many other points, all bearing on this important subject. We shall also consider the subject of picketing, strikes, &c.; their nature and results. And, in doing this, we shall, as a basis for our arguments, seek to be guided by the divine law—"Whatsoever ye would that men should do to you, do ye even so to them." This law is applicable to employer as well as to workman; both are to be guided by it in their dealings with each other. If we examine all good human laws, we shall find that this simple but divine law forms their essence; and it brings happiness or misery, just as it is observed or disregarded. We cannot too well remember the excellent words we have chosen for our motto, that "the divine law requires us only to take another's place, and to consider what our principles would require us to do under the other's circumstances. If one is a seller, he is to consider what—if he were a buyer—he would consider it right that a seller should do; if he is a master, what—if a servant—he would expect a master to do. By thus placing ourselves in the position of those with whom we have to do, we learn to be more just and merciful—to demand less and give more; in a word, to be more equitable." We are therefore to apply this as the standard by which we ought to measure the conduct of the employer to his workman, and of the workman to his employer. With such for our aim, and entering on the subject in a becoming spirit, we have no doubt that the result of our inquiry will be in every way satisfactory and profitable.

We will now proceed to consider the subject of capital and labour, and the law that regulates the rise and fall of prices. Capital stands to the employer in the same relation that labour does to the workman. If the demand for money is great, the value of it will rise accordingly; but if the demand for it declines, its value will decrease in a like ratio. For instance: in the month of May, 1866, money was in so great demand that the bank-rate rose to ten per cent.; but from the month of July, 1867, the demand has so decreased that the bank-rate has fallen to two per cent., at which rate it has since remained. It is the same not only with money, but with other commodities—such as tea, sugar,

4

coffee, corn, cotton, coal, or wood; for these things are only capital in another form. If there is a great demand for tea and sugar, the prices will naturally rise; and if the supplies are short they will rise the quicker. But if there is little demand for them the prices will naturally fall; and should the supplies be large they will fall the quicker, until they attain their own natural level. There are, of course, exceptions to this, when prices may advance more than is justified by the actual consumption; such as, for instance, when corn becomes scarce, speculators will buy up and hold for advanced rates. Or, again, when there is no great want of the article, they will combine, buy up, and hold, thus forcing the prices up to an unnatural level; for "bulls and bears" are to be found in commerce as well as on the Stock Exchange. In the first case the speculation is legitimate, because the short supply justifies it; and it serves a useful purpose in conserving a universal article of food by compelling the consumer—from its enhanced price—to partake more sparingly of it, or to seek for a substitute in another article of food until fresh supplies of corn arrive. In the latter case the speculation is not legitimate, because the quantity in the market does not warrant an advance in price, and therefore it partakes more of the character of gambling than of fair legitimate trading. Both these cases, however, are exceptions, and they only prove the general rule, that supply and demand—if left free and unfettered—will regulate themselves.

Now, labour may be regarded in the same way. When work is plentiful, wages, or the price of labour, will rise, and more so if the labourers are few; but when work is scarce, wages will decline, and still more so if labour is abundant. Let us give you an illustration of this: Before the discovery of gold in California and Australia, wages were moderately low; but immediately these great gold fields were laid open, the demand for every kind of workmanship gradually rose until the price of labour attained an almost fabulous height. Carpenters, joiners, shipwrights, seamen, masons, bricklayers, and diggers, all were wanted, and in proportion as that demand rose did wages increase; but when the gold fever gradually died out, wages gradually decreased. This law of supply and demand in its operations is altogether wise and good. Its admirable action is obvious. The low price of an article is an announcement to all concerned that there is more than enough of that article already in the market, and that it should cease to be produced, or be taken elsewhere. The high price of an article is, on the other hand, an intimation that more of the article is wanted, and it invites a supply from wherever the article is possessed. Nothing can be conceived more simple, clear, beneficial and beautiful than the divine law, and nothing more fully proves the truth, that the important subjects of capital and labour have not been left to the guidance of blind chance; but that they, like all other things in the economy of nature and of man, are under the control of wise and beneficent regulations, the safeguards of a Divine Providence.

Here, then, is a simple law in political economy, which every child may learn, but ignorance of which has led to many unhappy differences between master and workmen, as we will proceed to show. Now, then, what is the principle the trades-unions go upon in selling labour? It is this: They fix a *minimum* rate of wages for each workman—good, bad, or indifferent. The masters are at liberty to give more if they feel inclined, but the men are not at liberty to accept less, whether they feel inclined or not; and that minimum rate is oftentimes fixed by the unions in an arbitrary manner, irrespective of the state of trade with the masters, whether it is brisk or slack, or whether labour is scarce or plentiful. On this point we will allow the men to speak for themselves. Mr. W. Allen, Secretary of the Amalgamated Society of Engineers, in giving his evidence to the commisioners appointed to inquire into the organisation and rules of trades-unions, answers thus :—

Question 852. (Commissioner).—I think you said that a great number of the strikes arose from the resistance to a lowering of wages?—Yes.

Q. 853. When provisions rise, I suppose you sometimes ask for a rise of wages?—It has only been, as I have said, within these last twelve months that any material change has taken place, except in the north of England and in Scotland.

Q. 854. But when the masters endeavoured to lower the wages, did they ever state that provisions had fallen?—No.

Q. 855. They never gave that as a reason for wishing to give a less rate of wages?—As a rule, they assigned no reason at all beyond the simple fact that they want to reduce the wages inasmuch as the orders are not coming in so rapidly as they should; that trade is bad, in fact.

Q. 856. Then, the reason they give is slackness of trade?—Yes.

Q. 857. But you have never heard that masters have said that as the wages were raised when provisions rose, they thought they ought to be reduced when provisions fell?—No. I have heard that employers say, "When trade is good we have given you an advance of wages, and when trade is bad you ought to submit to a reduction."

Q. 858. But that position of the masters has been resisted?—It has been resisted, and is being resisted now at Blackburn. They propose there to reduce the wages two shillings per week, and the men are all on strike in consequence of refusing.

Q. 859. At Blackburn has there ever been a rise of wages in consequence of a briskness of trade?—Yes.

Q. 860. And the men, being pleased with that, have not yielded to the design of the masters to lower the wages when the trade is slack?—Precisely so.

Q. 861. So that you have one rule for yourselves and another for your master?—We keep what we can get, as a general rule.

According to this evidence, whether trade is slack or brisk, it makes no difference to the unions. If the masters attempt to reduce the wages, they strike. Now, can anything be more unfair and foolish than conduct of this kind? And is not this going against that healthy law of political economy—that wages, like capital and merchandise, should be allowed to regulate themselves according to the demand for labour? It is not only a violation of that law, but it is a violation of that higher law, which the Divine Law-maker has planted in the minds and hearts of all men—that law which says, "Do unto others as ye would they should do unto you." The men quietly take an advance of wages from their masters when trade is good; but when trade is bad with them, they show their gratitude by making it still worse, by ceasing to work. The masters have the sense to see that it is better to work for a small profit instead of allowing their hands and works to stand still, but the men answer, "Profit or no profit, we mean to have our full wages; and if you cannot give full wages we won't have any."

Again, the same witness thus answers :—

Q. 948. You have said that when an employer proposed to reduce wages, you generally resisted that on the principle that, having got what you desired to get, you would endeavour to retain it?—Certainly.

Q. 949. Do you, on giving advice on these matters, take into account questions of competition between different parts of the United Kingdom, or questions of foreign competition, and how far it is possible for employers to compete with each other, if certain wages are given?—We know nothing, in fact, of the circumstances of the employers' competition, and, therefore we are not in a position to deal with it; but as a rule I would say, if one employer runs into competition with another, we would decidedly object to allow the employer who has reduced the price of the article to reduce our wages to meet that.

Q. 950. But where the employers give you as a reason for a reduction in the price of the article that the demand is diminished, do you procure any information for your society which would enable you to judge whether such a statement is founded on fact, and whether, with the diminished demand for the article, the masters' wish to lower the wages is unreasonable?—We generally make inquiry, and we have, as a rule, means of ascertaining pretty accurately the condition of things before we give advice.

Here, then, the refusal to lower wages is not only damaging to the masters, but to the men themselves. Labour on the Clyde is cheaper than it is on the Thames; so that the Scotch shipbuilder, by offering to build ships for less, is able to carry off the trade of his English competitors by the English workmen refusing to lower their wages to their natural level; and the result is this:—The English shipbuilder loses the contract, and the workman his wages. And this will apply, not only to shipbuilding, but to every other branch of trade. Now, let us suppose that the principles of trades-unions in this respect were to become general; in short, that England was to become a huge trade-union society, maintaining the price of labour in the face of Scotch or foreign competition—what would be the result? It would be this: that as men buy in the cheapest market, they would, as a matter of course, leave the English goods, and buy the Scotch or continental goods. This artificial restraint on the price of labour cuts two ways. It harms both the master and the workman; and when the former sees that through these trade rules he has lost the contract, he will very naturally turn his capital to some other business where he is not hampered by them, leaving his obstinate workmen to reap the fruit of their folly, the charity of the public, or the tender mercies of the workhouse.

There is another important fact bearing on this branch of the subject, which English workmen do not sufficiently recognise. That is the competition between English labour and foreign labour. Owing to the great advances made by science, time and space are gradually becoming lessened—we might almost say, annihilated; and this is most materially affecting our commercial position. Countries that were formerly strangers in more ways than one are drawing closer to each other, and are becoming more firmly knit together in the great brotherhood of nations; and by these civilising agencies, that day is slowly, but surely, approaching when the prayer of the poet shall become a reality-

"And man to man the world o'er,
Shall brothers be for a' that."

Steam, railways, and electricity, are three of the great instruments employed by Divine Providence in bringing about this happy state. Our intercourse with Europe is every day increasing, while our continental neighbours are gradually following our example, becoming more and more "a nation of shopkeepers," finding that this is better for their health and purse than wasting their life and treasure in useless wars. And one of the results from this is, that British capitalists are taking advantage of the cheapness of foreign labour. Hence we have from Norway, Sweden, and Russia, large quantities of flooring boards, mouldings, doors, and even wooden spouts; and from Belgium, great supplies of manufactured iron, such as rails, steam-engines, &c. From a return of the Board of Trade, we learn that in 1862 the export of ironwork from Belgium was £1100, while in 1865 it had increased to £95,000. From the same source we also learn that £926,396 worth of

7

cotton goods were imported from the Continent into England in 1866. These, then, are a few of the results of advancing civilisation. They form a picture with two faces. The first, a happy one; the second, an unhappy one. The bright side shows that the influence of commerce, science, and knowledge are advancing, and that by them "the sword shall be turned into the ploughshare and the spear into the pruning-hook," and gentle Heaven-born peace shall one day reign in the place of the demon — War, that has so often desolated Europe. The unhappy side of the picture shows that these large importations of foreign goods are caused to a great extent by strikes and the high price of labour in England; so that the profit and wages on them made by our friend, the foreigner, are lost to the English employer and workman. To the British workman, therefore, we strongly recommend a careful study of this interesting, but unhappy side of the picture.

We now come to consider the question of the minimum rate of wages. Mr. George Howell, a member of the Bricklayers' Operative Society, thus testified before the Commission: —

Q. 1702. You fix a minimum rate of wages?—Yes.

Q. 1703. In the case of a clumsy, slow, and inefficient workman, who cannot do anything like the work of a quick and adroit workman, you do not make him work longer hours, or at a lower rate of wages?—No.

Q. 1704. The maximum is paid to the superior man, not by an additional rate of wages, but by allowing him an an hour's work which he does not in fact do?—That is frequently done; but I admit that in many instances more money is paid.

Q. 1705. In point of fact then, there is no encouragement for men of ability and skill to work more?—I very much regret to say that is true.

Q. 1706. You think it prejudicial, not only to the general interests of your particular trade, but to society?—Yes.

Mr. Coulson, member of the Operative Bricklayers' Society, answers:—

Q. 1479. Supposing two bricklayers to be engaged in the same description of work, the one being a very adroit workman, and the other a very slow and feeble man, have they the same rate of wages?—They have both the same rate of wages. Again, Mr. Howell answers:—

Q. 1718. What do you understand by the minimum rate of wages which you fix? Does it not mean merely this, as much as you can get for the trade considered as an average?—Yes.

Q. 1719. Your complaint of the masters, as I understand you, is that, after being compelled by the union to give as much to all men as the trade can get, they will not pay something more for skilled workmen? —Yes.

Q. 1720. Does the minimum rate mean the lowest rate which a skilled workman is paid?—Yes; which we think he should take under any circumstances.

Again, another of the men, Mr. R. Harnott, answers:—

Q. 1225. Supposing that an employer says, I should be glad to take John on at 5s. a day, which is rather above the contract, he being a very good workman, and I should be glad to take Thomas on, but not give him more than 4s., which is rather below the contract, he being a bad workman. Can he do that, supposing both of them to be union men?—So far as the society is concerned, it does not interfere with a superior man receiving what he can get, nor does it interfere with an inferior man who is not up to the average.

Q. 1226. But would they leave the employer to decide that, or would the lodge decide it for him?—The lodge, by the evidence of the men working in the same shop.

On this principle, therefore, a good workman is paid no more for his work than an inferior one. Of course the unions say the masters ought

8

to pay him more than the minimum rate, but the masters answer very justly, "You by your rule fixing a high minimum rate compel us to give an inferior workman more than he is really worth, therefore we are not able to give more to the superior workman. If you wish it, we will pay them according to their respective merits, by adding to the wages of the good workman what we deduct from the wages of the inferior one."

To see the injustice of the minimum rate, let us suppose for argument's sake, that a carpenter or mason, after he had finished his week's work and received his pay, went into a tailor's shop and said, "I want, sir, to buy two coats." The tailor immediately produces two, one of superior cloth and make, the other of inferior cloth and make. "There, my good fellow, there are two coats price three pounds each." The workman examines them and immediately fixes on the good one. "I'll take this one and leave the other," says he. "Oh no, you won't," says the tailor; "if you take the one you must take the other." "What!" says the workman, "do you charge the same price for an inferior coat that you charge for a superior one?" "Yes," answers the tailor; "and why shouldn't I? You sell your labour on the minimum scale, and I sell my coats in the same way; so the bad must go with the good." Now, the workman very naturally and very reasonably objects to pay the same price for an inferior article as he pays for a good one; and for precisely similar reasons the employer is justified in refusing to give the same price for inferior workmanship that he gives for superior. You pay more for a good lawyer than you do for a less able one; more for an experienced physician than for a less experienced; more for a highly talented than for a less talented artist; and so on through other professions; and this is as it ought to be. But the practical working of trades-unions is to deny this, from their overlooking the important fact that men are variously gifted. You will remember the beautiful parable where our Lord likened the kingdom of heaven unto a man travelling into a far country, who called his servants and delivered to them his goods. Unto one he gave five talents, to another two, and to another one; he gave to every man according to his several abilities. The master then departed, and the servants went and traded with their talents. On his return he required from them an account of his money. The one that had received five talents came and brought other five talents; he that had received the two talents made two more; while the third, instead of usefully employing the one talent, went and hid it in the earth. And the same lesson is conveyed in the parable of the nobleman, where the servant that had gained ten talents was made ruler over ten cities, and he that gained the five talents was made ruler over five cities. We learn, then, from this, that men are of different capacities, and according to that capacity do they receive, and according to their work are they rewarded. Now, trades-unions ignore all this. They try to make it out that there is little or no difference between the talents of their artisans; that each must do a certain amount of work and no more; and that all are to be paid alike. You will, therefore, see at a glance the difference between the Divine system and their system. The first is natural, perfectly suited to man's capacities; and for the free and full development of his powers, and that in proportion as he labours in that proportion is he paid. On the other hand, the system of the trades-unions is unnatural and artificial. It does not properly develope the faculties, but tends rather to cramp them by placing restraints on a workman as to what speed he shall work at. It says that the labourer must not do a bricklayer's work, or a bricklayer a mason's work, even although they are quite able to do it, and which, logically carried out, means that a labourer, a bricklayer, and a mason must continue a labourer, a bricklayer, and a mason to the end of his days; that piecework and apprentice work are to be limited, although the master is quite willing to take more of them; and, lastly, that each class of workman is to be paid alike. Now, all this is the merest folly and absurdity; and if you turn for a moment to Nature, you will find that the Creator there, too, protests against it, and corroborates in His works the teaching of His Divine Word. "There is one glory of the sun, and another glory of the moon;" and as "one star differeth from another star in glory," so one mind differeth from another mind in power, and one hand differeth from another in strength and skill; and Nature and Reason, as well as Scripture, repudiate the very idea of all these being measured by the same fixed standard, or rewarded on the same fixed scale. The talents of the artisan, like the talents in any other business, ought to be paid according to their real worth and intrinsic value. They ought to stand solely on their own merits, and not be dragged down by the skill of a second, third, or fourth-rate character. And when they are not so considered, a serious injury is done to their owner by taking from him that incentive which is ever urging him on to rise higher and higher in his profession. He says to himself, "What is the use of all my efforts to improve? I have no encouragement; for I am paid no more for my work than my mate here, who seldom or never makes any exertion, and who has little or no energy and activity in him." It is not the master's fault the good workman is not paid more. It is the fault of the union in fixing a minimum rate; and until that mode of payment is abolished, the really good and deserving workman will be underpaid. Besides, the minimum rate is unfair to the inferior workman as well. When, for instance, trade is brisk, the employer takes on say 500 men; but when trade becomes slack he requires to pay a certain number of them off, and who amongst them does he discharge? Why, of course, the inferior workmen, and he keeps on the good ones.

Mr. Geo. Smith, of the firm of Smith and Taylor, contractors for the Foreign Office, thus speaks in relation to this point, in his answer to Question 2823:—"I think that the greatest mischief, however, arises from the fixed minimum. It throws a vast number of men out of work who would be receiving fair wages, because in every establishment there are men, whom I will not call inferior, but who are physically incapable of earning as much as others; for instance, a man may be so from age. The consequence of our being obliged to pay almost a uniform rate of wages is that the moment there is a reduction, the shops or the works are gone through, and all these poor men are picked out and discharged. I mean those who are receiving a greater rate of wages than they are fairly entitled to."

Q. 2851. I think you stated that the effect of a uniform rate of wages was injurious to the skilled workman by taking from him his zeal in the execution of his work, inasmuch as he got insufficient remuneration for it?—Yes. I do not think there is that incentive which he would otherwise have; and it is also a very great hardship upon the less skilled, those who may happen to have met with an accident, or who are not so active as they were, because they are as it were a sort of outcasts, who go from shop to shop, and ultimately come to a workhouse. I believe that all these attempts to carry out an artificial system are injurious alike to men and masters.

Q. 2854. So that there is first an injurious operation upon the superior workmen, because they are prevented from earning according to their worth; and at the same time there is an injurious operation upon inferior workmen, because in times when work is scarce they must be the first discharged?—That is my view of it. It is far better that all should be employed, than that a certain number should be wandering about subsisting on alms.

10

There is something delusively attractive in the idea of a minimum price of labour being fixed, inasmuch as it may be said, with some show of reason, if you employ a man at all you must pay him not less than such or such an amount necessary to his comfortable subsistence. But the fallacy of such an idea appears at once, when we reflect that the natural—that is, the Divine—law will give a man what the state of trade will allow, and anything more than that would cause the trade to stop by hastening the ruin of the employer, and therefore the stoppage altogether of the wages of the workman. Low wages warn the operative away from the particular branch of labour in question, which in time will produce a rise, because the demand will continue and the supply decrease; but an artificial elevation of the scale of wages invites more labour even when the trade is fully stocked, and hence tends to perpetuate and increase the downward tendency. We can thus see in their operations the essential difference between the Divine law and the artificial Union-made law. The first, like all God's laws, is not only a law of wisdom, but a law of mercy, for it acts like a beacon to warn the workman away from danger; the second, like a false light, luring and tempting him into danger.

Look at this question how you will, and from any point you choose · turn it inside out, and examine it with microscopic closeness and keenness, and you will come to the same conclusion. The idea of fixing a minimum rate was conceived, in order to get by fair or unfair means as much as possible out of the employers; but like everything begotten of selfishness, it has outwitted itself. The mind's eye of the individual who originated the idea must have been afflicted with a squint: he could not have seen straight; and in attempting to cheat others, he has cheated himself. By the inexorable law of the trades-union, the poor workman is bound to take according to the minimum principle. He has no alternative. All are to be paid alike, good, bad, and indifferent. Classical tradition tells us there lived once upon a time a famous robber, named Procrustes; and this robber had a bed on which it was his cruel practice to lay all his unfortunate captives. If they were too short for the bed, he strained and racked their limbs until they attained the required length; and if they were too long for the bed, he cut and maimed and mangled them until they exactly fitted in. The trades-unions of England are like unto this robber; for in them they have the bed of the minimum wage, and on which it is their cruel practice to lay all their unfortunate members, and by which they are measured and mangled by this modern Procrustes

The endeavour on the part of the unions to fix the price of labour according to an arbitrary rule is very similar in character to that made in olden times, when the justices of the various counties fixed the labourers' and workmen's wages. On this subject we have some very curious and interesting information. About the beginning of the year 1685, the justices of Warwickshire, in the exercise of a power intrusted to them by an Act of Elizabeth, fixed at their quarter sessions a scale of wages for the county, and notified that every employer who gave more than the authorised sum, and every working man who received more, should be liable to punishment. The wages of the common agricultural labourer, from March to September, were fixed at the precise sum mentioned by Petty—namely, 4s. a week without food. From September to March the wages were to be only 3s. 6d. a week.

Again, "The magistrates of Suffolk met there in the spring of 1682 to fix a rate of wages, and resolved, that where the labourer was not boarded, he should have 5s. a week in winter, and 6s. in summer."

Again, "In 1601 the justices at Chelmsford fixed the wages of the Essex labourer, who was not boarded, at 6s. in winter, and 7s. in summer."

11

The wages when so fixed were announced by proclamation at the market-places, and were to continue until the justices thought proper to alter them.

About the year 1610, a bailiff of husbandry received 52s. a year, or 1s. a week; an agricultural man-servant of the best sort, 50s.; a common servant, 40s.; a middling servant, 29s.; a boy under 16, 20s., and so on. A mower received 5d. per day with his meat; a man reaper, haymaker, hedger or ditcher, 4d; a woman reaper, 3d.; a woman haymaker, 2d. If no meat was given, these sums were to be exactly doubled, except that the woman haymaker was to receive 5d. instead of 4d. A master carpenter's wages was 8d. per day with meat, or 1s 2d. without; a chief joiner or master sawyer 6d. with meat, or 1s. without: bricklayer 5d. with, or 9d. without, meat; and tailors 4d. with, or 8d. without.* These were the rates of wages at that period as gathered from an account printed in the *Archæologian*, and fixed by the magistrates for the county of Rutland, and which continued from then nearly down to the breaking out of the Civil War. One can scarcely conceive how the poor labourers and workmen existed on such miserable allowances; and their condition was intensely aggravated by an Act of Parliament passed in the reign of Elizabeth, in which it was declared that "The givers of greater wages than what were allowed by the justices were liable to ten days', and the takers to twenty-one days' imprisonment."

Now, this savage and inhuman system has long since been abandoned. Justices of the present day would never think of fixing the price of the workman's labour, and still less would they think of putting the man in prison because of his taking a shilling or two more; but the trades'-unions wish to do something similar. The wages are by them to be fixed by an arbitrary and artificial law; and when so fixed, the men may take more, but they must not take less. English employers have long ago seen and acknowledged the folly of such a proceeding; and surely English workmen will never be so foolish and insane as to adopt laws and rules which experience has proved to be utterly worthless.

Before leaving this part of the subject, we would like to compare the price of labour of this day with that in times past. It is very usual to hear it said that all classes of the community are, and have been advancing, except the labouring portion. The middle and higher classes, we are told, are increasing in wealth, but the poor working man still remains at starvation's door. There never was a greater fallacy foisted on the public than this; and the slightest reference to history proves the contrary. Compare, for instance, the rates of wages as quoted above with those at present received, and, making allowance for the depreciation in the value of money, it will be seen that the wages of the working classes have vastly improved. And this improvement has not been owing to strikes, but to the gradually increasing wealth and prosperity of the country, the results of Free Trade and wise legislation. "I appeal to every man," says Mr. Bright, speaking on the abolition of the Corn Laws, "who knows the condition of the country, if it is not the fact that in a majority of cases the rate of wages has been advanced from 20 up to 50 per cent. There is not an agricultural labourer—and he is the poorest paid labourer in the country — whose wages have not been increased, whose comforts have not been made more secure, by the abolition of the Law which was supposed to give ample protection to that interest." † Unfortunately, while their wages have been increasing, their education, moral and intellectual, has not been equally attended to, and hence we have much discontent. Some among them have taken advantage of this state of things, in order to sow the seeds of discord

* "Pictorial History of England," vol. iii., p. 659. Also "Eden on the State of the Poor," pp. 125—127.

† Speech of Mr. Bright at Birmingham, October, 1868.

between the men and their employers. These parties are devising all sorts of plans and schemes for the benefit of the working man, except the right kind, and diverting his attention from the quarter where improvement ought to begin. Richard Cobden, one of the best and truest friends the working man ever had, says on this subject, "I wish to see the great mass of the working classes of this country elevate themselves by increased temperance, frugality and economy. I tell you candidly that no people were ever yet elevated except through their own advancing wealth, morality, and intelligence; and any one who tells the working men of this country that they may be raised in the social scale by any other process than that of reformation in themselves, is interested either in flattering or deceiving them. If you trace the rise and progress of the middle classes in this country, you will see that in proportion as they became intelligent, as they accumulated wealth, and as they improved in morality, did they rise in political and social importance." Mr. Cobden wished to see temperance, frugality, and economy practised amongst the working men, while Mr. Allen, of the Amalgamated Engineers, shows how highly he appreciates these virtues by spending £40,000 of the money of the Society of Engineers the first six months of the lock-out of 1852; and, from his own confession, the men were even worse at the end of this strike than they were at the beginning.*

Beware, then, Working Men, of your advisers. It is not by warring with your employers, and casting your hard-earned savings into the bottomless pit of strikes, that you can improve your position. Your little will be made to go further by "temperance, frugality, and economy;" and the comforts of your homes will be sweetened by contentment. Remember the advice of your true friend, Richard Cobden —advice which he embodied in his own pure and unselfish life—and whose "unadorned eloquence" pleaded your cause and cheapened your bread, that it is by the practice of honesty, industry and intelligence in all your daily duties, that you can better your condition; thus rendering your services indispensable to your employers, who, sooner than part with faithful servants, will rather part with higher wages for you to remain. It is in this way, Working Men, you can truly raise yourselves in the scale of social life.

We come now to another evil connected with this branch of the subject, and that is the unions not allowing a man to accept less wages if he is willing to do so, and forcing him to support the union. The union acts the part of the tyrant to the employer in not allowing him to give less, and it performs a similar duty to the workman by not allowing him to take less. Mr. Trollope thus answers

Q. 2912. I think you said that there was a disinclination on the part of the unions to allow inferior workmen to take less than a certain rate of wages?—They cannot do it; they would be worried out of their lives.

Q. 2913. If men were to come to you and say, "We are willing to work for 4s. 6d. a day," while the standard rate was 5s., they would not be allowed to do it?—At the present time we should not entertain such a proposition; in fact we shouldn't be able to do it. I have known a case in our shop where an old joiner has been at work, and we have been paying him less wages, and he has said, "I am sorry, sir, I cannot stop, but the men worry me out of my life. I know I am not worth what they are; but I am not allowed to take less wages."

Q. 2914. And if he persevered in working with you without the consent of these men, they would probably strike?—Probably they would strike, but they would first of all try to worry him out.

Q. 2915. And in the event of their failing to do so, they would strike?—Yes.

* See evidence before Royal Commission.

Q. 2916. Do you know anything about picketing?—In 1859 our works were picketed morning, noon, and night. Several of the men who came up from the country said, "I am sorry I cannot stop; but when I go out the men follow me, my wife is annoyed, and my children get hooted at in the street; and therefore I really cannot stop."

Q. 2917. So that, in short, although there is no actual physical violence towards the man, there is that sort of social violence which prevents him following his employment?—Quite so.

This is the evidence of the masters, and it is corroborated by the men themselves—in fact, their rules sanction it. The Manchester Operative House Painters' Alliance have a rule in the following words:—"That any person when asked to join the association, being obstinate, and causing delegates, deputations, or any other expenses to be incurred by their obstinacy, shall pay all such expenses, together with the amount of the entrance or re-entrance fee, as the case may be."

Rule 23 of the Bradford Plasterers' Union says:—"That no plasterer be allowed to work in any shop more than six days without giving satisfaction to the shop steward that he is a member, or will join the society." And we will here give you an instance, showing to what a pitch of audacity these plasterers had risen in enforcing these tyrannical rules. A Mr. Howroyd, of Bradford, had four sons working with him. They did not belong to the union. The union men thought they ought to subscribe to its funds, and for this purpose they sent the following letter to the father, dated 7th January, 1867:—

"MR. C. HOWROYD,—We, the operative plasterers of Bradford, do hereby give you notice that all your sons that are working as plasterers which are above twenty one years of age, are requested to join the society on or before Saturday next; and failing to do so, all our men will cease work on Monday morning next, and not return again unless you pay all the expenses of the strike.—We remain, yours, "THE OPERATIVE PLASTERERS OF BRADFORD."

Mr. Howroyd's sons did not wish and would not join the union, and their father would not force them to do so; therefore the men struck work. In relation to Mr. Howroyd's case, we may also notice that one of the reasons given why the men struck was that he did not obey Rule 27 of the Bradford plasterers, which is in these words:—

"That no member of this society be allowed to work for any firm consisting of more than two masters, unless every partner's name is publicly stated in full."

Imagine the clerks or the warehousemen of the City, or Messrs. Barclay, Perkins, and Co.'s draymen going to their employers and addressing them thus:—"Gentlemen,—Unless you put up on your signboard in front of your premises the names of each of your partners, we shall be compelled to strike work." Here is another case which occurred on the 1st March, 1867:—

"Mr. Foster,—Sir,—You have three men in your employ that do not belong to any society. I am requested to inform you, that if they do not become members of a society on or before the 11th of March next, your shop will be called a black shop, and no society man will have to work in it.—I remain, yours respectfully,

"J. HAYES, Secretary

"Blackpool Operative House Painters' Association."

At Bath a similar case occurred to Mr. Joseph Bladwell. A workman went to him, and said, 'You have got a non-society man in your employ; if you don't discharge that man, I will call off all the rest of your men, and none of them shall work for you." This time the union reckoned without its host. Its emissary was taken before the magistrate on the charge of interfering with the master's business, and he received a month's imprisonment. Now we could fill whole sheets with cases of this kind, and not confined to one particular trade or locality, but to almost all the trades, and from every part of the country, from Manchester, Liverpool, Bolton, Derby, Yorkshire, Scotland, and even America. In some cases the men's tools have been stolen, the masters' property destroyed, and even murder itself, as in Sheffield and Manchester, has been committed. We hope, and we are quite willing to believe, that the Sheffield and Manchester outrages are exceptional, and that they ought to stand apart and be judged by themselves; but at the same time we cannot hide from ourselves the fact openly staring us in the face, that where trades-unions have been planted and allowed freely to grow, they have invariably developed the same lawless spirit, manifested the same disregard for the rights of others, and produced the same evil fruit. "Men do not gather grapes of thorns, or figs of thistles;" and as we are commanded to judge the tree by its fruit, we are driven to the conclusion, that not only the tree, but the very seeds of trades-unionism are evil. Here then are cases where a man is willing to take less than the minimum price, but the union steps in, and says he must not. Is this freedom? Is this the liberty of which English working men are so fond of priding themselves, and about which they so continually talk and preach? True liberty allows everyone to think and act for himself, without any compulsion of this kind. But this is the modern Procrustes over again. Not content with plundering the workman of his wages by his absurd rules and strikes, he is now robbing both him and his employer of their liberty. This spirit of intimidation is utterly foreign to every principle of right and justice, and ought to be held in abhorrence by every honest man. We have no right to interfere between a master and his workman by saying to the one, "You must pay so much," and to the other, "You must accept so much." That is a matter entirely between a workman and his employer, just as his religion is a matter between him and his Maker; and trades-unions are no more justified in interfering in the one case than a body of religious inquisitors would be justified in interfering in the other. Happily for us we are not left to the unions for light on this subject. We read—"Then came the Publicans and said unto Him, Master, what shall we do? and the Master replied, Exact no more than that which is appointed you. And the soldiers likewise demanded of Him, saying, And what shall we do? And He said unto them, Do violence to no man; neither accuse any falsely, and be content with your wages." * These words were uttered 1800 years ago; and yet so fresh and appropriate are they for us at this very moment, that they read as if only spoken yesterday. To all honest working men then we say—Abandon those false friends and teachers who are wasting your substance and destroying your character, and follow the advice of Him whose wisdom never fails.

Let us now consider another important part of this subject—namely, the rule by which no bricklayer shall be allowed to do the work of a mason, and no labourer the work of a bricklayer, although quite competent for it. We give the evidence of the men first:—

Q. 2245. Suppose that I, as a master, were to put on certain work three or four labourers whom I consider fit to lay bricks, would your members say that, in their opinion, these men were not fit to be bricklayers? Certainly; the bricklayers would know whether they were fit to be bricklayers.

Q. 2246. But, even though these men had satisfied the master, your members would still persevere in their objection to work with them? If labourers were put to bricklayers' work, they would refuse to work with them.

Q. 2247. That is to say, they would not allow the employer to judge of their capacity? They do not interfere with the master.

Q. 2248. Except by striking?—Yes.

* Luke iii., 14.

15

Q. 1285. You do not allow a bricklayer to lay a stone at all—do you?—No.

Q. 1286. Even although he can do it?—I do not know about his being able to do it.

Q. 1287. But, supposing a bricklayer can lay a stone, would you prevent his laying it?—Yes; if he professed to be a bricklayer.

Q. 1288. So that if a bricklayer were to lay stonework, and there were other work to be done on that particular job by stonemasons, they would make him undo that work before they would go to their work; and that would be consistent with your regulations, would it not?—Yes; but on the other hand, I may remark, that our men do not interfere with brickwork.

The following letter was addressed to Messrs. Worthington and Challoner, master house painters, Blackpool:—"Blackpool, April 18, 1867. Sirs,—It having been brought before the meeting on Monday night that your book-keeper has been doing some writing and painting, you are requested not to let him, he not being a painter. By doing so, you will oblige, yours respectfully, The Committee Blackpool Operative House Painters."

Here is another case, from Sheffield:—"Sept. 12, 1864. To Mr. Reynolds. Dear sir,—I am authorised by our Lodge to state that all our members are willing to turn out at the request of this Lodge, on account of your employing persons to interfere with the work of the bricklayers, contrary to our rules," and so on.

Here is another case from Ashton-under-Lyne:—"Mr. Geo. Colbeck, a joiner and builder, sent in December, 1865, a joiner, bricklayer, and two labourers, to make some alterations in a house; a door was to be removed half the width of itself. The bricklayers built up the part requiring it; the joiner, having to stand idly by while this was doing, proceeded to pull out the few bricks which had to be removed. The bricklayer struck work and left the job. For Mr. Colbeck having infringed the rules by allowing a joiner to pull out bricks which must be done by a bricklayer, he was fined £2, which he paid." Now, we could quote whole pages of instances of this kind from the evidence before the commission; some of them, from their arbitrary cruelty, raising one's honest indignation; others, from their utter absurdity, his merriment and laughter. Here is one from Nottingham. Mr. Murdy, a master plasterer, writes:—"Two years ago, a respectable man, who had formerly been in business as a bricklayer and plasterer, but who had been reduced by circumstances to obtain his livelihood as a journeyman, came to me for employment. Feeling sorry for the man, and knowing him in better days, I gave him employment. After he had been at work a few days, the trade-unionists, in my employ, informed me that I must discharge the man, as they had passed a resolution 'That as he professed to be as much a bricklayer as a plasterer, he should not be allowed to work as a plasterer, and he might go somewhere else and work as a bricklayer.' I remonstrated with them, but to no purpose. The poor man tried all he knew to retain his place of employment with me. He offered to join their trades-union, and become one of their body, if they would allow him to work for his living. He pleaded to them that he had a wife and a large family dependent upon him for bread. But it was no use; they were determined that he should throw himself out of employment, otherwise the whole body of men would strike against him. The result was, I had to come to the understanding that he should leave my employment." "Mr. William Wildsmith, of Manchester, in Nov. 1866, had a dispute with some of his bricklayers, two of whom waited on him and read to him the laws of their Society, and concluded by coolly asking him for their day's wages, amounting to 7s. each, and 5s. each for first-class railway fare from Manchester, for coming to give him what they called their orders, which he had to pay." A master brickmaker, for having worked in his own brickground, had the following notice sent to him by his workmen, the Lancashire and Cheshire Operative Brickmakers:—"And you must, if you take an active part in your brickground, contribute to this Society, and you are not to employ any non-society men in casting clay." (*Q.* 3428.)

The unions have similar objections to work done by machinery. In Question 3209, an instance is given of an employer having the walls of his building, and many thousands of bricks, destroyed by gas-tar, because the bricks were made by the machine. A reward of £20 was offered, but without effect. In Question 3217, we have another instance. Mr. Israel Brooks, a master mason of Ashton-under-Lyne, received a quantity of sawn base from Kerridge, near Macclesfield. "My men refused to fix it, being polished, as it was against the rules, and struck work accordingly. After standing out against what I considered this injustice three weeks, I was forced to submit to have the polished part defaced, so that they might polish it again by hand; and the men declared after all this expense, it was not so good as when it came from the quarry."

Another instance of a similar kind is given in Question 3362, where carved granite was brought to the men, but this time instead of defacing it, the workmen stood over the stone and did nothing, and for doing nothing, the master was compelled to pay as if all the work had been carved by them, although they had not the skill to do it. Now all this is the very essence of human folly and nonsense; indeed when reading it, one can scarcely believe that men could be guilty of such stupidity. Fact is sometimes stranger than fiction; and if these facts had not been proved from undoubted evidence, they would have defied belief. Well might Mr. Gladstone say that such things were worthy a nation of savages. The practical working of trades-unions, as thus manifested by their own naked monstrosity, is their own condemnation. Comment is unnecessary; but we would call our readers' attention to this, that all these terrible things do not spring from nothing; on the contrary, they are the results of hidden but actual causes. If you plant a certain seed, that seed will produce a certain tree, and that tree, certain fruit; and in the same way if you inseminate certain principles into the minds of men, these principles will produce certain theories, and these theories, when practised, certain results. These, then, are the results produced by trades-union principles; and if these principles become dominant, the same frightful evils will spread all over the country. We will not, however, indulge in prophecy; the picture is already bad enough without investing it with the horrors of the imagination; but we will proceed to a fuller dissection of the union system.

From this evidence there exist three evils in connection with the system. The first is; the unions seek to establish monopoly of labour;*

* This monopoly is carried out in the brick-making trade in Manchester in the most systematic, and, at the same time, the most unjust and tyrannical manner. A ban is placed on machine-made bricks, and the neighbourhood is mapped out into districts—Ashton, Stockport, and Manchester—so that brickmakers, or builders, or carters, are not allowed to contract for or convey bricks out of one district into another; and to carry out this villanous system, a staff of inspectors, or, in other words, custom-house officers, are employed and paid by the Union. It is the duty of these inspectors to watch all brick-carts, so as to prevent Stockport bricks being conveyed to the Manchester district, and Manchester bricks to the Stockport district. One of these inspectors, we are informed, had not the honesty said to be proverbial among thieves; for he was not above receiving

the second, dictation and tyranny to the masters: and the third, injury to the men themselves. With regard to the first, the unions wish to keep out strangers, in the shape of non-union men, machinery, apprentices, and also by not allowing their fellow workmen to rise in the several grades of their business. By this means they wish to confine the trade to themselves; for, they reason, the more work there is and the fewer hands there are to do it, the higher will be the wages and the more certain the employment. Here again you can easily detect that serpent of selfishness in trade and commerce, whose head was bruised when the corn laws were abolished. If trades-unions are necessary for workmen to keep up the price of labour, why not have them for grocers, bakers, and butchers? These parties are as much entitled to protection in their business, as the members of trades-unions; but if they were to combine to raise the price of provisions, the very first that would cry out against it would be those very men who are now seeking to establish protection for labour. The answer to this question was given when the corn laws were abolished; and a monopoly in labour can now no more be tolerated in this country than a monopoly in food. It is quite true, as the defenders of trades-unions observe, that there are other professions, such as barristers and brokers, where similar restraints and restrictions exist. The answer to this is short and conclusive. One evil does not justify the existence of another; it only proves that both should be removed. While, therefore, we disallow the workman the right of monopoly, we grant him the liberty of getting the very highest price he possibly can for his labour, so long as the means he employs for that end are fair and legitimate. And here let us examine for a moment the principles which guide the unions in obtaining higher prices. In a lecture on Capital and Labour delivered in their defence, Mr. Ernest Jones asks, "Is it wrong for the manufacturer to withhold his goods from the market in view of better prices? If not, neither is it wrong for the workman to withhold his work in view of better wages. Is it wrong for corn merchants, or coal merchants, or iron merchants, or cotton merchants to *arrange together* not to sell their wares under a certain price? If not, neither is it wrong for labourers, or miners, or spinners, or weavers, to *arrange together* to sell their labour under a certain wage." In answer to this, the workman, like the manufacturer, is at perfect liberty to withhold his labour, in the hope of getting a better price; but neither workman nor manufacturer is justified in resorting to *combination* to obtain that price. The two cases in these questions are quite distinct. The first is fair and legitimate; but when, as in the second, we resort to combination, or as Mr. Jones mildly terms it, "arranging together," which in plain English means commercial conspiracy, then we overstep the bounds of legitimate trading, and we destroy that honesty of dealing which ought to exist between buyer and seller. Let us suppose, for instance, that the corn merchants were to unite as one man to raise the price of corn beyond its natural level: their customers, the public, would denounce such con-

---

and pocketing a bribe from the carters, or, as one of the Commissionors happily characterised it, "he cheated his employers into doing right." From the evidence placed before the Commission as to these districts, the most frightful atrocities are revealed. Needles have been placed amongst the clay, large quantities of bricks have been destroyed, bottles of gunpowder have been thrown into the houses of obnoxious and non-union employers, while they in self-defence have been compelled to watch, armed, at night to protect their houses and property; and on one of these occasions two policemen were attacked by a number of ruffians, and one of them brutally murdered. All these terrible doings are clearly traceable as trade outrages, being testified to by several witnesses.

duct as being nothing else than extortion. The combinations of trades-unions is often compared to that of railway companies and joint-stock companies. Well, let us see how far the comparison holds good. It is quite right that a number of individuals should combine to raise the necessary capital to construct a railway; but after it is completed it would be very unjust of that company to combine with other railway companies on the same route, and then coolly to inform the public, "You are hereby informed that we have combined with our friends over the way, and raised our fares from 1d. to 6d. a mile." The combination of the individuals to form the company and raise the necessary capital, is a means for a great good; but the combination of the separate railways for the purpose of raising the fares is a means for great evil and wrong.

Again, let us suppose the joint-stock banks were in some way or other to unite to raise the price of money from its present natural rate of 2 per cent., and then to tell their customers, the public, "We, the Joint-Stock Banks, have all combined, and if you want any more money you must now pay 10 per cent." The thing is manifestly unjust. There is then no harm in a number of individuals combining to form a company, and carry on business in a fair, honourable, and legitimate manner; but there is a vast amount of mischief done when these companies unite as one body to raise their fares, or the price of their money or goods. The combination to form the single company is necessary and lawful; but the combination to unite all the separate companies is a totally different thing—a gigantic monopoly, and therefore most undesirable.

Now, it is exactly so with unions or companies formed for the selling of labour. If the unions simply sought the privilege to form so many separate companies, each of these acting independently of each other; each being at liberty to take a low price without being called in question by the others; each competing with the other, as in other trades, to keep down the price of labour instead of raising it, there would not be the slightest objection to them, but the unions want more than this. Their object is to form different societies, but all to act as members of one common body. Hence the carpenters of London have at times singled out one large building firm and demanded from it an advance of wages. If it is refused, they strike, then picket, and are supported while out of work by the different unions throughout the kingdom. The building firm has thus to struggle not only against its own men, but it has pitted against it the full power of this gigantic combination. Herein lies the sting of unionism; and the closer one examines it the less one likes it By a combination thus formed, and having ramifications and branches throughout the country, we fully believe that there is no single building firm or railway contractor, however powerful, but could be overpowered by a combination of this kind; and for that reason, therefore, we say of it, as we said of the railway and bank combinations, it is a gigantic monopoly, and therefore most undesirable. Let our law-makers then pause before they legalise a combination of this nature; for if such is sanctioned it will beget another great evil, viz —the combination of employers to resist that of the men. The conclusion we arrive at then is:—workmen are not justified in combining to raise wages; and employers are not justified in combining to lower them. No doubt this "rigging" the market, as it is called, is to some extent practised in commerce and on the Stock Exchange, and there is no written law to prevent it; but gambling in these quarters does not justify gambling amongst workmen and their employers. The natural course is the right one; to allow the prices to rise and fall just as they are influenced by the supply and demand.

Mr. Jones goes on to say:—"The manufacturer, while his merchandise is lying idle waiting for a better price, lives on his accumulated money capital. That capital is at his bankers. The trades-union is the bank of the working-man, and enables him to hoard his idle labour even as the capitalist hoards his idle goods." The truth of this we must all admit. There is nothing whatever to be said against the workman hoarding his little savings; on the contrary, it is highly commendable; and if he cannot find a purchaser for his labour at *his* price, then he is quite entitled to live on his savings until he does. Every workman is entitled to the privilege of going to this extent; but here his privilege ends. He ought not then to resort to combination, to picketing a master's premises, or intimidating by threats or violence others who are willing to serve that master at lower wages. This is at variance with all law, human and divine. When the offer of his labour is declined, then the next thing for him to do is to try other buyers; if accepted at his price, well and good; if declined, then he may shrewdly guess he is asking too much, and that it is better to accept what the market can fairly give, and live in the hope of things improving. And this is what the merchant does. He offers his tea or cotton, or whatever the goods may be, at a certain price. If the price is above what the buyer is willing to give, the goods are declined. After that, what does he do? Does he follow the advice of Mr. Applegarth,* and picket the buyer's offices? Certainly not. He then either determines to hold the goods or consider the advisability of taking the price offered; and, perhaps, while so considering, a second seller will step in and, by fair competition, cut him out of the sale by selling goods of the same, or similar, quality at the required price. Yet all three parties are as good friends after the transaction as they were before it. And this is done not only once, but thousands of times every day amongst the mercantile classes. Why, then, cannot the buyers and sellers of labour act in the same free, friendly, and charitable spirit?

Although, therefore, we contend most strongly for freedom of trade to the employer, we do not, therefore, advocate employers paying their workpeople badly. We think the fact cannot be denied that there are amongst employers many selfish men who have no sympathy with their workmen, and whose duty goes no further than the letter of the civil law dictates, forgetting that there is, above and beyond this, the moral law, or law of the conscience. Such employers grind down their workpeople to the lowest possible wages, in some cases scarcely enough to keep soul and body together. They amass large fortunes in their lifetime; and when they die, finding they cannot take their wealth with them, they pay, as it were, to enter heaven by bequeathing their money for the erection of churches, chapels, and hospitals, thus giving to man proofs of their piety and charity, but which in the sight of "Him who looketh on the heart" are only enduring monuments of their hypocrisy and meanness. In our opinion it would have been

* Not content with refusing the wages, the Union, like the dog in the manger, will not allow others to take them. Mr. Applegarth says, as to picketing, Ques. 6736: "I say it is perfectly justifiable for men to appoint other men to wait at a shop door and say to those who come, 'The men were dissatisfied with the terms upon which they were working at that place, and if you go in you will undersell us. Now, we beg that you will not do that;'" and he further adds the important information that if there were no picketings, strikes to a great extent would be useless. Although Mr. Applegarth thinks it right for the *men* to picket, yet in Question 7337 he gives the Commissioners several instances of picketing on the part of the *masters*, which he complains of as being very unjust. Cannot Mr. Applegarth see that if it be wrong in the one case it must be equally wrong in the other, and more especially when, as in the case of the men, the advice is accompanied by certain significant winks and gestures?

much truer charity to have spent that money in the better payment of their badly-paid workpeople. And one can readily believe that many honest workmen, groaning under such oppression and disgusted with such behaviour, have betaken themselves to the union, forgetting, in their haste and indignation, that they were raising up a greater evil than that they sought to destroy. We do not, therefore, defend conduct of this kind; on the contrary, we think it is for the advantage of the workman to do his work well, and for the master to pay him liberally. Selfishness on the part of an employer is, however, no valid argument against freedom in labour and trade, any more than his hypocrisy is against genuine charity and true religion. We must not confound the one with the other, but endeavour carefully to discriminate between both, and see that all laws and rules laid down between employers and employed, should be founded on truth and justice; and more especially should we do so, as there are hundreds and thousands of employers in this country, animated by the best and kindest motives towards their workmen.

With regard to the second evil, namely, the dictation and tyranny of the men to the masters, it is greatly to be deplored; but from the evidence given there can be no doubt that it prevails to a large extent. The unions dictate what work is to be done, and what men are to do it. Now this is reversing the proper order of things; for the men stand in relation to the masters as the hand of the body does to the head, and it is for the head to direct and the hand to obey. Besides, if the master pays for the work, it should be his business, and his business only, to say who shall and who shall not do it; and if he is content with the manner in which it is done, whether by apprentices or by non-union men—by labourer, bricklayer, or mason—whether by hand or by machinery—why, in the name of common sense, should the men find fault? It is the men's duty to do their work, and leave the master to do his; and if he does not do it properly, he, not they, will be the loser, and he will soon find out his mistake if he has not done it rightly. The men might with equal justice, after the master had paid for his dinner, go and tell him that a certain cook must cook it for him; or, after he has paid for his cloth, that a certain tailor must make his coat.

The third evil referred to is the great injury the men do to themselves by these rules. They say that a man shall only learn one branch of the trade: that if he is a labourer, he must not do a bit of bricklaying or mason work; and that if he is a plasterer, he must not turn his hand to painting or carpentering. Now, what is this but to stunt the growth of the man's faculties, and to deprive him of the means of earning his livelihood? It is to exchange the Divine law of liberty for the Hindoo law of caste. For if trade is bad in one thing, he can, if he knows more than one branch, turn his hand to another, and thus, as one door shuts another door opens for him. But to this trades-unions say, "No; he must be a bricklayer, a mason, a painter, or a carpenter," and such he must continue. Again, how often do we meet with in life instances of young men being apprenticed to trades quite unsuitable for them; or, as the late President Lincoln aptly called it, where the square peg is put into the round hole, and the round peg is put into the square hole; where, for instance, a man whose taste is for carpentering is put to making boots and shoes; another, who would make a very good shoemaker, finds, by mistake, he is on the tailor's bench. Even these mistakes are not confined to the lower classes; they are to be found in the higher. Let us suppose a case, and such a case is not beyond the bounds of probability, of a young man, pious, amiable, and of a noble house, the son of one high in the Church and State; he tries the law, but finds he has not sufficient application to master its studies, or ability enough to hold a brief. The profession is not to his taste, so he leaves it and enters the army. Here, in time of peace, when the soldier's duty consists in acting the gentleman, he gets on very well; but "wars and rumours of wars arise," and he is ordered abroad. Alas! it is found he cannot go. No one for a moment doubts his courage, for he is as brave as a lion; but then his physical constitution is unable to endure the fatigues and hardships of campaigning; so he doffs the red coat and dons the black—he quits the army and enters the Church. Here he continues, learning to labour and to wait, until a merciful Providence, through parental solicitude, kindly interferes and helps him into a bishopric; so that he who had made a bad lawyer, a feeble soldier, and only a moderate preacher, at last makes a most amiable and excellent bishop. How great, then, would have been the loss to his country had the early prospects of this promising young man been blasted by the arbitrary rules of an aristocratic trades-union! Like the bishop, then, the middling labourer may be a very handy and intelligent man, and make, perhaps, a better bricklayer than the bricklayer he is under, or the bricklayer may be better suited to the work of the mason than the mason himself—and the same holds good with reference to plasterers, painters, carpenters, joiners, and so on; but the working of these arbitrary rules of trades-unions prevents a man from rising, by forcing him to that for which he is naturally unsuited. This, then, is a great mistake, and a serious injury done to the men themselves. And here we wish especially to point out the beauty and excellence of the divine laws in relation to this part of the subject. The law of heaven is that each one follow the bent and inclination of his own peculiar talent and genius, because in heaven restraints and restrictions are unknown; there the most perfect freedom reigns, and by this the individuality of the angelic mind and character is most fully and beautifully developed. Let us, then, try to make earth like unto heaven, by sweeping away all monopolies and unnecessary restrictions, by giving to every one—high and low, rich and poor, in every profession—equal opportunity for the development and reward of his talents. As the divine parable expresses it, let him that hath the ten talents be made the ruler over the ten cities, and him that hath the five talents the ruler over the five cities.

Let us now briefly consider the subject of the limitation of apprentices. We need not quote instances to prove that rules against apprentices exist. They are acknowledged by many if not by all the societies; and the reason given by Mr. Potter is, that where there is no limitation of apprentices, inferior articles are manufactured, work of every kind deteriorates, it drives skilled workmen out of England to foreign countries, where their superior work supplants our inferior labour to the injury of our export trade; and slop-work, the sweating system, and many other so-called evils are the result. As regards the slop-work trades, we would only say in extenuation of them, that they, especially those of the eastern districts of London, have, like that of the costermongers, been called into existence to supply the want of a certain class of the community; and if any one is to blame for the deterioration of the work and wages, we think it is the public who have deteriorated the prices; and this sin, so terrible in the eyes of Mr. Potter, may readily be forgiven, considering the poverty of those who have committed it. But, supposing you apply Mr. Potter's remedy by lessening the number of apprentices, you would only aggravate the evil by making the goods dearer, and therefore less accessible to the poorer classes; and when the price rises, the demand decreases, and the employment with it. But in these trades, and in all other trades, we think the master should be at liberty to employ as many apprentices as he thinks proper. If he is quite willing to purchase the inferior labour, why should another step in and say he must not? On the same principle you might refuse to allow a man to purchase an inferior house or inferior clothes. Many are able but not willing, and many are willing but not able to purchase superior houses and clothes; and whether they are able or not, they ought to be the only parties to decide whether or not they shall buy. And if you allow inferior houses and clothes to be sold, on the same principle you ought not to step in and say to the seller, "You must not employ inferior labour to make them." The very reason of these things being cheap is because the labour that produces them is cheap; and if you take advantage of the cheapness of the article, by all means allow the maker to have the advantage of the cheap labour. The employment of apprentices, like all the other laws on this subject, should be quite open and free; and, when they are so, the matter will regulate itself. There is no fear that the master will employ too many apprentices; for no master can with advantage to himself or to the public he serves, employ too many lads. If he does so he runs the risk of being cut out of the trade by the inferior quality of his work. It is for his own interest, then, to employ a limited number; and, as we give the public the liberty of deciding what quantity of apprentice-made work they will buy, so we ought to give the employer the liberty of deciding his own quantity of apprentice labour. If you curtail the liberty in the one case, be consistent, and curtail it in the other.

This reason of the deterioration of the work is only the *apparent* reason of trades-union opposition to apprentices; it is not the *real* one. The true reason is given in answer to question 1102.—And why do you interfere with the master who determines to use the labour of his apprentice in preference to that of his journeyman?—Because it generally leads to annihilation of a regular journeyman's usual labour.

Q. 1103. That is to say, it diminishes the wages of the journeyman? —It does.

Q. 1104. Why should you interfere with me, when building a house, so as to prevent my building it at a cheap rate?—The question certainly is rather a powerful one. I can only say that the custom is adopted throughout the trade.

The secret of trades-union opposition to the employment of apprentices is to be found in these words, "it diminishes the wages of the journeyman": that is where the shoe pinches. Had it not done this we should never have heard a syllable about the deterioration of the work, or of slopwork, or the sweating system, &c., &c. The men wish to keep out apprentices in order to keep up their own wages. This again is another instance of the spirit of selfishness running through all these trade rules. The master should be allowed to buy apprentice-labour, and apprentices should be allowed to sell it. And why? Because the apprentices have to live as well as the men. "Self-preservation is the first law of nature;" and although an overabundance of labour may at times cause a scarcity of work, it is rather too much to tell the apprentices to keep out of employment and starve, so that the men may keep in employment and live.

By limiting the number of apprentices you produce another evil; and this was pointed out to Mr. Potter and his friends when they waited on Mr. Gladstone. He said, "All attempts to limit the labour of boys are attempts to limit the aggregate of the wages fund of the country." The cry at present is, The working classes have too little to live upon. The trades-unions, by their short-sighted conduct, are making that little less. Now, what would you think of the man who continually complained of illness from want of food, if you saw him try to cure his complaint by still more lessening his daily nourishment? The working-man of England is such a man, afflicted with such a complaint, and, having called in the doctor in the shape of the trades-union, he is having this extraordinary prescription administered to him in the vain hope of making him strong. What would you think of such a man? Would you not say the patient is a fool, and his doctor a quack?

We will now briefly consider another rule of the union—that of "chasing." Chasing means a workman working too fast, doing his work too quickly; in short, being active and expert in the execution of his duties. Now, the unions have a strong objection to this; and the reason given against chasing is, that an active workman may put himself to inconvenience, and also to distress (Quests. 1145-6), and he may also distress his fellow-workmen by working too hard. How remarkably solicitous these rules are framed for the welfare of their members! One would hardly have thought the union would have taken any notice of a trivial matter of this kind. But it does; for the men being so remarkably willing to work, and their constitutions so remarkably delicate, it is necessary to prevent them from over-exerting themselves. We would beg to suggest to the unions that they should have a rule laying down what quantity their members should eat and drink. The unions have great doubts as to one workman being able to do more than another; but in the matter of eating and drinking we think they must admit the men have different capacities; so that a rule is more required to regugulate their appetites than to regulate their work. For our own part, we do not think a rule is required in either case. The man, in both cases, is the best judge how much he is fit for; and it is a piece of presumption for the unions to meddle in such matters at all. The above is the *apparent* reason for the rule against chasing; but we will now give you the *real* one, and we quote it from the fifth rule of the Bradford Labourers' rules, which is in these words:—"You are strictly cautioned not to overstep good rules by doing double the work you are required (by the society), and causing others to do the same in order to gain a smile from the master. Such foolhardy and deceitful actions leave a great portion of good members out of employment the year round. Certain individuals have been guilty, who will be expelled if they do not refrain." (Ques. 3120.) And the Leeds bricklayers have a rule inflicting a fine of 1s. on any member carrying more than eight bricks (3122). Here, again, you are let into the secret of the matter. The union does not care a straw about its member's delicate constitution, or his hurting himself by over-exertion; but when he does his work quickly he lessens the labour, and therefore the amount of the wages, and it is here that the shoe pinches. Here, again, you have the spirit of selfishness showing itself; the work is to be spun out as much as possible, so that the master may be made to pay to the very uttermost farthing.* His interest is not consulted,

* The following instances will show how the work is spun out:—By the rules of the Bricklayers' and Brickmakers' Unions of Manchester, it is enacted "That any man found running or working beyond a regular speed, or trying to run off or take advantage of their fellow-workmen is to be fined the sum of 2s. 6d. for the first offence; for the second offence, 5s.; and for the third offence, 10s.; and if persisted in, to be dealt with as the committee thinks proper. Any man working short-handed, without man for man, will be fined or punished as the committee think proper." It is also enacted by this union, that the walking rate of the men be three miles an hour, doing more entails a fine. Here is a case that caused a strike. The master at Birmingham was doing some heavy work at the waterworks on the canal side, he put some planks across from the boat, and set some labourers to wheel the bricks from the boat right down to the place where they were to be used; but the labourers struck against it, and said that they must all be unloaded in the ordinary way, and stacked on the canal side, and then carried by hod down to the works. (Question 3460.)

and the true interest of the men is overlooked; both must give way to the will of this Prince of Selfishness. Now, what is this rule but an artificial barrier, raised to prevent the workman actively and faithfully doing his work? And what is the tendency of all such rules but to make the man an eye-servant by raising in his mind the idea that his interests are opposed to his master's interests, and the less he attends to his master's interests the more he attends to his own? Nothing can be more inimical to both parties than teaching of this kind. An honest and intelligent workman will endeavour to work cheerfully at his duties, and attend to them as thoroughly as if he were himself the master; and a good employer will show his appreciation of such services by advancing and rewarding his servant. The interest and success, then, of both parties are bound up together; and anyone who steps in to separate the two is an enemy to both. No doubt there are selfish men among the employers, who would not recognise faithful service from their man beyond the mere payment of his weekly wages; but even in such a case the man's duty remains the same, and, whether it is recognised or not, he is none the less bound faithfully to discharge it. Again, sometimes the workman may have a doubt as to how far his duty goes; such, for instance, in what consists a fair day's work for a fair day's wage; and perhaps he cannot rely on his master's giving a just decision in the question. What, then, is he to do under the circumstances? Is he to surrender his own manly independence and freedom, and his own right of private judgment by appealing to the tribunal of a secret and irresponsible body? By no means. Every honest and intelligent workman is to ask *himself* what is right and necessary in such a case. He is to be guided by the light that is *in* him, and according to that he is faithfully to act. For a trades-union, then, to step in and thrust its opinion and will on the workman, is to usurp the place of that conscience which God in his mercy and wisdom has planted in his heart.

We now come to consider one of the greatest evils, we might term it the greatest curse, arising from the action of these unions, and that is the antagonism it has raised between the employer and his workmen. It is indeed painful to read the evidence on this part of the subject. Formerly the man went direct to the master and arranged the matter: if they did not come to terms, there was nothing more than a simple refusal between the two. If they agreed and the man went to work, the master naturally began to take an interest in his workman; and the longer he continued in his service the more that interest increased, the master treating the man with greater kindness, and the workman returning that kindness by respect to his employer and attention to his interests. Now all this has been very greatly changed. The union has stepped in between them; and, as the workman cannot serve two masters, he has, in an evil hour, cast off allegiance to his employer, preferring the allegiance of the union. Hear what the employer says:

Q. 2904. As far as I and my partners are concerned, and I believe the generality of the masters, we are very desirous to consult the comfort and the advantage of our workmen. But that feeling, unfortunately, has changed of late years. In my younger days, there used to be some sort of attachment between master and men, but that has entirely gone, and I say that it is the unions, and nothing else, that have brought that result. In the strike of 1859, men came to us who had worked at the place for thirty or forty years, and said to us, "This is the saddest day that ever happened to us in our lives, but we must go; we are bound to go."

Q. 2905. You believe that the effect of these unions is to loosen the tie between the workman and the employer?—That experience has most decidedly been ours. When we find that we cannot come to these men and talk to them as friends; that, in fact, their individuality is lost because they are members of a union, it seems to me a most distressing thing. I have talked to them and argued with them; but in two or three days' time it is all obliterated by the influence of the union.

Q. 2906. In the absence of any operation of the unions when masters are left to themselves, the tendency is for each master to keep his men as long as he can, is it not?—Most decidedly.

Q. 2907. On the other hand, when the union tries to screw every penny out of the master that it can, it destroys the very essence of the relation which should subsist between them?—Yes.

Q. 2908. It is, then, natural for the master to say to the men, "Having got out of me everything you can, you have no right to ask me on social considerations for anything more?"—Yes. I have heard many masters say, "I used during the winter to keep on as many men as I could, even at my own personal loss; but now these men care nothing for me." In fact, the effect of the unions is to induce them not to do anything for my benefit, and, therefore, why should I put my hand into my pocket for them?

Q. 2909. That feeling is general throughout the trade?—I am afraid it is.

Q. 2910. And it is increasing, you fear?—Yes, every one notices that there is a difference in the very behaviour of the men: some hardly address you with ordinary civility.

Q. 2911. So that the effect of these unions is to increase the antagonism between the interest of the master and the interest of the workmen?—I think so.

This is the evidence of Mr. Trollope, and, if necessary, it can be corroborated by other witnesses; but this is unnecessary: it tells its own tale, and is so simple and so painful, that no one can for a moment doubt its truth. The employer and his workmen were intended to be somewhat in the same relation as husband and wife; and when they had any little differences, as no doubt they would—for "mistakes will occur in the best regulated families"—they ought to have arranged them between themselves; but the wife in this case, in order to correct her husband, has abandoned her legitimate protector and taken up with a stranger and intruder, and since then there has been nothing else but rows in the house. Whatever, then, may be the employers' faults, they do not justify the course taken by the men; and the first step towards restoring peace and harmony to the house is to turn the intruder out.

Having said thus much on the rules of the unions, we now proceed to the subject of Strikes; and as a fitting introduction to it we will give you an idea of how the money of the workman is not merely spent, but squandered. In question 1998-9 we are informed by the secretary, Houlsey, that "a Delegate meeting, to revise the laws of the Operative Bricklayers' Society, cost £1088. Each delegate was paid 11s. 6d. per day, and their lodgings and railway fares, and 6d. each for refreshment, and that meeting occupied a fortnight." Thus, reckoning seven days to the week, each delegate received £4 per week, besides the expense of his lodgings, railway fares, and refreshments. If the revising of the rules cost £1088, what did the framing of them cost? However badly the men may be paid, the delegates, in this instance, have taken very good care not to allow their services to go unrewarded.

In question 6522 we find that Mr. Applegarth's printing and stationery account, for the year 1866, amounted to £1052 17s. 8d., and that the delegates were paid £567 16s. 2d. In explaining this item of stationery, he says:—"*Q.* 7276. It will be remembered that the item of printing and stationery was pointed out as not economical. Now, I may state, in that very item which was selected, there were 20,000 copies of rules, which will produce a *profit of at least one hundred per cent.*, when sold." This is a most extraordinary defence, and comes with a very damaging effect from a unionist like Mr. Applegarth. His complaint is that the masters' profits are too large, and the men's wages too low; but when *he* becomes a sort of master printer and stationer, he sells his goods at the very high profit of at least one hundred per cent., and this, too, to his dear friend and brother, the poor, badly-paid working man. We very much fear Mr. Applegarth's love for the working man is, after all, like a certain lady's "everlasting beauty"—only skin deep, and rises in price in proportion as it decreases in usefulness. We would ask you, then, Working Men, now that you have got the Union, with all its expensive machinery, don't you think, like Dr. Franklin, you pay too dear for your whistle? And do you not also think that you could employ your hard-earned wages to far greater advantage amongst your wives and families, instead of wasting them in the purchase of Mr. Applegarth's high-priced stationery, or in providing food and refreshments for Mr. Houlsey's hungry delegates?

In question 964-971 evidence is given of a strike taking place because the men had an objection to work with the foreman, although he had only been a few hours in the shop. In question 961 we have the opinion of one of the men as to the worth of strikes. Mr. William Allan:—

Your experience then is that strikes have not benefited the men? They have in some instances.

Q. 962. As a rule not?—As a rule, they have been much the same at the finish of the strike as they were before it commenced.

Q. 963. With all the misery of the strike meanwhile?—Yes.

Again, in questions 3595 to 3600, We have evidence on what trivial grounds they will sometimes take place, brought about, too, by the most worthless class of the operatives. In this case, the men's time was divided into thirds, and the union determined to have it divided into fourths. The result to a steady, industrious man would not have made 2s. 6d. or 5s. either the better or the worse; yet on this trifling matter a strike took place, lasting four months, and in the end the men had to yield. This strike took place in the winter, and in the following spring trade had so greatly improved, that the men gained the advance they had before struck for, thus showing the absurdity of strikes; for it is trade, and not strikes, that regulates wages.

In question 4326 and following, we have the history of the great strike in Manchester at the New Gaol and Assize Courts. The men determined to remove an obnoxious foreman; the offence of this man being his placing an efficient gang-leader in the place of an inefficient one, which was simply doing his duty, and because of this the men struck work. They continued on strike for many months, spent £2600, besides losing all their wages, and after all they were most ignominiously beaten. One of the men (O'Neil) was examined on the subject of this strike. He was Secretary of the Manchester Lodge; and, though occupying that important office, he had a memory like a sieve—it could hold nothing; and out of him the truth was as difficult to get as it was to draw mercy from Shylock. After severe examination he admitted the fact that £920 was given towards this strike by the executive of the Union *without even consulting the members*, and also that the London Society sent their Secretary (Coulson) down to Manchester, who bribed the men to leave their employment. This strike was characterised by compulsion, picketing, intimidation, violence, bribery and corruption in their worst features, and stands out in the history of trades-unions as one of the most gigantic attempts at tyranny they ever made, as well as one of their most complete failures.

From Question 615 we learn that it cost the Amalgamated Society of Engineers £40,000 the first six months of the lock-out of 1852, besides the loss of the men's wages, and thousands of pounds were lost both to men and masters in the great building strike of 1859. Besides these, we have on record the strike in the building trade in the Eastern Counties in 1864; also the lock-out in the iron trade, arising out of the North Staffordshire strike. According to the *Times* of March 22nd, 1865, out of a population of 133,000 in the counties of Staffordshire and East Worcestershire, 33,730 workers were thrown out of employment; and out of a population of 59,128 in the North of England, 14,782; thus in this strike alone nearly 50,000 of the working classes were deprived of their livelihood. The coal strike and lock-out of 1858 is still remembered in West Yorkshire, and the Preston strike of 1853 in Lancashire; and, lastly, the colliers of Lancashire have again been out for three months this year; and after much violence and rioting, to quell which the police and military have been called in, as well as to protect non-union men, the union men have been compelled to yield and agree to the masters' terms. To carry on this strike it is reckoned, according to the *Morning Post*, 8th May, 1868, that £15,000 to £20,000 have been drawn out of the funds of the unions, or by levies on and contributions from the working classes, which sum has been uselessly spent; and, reckoning £1 a week as wages for each man out for the three months, this strike has caused the loss of wages to the amount of £240,000, besides the loss to the employers and those indirectly connected with the trade. But this is not all. Some of the men engaged in the rioting have been convicted and sentenced to five years penal servitude, and many of those remaining have sullenly returned to work, exasperated and embittered against their employers. These then are some of the deplorable and disastrous results of this strike, and it is the unions, working on the ignorance of the men, that are to be held responsible for them.

But, further, we learn from "Chambers' Information for the People," "Popular Statistics," p. 553, how wages have been gradually rising by the natural growth of trade and intelligence.

Before the early part of the reign of George III. the wages of English workmen, as given by Young in 1768, were:—

	Per week
Iron works at Rotherham	10s. 0d.
Cutlery at Sheffield	13s. 6d.
Clothworkers at Wakefield	10s. 0d.
Colliers	11s. 0d.
Colliers at Newcastle	15s. 6d.
Cotton workers at Carlisle	9s. 0d.
Shoemakers and pinmakers at Warrington	8s. 7d.
Fustian-workers, &c., at Manchester	7s. 1d.
In the Potteries (men)	9s. 6d.
Agricultural labourers	from 5s. 10d. to 8s.

By the gradual increase of trade up to 1839, Mr. Jellinger Symons gives the wages of the same classes as follow:—

	Per week.
Sheffield workmen	from 25s. 0d. to 40s. 0d.
Manchester workmen	from 13s. 0d. to 30s. 0d.
Colliers	21s.
Agricultural labourers	8s. 6d. to 10s. 6d.
Flax-workers at Leeds	17s. 0d. to 19s. 0d.

Thus, by the gradual operation of trade and intelligence wages were doubled, and in many cases trebled, in seventy years. The wages at the present time are still higher all over England. Bread is about the same price, or rather cheaper. Potatoes, much cheaper. Tea, coffee, and sugar are much reduced in price; and clothes are immensely cheaper and better.

All this by peaceable change; and if the intelligence of the people had been as further increased by education and thoughtfulness on their own parts, these results would have been more favourable still.

The contrary results of strikes may be strikingly exhibited by the following statements extracted from a paper read by Dr. Watts before the Statistical

28

Society at Manchester, September, 1861. Dr. Watts is one of the most zealous and enlightened friends of the working classes :—

The great Preston strike lasted 38 weeks ; the one at Padiham, 29 weeks ; Bolton, 6 weeks ; Ashton and district, 6 weeks ; Clithero, 6 weeks ; Blackburn, 3 weeks ; the strike of the London builders, 26 weeks ; and the late Colne, 50 weeks.

The amount in dispute in a strike is seldom more than 5 per cent., often not so much.

Let us assume that the dispute is for 5 per cent. and that it succeeds. We shall even then see that he who wins loses. The following table shows the length of time it would take to make up for the loss of time in a strike :—

The loss of	1 month	requires	$1\frac{3}{4}$	years.
„	2 months	„	$3\frac{1}{4}$	„
„	3 „	„	$4\frac{4}{5}$	„
„	4 „	„	9	„
„	12 „	„	19	„
„	$12\frac{1}{2}$ „	„	20	„

But if the strike does not succeed, the bitterness and loss are sustained with no compensation ; and here is a list of unsuccessful strikes and their results, which occurred within $2\frac{1}{2}$ years, with the exception only of Preston, which we include on account of its striking character :—

	Number of Persons.	Weeks of Strike	Amount of Loss.	$12\frac{1}{2}$ per Cent. Loss on Capital.	Subscriptions to Support.	Total Loss.
Preston........	15,000	38	£427,500	£92,625	£106,875	£627,000
Padiham......	800	29	17,400	3,770	4,350	25,520
Clitheroe......	3,000	6	13,500	2,925	3,375	19,800
Blackburn....	40,000	3	90,000	19,500	22,500	132,000
Ashton.........	22,000	6	99,000	21,450	24,750	145,200
Colne......	1,500	50	56,250	12,187	14,062	82,499
Bolton.........	12,000	6	54,000	11,700	13,500	79,200
London Building ...	10,000	26	325,000	53,125	81,250	459,375
			1,082,650	217,282	270,662	1,570,594

This is the money loss. Then the loss in morality by the men getting into loose, bad habits, and the loss in good will between men and masters are beyond all price.

David Chadwick, Esq., then Treasurer of the borough of Salford, says in a paper read before the Statistical Society, December, 1859, "On the rate of Wages in the Manufacturing Districts, 1839 to 1859 :"—

"The wages of all classes of factory operatives appear to have increased from 10 to 25 per cent. during the last 20 years. This is mainly owing to the improvements in the construction of the machinery, and the increased speed at which it is worked."

Mr. Henry Ashworth, of Bolton, has estimated that before 1842 the net earnings of the operative spinner amounted to 20s. per week; in 1859 to 30s. 10d., with some slight help by a boy, although the price for working 20lb. of yarn is reduced from 4s. 7d. to 3s. $11\frac{1}{4}$d.

Contrast these statements, all founded on exact figures. By peaceable means there had been a gradual constant rise of 2s. or 3s. in the 20s. By violent means there was wasted in unsuccessful strikes in two years and a half a million of money by the working men.

Take another illustration.

The Staffordshire colliers turned out in 1864 against a reduction of 3d. a day arising from the state of trade. They remained out four months, and then went in at the masters' terms. This strike cost as follows :—

6932 men at 22s. per week for ten weeks	£76,252
3068 youths at 9s.	13,806
Employers' loss £100 per man at 15 per cent... ..	23,076
Shopkeepers' loss on wages and profits at 15 per cent.	16,104
	£129,238

without reckoning subscriptions to the men, or loss to the public in extra price of coals.

This irrational and suicidal conduct on the part of the men has been committed under the influence of the trades-unions; and we can only find a fitting parallel to such infatuation in the annals of Eastern Paganism. It is like the conduct of the poor deluded fanatic, who, under the influence and at the instigation of his priests, insanely throws himself before the wheels of the car of Juggernaut. Beware, then, Working Men, when certain Professors of Political Economy, whose zeal in your cause surpasses their wisdom, tell you that you have the right to strike. You have the right to strike, just as you have the right, or rather the power, to commit self-destruction; but only *madmen* would exercise such a right.

And what does all this painful evidence prove? Does it not show on the very face of it that strikes will do more to impoverish the men than any lowering of wages the masters may attempt? The Recorder of the Wigan Borough Sessions, in addressing the jury on the subject of the above riots, says:—"I believe there never has been a strike, certainly not one of long continuance even when successful, which has not resulted in more poverty, privation, and suffering among the men and their wives and families, than if the turn-outs had continued to work at even far less wages than they had rejected." And with such facts before them, any one open to conviction must come to the same conclusion. In many instances strikes have occurred for the sake of 2s. or 3s. a week; and, to gain this paltry sum, the men have thrown away their weekly wages for months together; and, after all, they have not got what they struck for; and even in those cases where they have got it, don't you think they have reversed the good old proverb by giving the herring to catch the sprat? In return, then, for the immense expenditure of strikes, the men gain nothing. Looking at the subject merely as a matter of pounds, shillings, and pence, the folly of the thing is self-evident; and there is not only the loss of the money to be considered, but there is something far worse in the misery and desolation caused to their poor wives and families and homes during its continuance. Trades-unions will answer to all this, that if we put the gain on the one side, and the loss on the other and strike the balance, the difference will be found to be in their favour. Now, admitting this for argument sake, although we have proved exactly the reverse to be the truth, we ask, has that balance been honestly come by? The mere success of a thing is no justification of the means employed for its accomplishment. A man, for instance, may have made a good round sum at smuggling or burglary; but his success is no defence of the profession by which he has amassed his ill-gotten gain; and it is far better for him to remain a poor but honest man, than to become a rich and successful rogue. We must, before considering the success of the thing, consider the legality of the means employed; and we think it has been clearly shown that a workman is not justified in resorting to combination, with or without violence, to raise the price of labour, any more than the merchant is to raise the price of tea or cotton. Then again, admitting that the means employed are fair and honourable, the next question is, have they received a fair return for all the money expended, and misery endured in the strife? When a merchant engages in a speculation, he always looks for a fair profit on the amount risked; and viewing strikes in this light, the results from them are by no means worth what is ventured for them. Again, supposing when trade is good, a strike takes place and a rise is obtained, trade afterwards may become bad, and that rise is again lost, so that the union cannot hold the gain they seek. The union, in this respect, is like Sisyphus. He, according to classical fable, was, by the gods, as a punishment for his crimes, condemned to continually roll an immense stone up a steep hill, and just as he was on the point of reaching the summit, and hoped his labour was finished, the stone would elude his grasp, and roll down again. Like him, the union is condemned—we will not say for its crimes, but most certainly for its folly—to be continually rolling up the wages, and just as the modern Sisyphus gains the summit the wages roll back again. The unions may do all they possibly can to keep up the price of labour; but when trade becomes slack they cannot maintain it; and the reason of this is evident, when we bear in mind the important fact that the unions cannot create trade, any more than the Bank of England can create commerce; and it is for them to follow the wise example of the Bank, and allow the wages, like the gold, to rise and fall according to the supply and demand. It is possible they may not by this method gain so quick and so high a rise as in some cases they gain by means of a strike; but the rise is more certain and permanent, because the cause is more natural, and the men are in pocket all the money they might foolishly spend on the strike, and save themselves and their families all its miseries, besides having the pleasure, which every honest workman must desire, of knowing and feeling that he has acted honourably. Another very important consideration to bear in mind is this—that strikes unsettle and disturb the financial arrangements of the employers. The great railway and building operations of the country are carried out by advances from bankers and financial agents. There is no body of commercial men who are so willing to employ capital at a moderate rate as bankers. They are far more content with a small profit and safe and sure investment than a large profit and hazardous trade. At the same time, no commercial men are so sensitive to risk and danger; and if they have the slightest suspicion that there is anything wrong, or that the work is liable to be stopped at any moment, then they do not care about advancing money to carry it out, so that when the capital is stopped, employment is stopped and wages also. Mr. Rupert Kettle, in his evidence (Question 7185), says: "But there is another person who is disturbed, and whose influence on the trade is paramount, and that is the banker, who finds the money for the builder. I am confident that when the trade is so disturbed that a banker cannot be quite certain whether a strike will not interpose to prevent the builder having the architect's certificate, showing that he is able to meet the state of his banker's account periodically, he would hesitate to advance money to carry on the business. In all things in fact, security increases confidence, and so promotes investment; and if you do anything to disturb security and bring about such a state of things as exists in the building trade, you disturb the confidence of people in it, and divert the investment of capital." Here, then, is another very powerful reason proving that the interests of labour and capital are identical, and that both should work harmoniously together. The more capital there is invested in a trade, the more will the operations of that trade extend, the greater will be the demand for labour, and the higher the wages; but strikes, by driving capital from the trade, bring loss on the employers and poverty on the men. And, lastly, strikes ought to be condemned because they are in their very nature a species of social civil war. War, whether of a national or social kind, can very rarely be justified, more especially when waged between citizens, and workmen and employers. Whatever differences arise, they can be arranged by arbitration. The spirit of our age is to settle disputes, not by blows, but by argument; not by physical or social violence, but by dispassionate and enlightened reason. This is one of the most pleasing signs of our times; for it shows a free people, the makers and supporters of free laws; and it is the duty of every one to strengthen this growing sentiment by discountenancing all attempts of individuals resorting to physical or social violence to redress public or private grievances. We have a free press, public teachers, and a Parliament that is becoming more and more the reflex of public and national opinion, before whom we can place our disputes, and by whom we can have them fairly discussed. Let us then appeal to them, and calmly wait a righteous judgment.

We have thus far endeavoured to show that the objects the unions seek are many of them not lawful, and, where lawful, the means employed are not legal or right. We have seen that picketing, intimidation, violence, and strikes, are contrary to law and order; that the minimum rate of wages, limitation of apprentices, and the rules relating to labourers, bricklayers, masons, painters, &c., also as to chasing, are all artificial restraints, tending to stunt the growth of the men's faculties, to retard their advancement, and lessen their employment and wages, as well as to injure trade, and defraud the employers and the public; and, lastly, they are at variance, both in the letter and the spirit, with the Divine law of "doing unto others as ye would they should do unto you"—that law which ought to be the guide alike of individuals and nations in all their dealings. Let not, then, trades-unions for one moment suppose that their ideas are for society's good. What mankind requires is more freedom in labour, commerce, and religion. The world is moving onwards, not backwards; and they might just as well expect the strong man, proud in his strength, to return again to the swaddling clothes of his babyhood, as to expect this country to adopt the artificial restraints and restrictions of the shortsighted, narrow-souled system of trades-unionism. The question, then, arises, what good are these unions? or are they a good at all? They were originally benefit societies; but, by having trade rules grafted on them, they have been taken out of their proper place, and applied to purposes for which they were never intended, and thus have become an evil. There is no such thing in existence as pure evil. All evils are simply good things abused or perverted. Fire and water, for instance, are very good things in their proper place, but very dangerous things out of their place; and trades-unions, as at present conducted, are a striking illustration of this truth. They originated as benefit and friendly societies in the strict sense of the term, and were designed to be the savings' bank of the workman, to which he could apply for help when out of work, or in times of sickness, or from accident, and, when he died, where his family could receive assistance. From these most useful and commendable objects they have been turned, and applied to purposes altogether beyond their proper sphere, and foreign to their true nature. Mr. Allan, of the Amalgamated Engineers, thus speaks on this point:—

Q. 7573. Trades-unions are not now the same as they were formerly. A simple trades-union for the protection of trade, in my opinion, is calculated to bring about a great deal of unpleasantness and ill-feeling between employers and workmen; but when we have a system of benefits attached to it, it brings a better feeling to bear on many of the members, and, in many instances, no doubt they adopt a wiser course than they would adopt if they were simply a trade society.

Mr. Allan by this fairly admits the evil of the trades-union; and what does he do to cure it? Does he do as every sensible and clear-headed man would, eradicate the evil and leave the good? No, he does not; but he mixes the good and the bad together, thus defiling and polluting the good of the friendly society with the evil of the trade rules. The evil of the unions is admitted, and the cure is very simple, namely, by confining them to their proper place, and so preventing their interfering with trade. So long as they are kept there we admire and commend them, and heartily wish them and their members every success, and that same feeling which prompts us to support them as a good prompts us to expose them when an evil. As benefit societies they can be the means of great use, and are therefore justly entitled to legal protection for their funds against dishonest officers. The sympathies of all true Christian men are with the working classes. Their labour, however humble, is honourable, and their poverty is no disgrace; for

"The honest man, tho' e'er so poor,
Is king o' men for a' that."

We wish to see their condition improved in every respect—better paid, better clothed, better fed, and better educated, and the worthy amongst them rise to positions in life corresponding to their merits and abilities. All this and a great deal more we sincerely desire; but we wish to see it brought about by lawful, not unlawful, means. In conclusion, we would beg to offer a suggestion, or rather to support a suggestion that has already been made to the Legislature. In reading the evidence laid before the Commission, one is appalled by the terrible revelations made, especially in relation to the state of immorality of the lower orders in Sheffield and the Manchester districts. They reveal a state of society which neither the higher, the middle, nor the respectable part of the working classes ever believed to exist. However painful it is to make this discovery, it is far better that these terrible evils should be laid open, than allowed to lie hid like a social ulcer, poisoning and corrupting the life's blood of society. These men are controlled by no conscience, guided by no principle of religion, honour, or duty, and are scarcely moved even by the stern, iron rod of the law. And what is the cause of such a state of things? They are for the greater part the sins of ignorance, and can in the course of time be removed by education. Let, then, a system of National Education—compulsory, if necessary—be provided, so that every child may have the means of instruction placed within its reach, an education that will teach him not only to read and write, but, what is far more necessary for the duties and difficulties of life, to think, understand, and act for himself as an intelligent, conscientious, manly, and independent being. This duty has been too long delayed by our Government, and we are being punished for its neglect by having the children reared in the midst of all this vice and darkness, becoming "rotten before they are ripe." Spread, then, the blessings of Education far and wide amongst our labouring population; and before its heavenly light those black, thick clouds of ignorance, settled on the minds of our people, and big with so many curses, will gradually roll away, and a brighter and happier day dawn on our land.

PRINTED BY RANKEN AND CO., DRURY HOUSE, ST. MARY-LE STRAND, LONDON

TRADE UNION BILL, 1871.

A LETTER

BY

Mr. W. P. ROBERTS,

(OF LONDON AND MANCHESTER, SOLICITOR.)

TO

MR. GEORGE POTTER, MR. WILLIAM ALLAN,

MR. ALEXANDER M'DONALD,

AND OTHERS INTERESTED IN TRADES' SOCIETIES.

Printed for Presentation. Additional Copies—Twopence each—may be had by applying to the Bee-Hive Office, Bolt Court, Fleet Street, London, E.C.

1871.

A LETTER, &c.

61, Great Russell Street, Bloomsbury,
March 18, 1871.

Gentlemen,—

Permit me to address you on the subject of "The Trades Union Bill" now before Parliament. This Bill is of the greatest importance to the working classes, and it is imperative that they should well consider whether they will give to it their sanction; and thus to some extent add to its weight and power; and, moreover, preclude themselves from future opposition or comment—or whether they will treat it with such an amount of guarded opposition as will allow of their speaking of it as not *their* Bill but that of the Government; so as the more readily to allow of its being qualified and varied on some not distant opportunity. There is at present, with all classes, a strong tendency to regard the proposed Bill as one specially demanded by the workmen themselves—as if it had been prepared in their interest; and this belief should not, perhaps, be permitted to grow into permanency—at all events, not until the question has been well considered.

My own opinions are entirely against the Bill. It not only gives nothing—certainly very little—to the workman, but it deprives him of many advantages which he accidentally possesses under the present law. I think, if you will give me your quiet attention for a short time, that I can succeed in making it clear that the Bill is not favourable to your interests, and that you ought to give to it your most determined opposition. There are one or two general points to which I will ask your observance before I comment on the Bill itself—its particular clauses.

First.—Bear in mind that any diminution of the legal right of workmen, to combine together for mutual protection, is more or less destructive of their power to better their position and hope of improvement. The motto on their shield should be that, *whatever one man alone may lawfully do, several may lawfully do in combination.* Until, indeed, this doctrine, of which several of the superior Judges have expressed their approbation, is recognised as a legal axiom, your position as Trades Unionists is not only unsatisfactory but dangerous.

Secondly.—Consider the generally acknowledged fact, that of late the tendency of circumstances has been to make the Courts where the law is administered by magistrates, either at their own houses, or in their clerks' offices, or in Petty Sessions, or at Quarter Sessions, more severe against working men than formerly prevailed. As lately as thirty years ago the

4

magistracy was composed of more educated men than now—of more solid social rank and wealth; and though many acts of cruelty were committed with their sanction, still their conduct towards working men frequently exhibited a sort of "live and let live" kindness and sympathy. Of late years there has been a change for the worse. The magistracy of the last few years has been very much selected for political considerations from the manufacturing and trading classes. There was a *Derby* batch, for instance, and then as a counterpoise, there was a *Gladstone* batch. And it has come to pass—no matter how—that a workman has to sustain a more keen and subtle application of the law's discipline and correction, —more severe and artistical—than formerly. By some friends of the working classes, myself among them, it is therefore considered desirable that the power of the present magistracy should be restrained rather than extended. The present Bill extends that power. The loss of a jury is a heavy loss to a working man. For him the Quarter Sessions is a most unsatisfactory tribunal. This, however, is a consideration for the future.

There are some other general remarks I should like to make; but I will work them in as I go through the Bill. I will go through all its clauses in their order. My figures refer to the same figures as appear in the Bill.

1.—The title is "The Trade Union Act, 1871." You may as well remember that this is not the only act on the same subject—workmen and their employers. There is one especially that is popularly known as *Lord Elcho's Act*, which, as working men become more alive to their rights and interests, will have to be considered before very long.

Then we have the second clause. Its most clever ingenuity ought to be well studied that it may be fully perceived. With an appearance of doing much, it really does nothing, or next to nothing. It has a sentiment indeed, though but little substance. It says that "*the purposes of a Trade Union shall not by reason merely that they are 'in restraint of trade' be deemed to be unlawful, so as to render any member of such Trade Union liable to criminal prosecution for conspiracy or otherwise.*" Now, one would believe from reading this that at the present moment "restraint of trade purposes" are illegal, and subject the men to criminal prosecution. But in truth, there is not the slightest ground for such a belief. Purposes "in restraint of trade" are *not* illegal; nor do they subject to criminal prosecution for conspiracy or otherwise. Nor have such purposes been illegal since the repeal in 1825 of all the old combination laws. This is expressly stated in the judgment in the case of "Hornby *v.* Close."

That case, Hornby and Close, merely amounted to this. In order to punish a defaulting Treasurer, the society, the Boiler Makers, prosecuted him (I acted as the attorney of the society) under a clause in the "Friendly Societies' Act," which gave a summary, cheap, and quick mode of dealing with such offences. I had several times before made use of this clause, and successfully. The Court of Queen's Bench, however, held that the society was not in a position to avail itself of this special remedy, because its rules were (some of them) *in restraint of trade.* The special restraint most relied on for the defaulting treasurer, was that members of one branch of trade were not allowed to take work in a

5

higher branch until they had worked as learners in such higher branch, and so elevated themselves to its emoluments and privileges. The judgment, however, expressly decided that in the ordinary sense of the word "unlawful," the liability to a prosecution, the society was not unlawful; the decision was merely that the society was *not within the scope* of the "Friendly Societies' Acts," and therefore was not entitled to the benefit specially given by those acts. So far, therefore, as this second clause is concerned, it cannot be said to have much practical utility. There is, however, something, though but little, to be said for it. In a subsequent part of the proposed Bill there is a power to *register*; and then, after registration, the society would have, *but only on condition of its being first registered,* the advantage of a summary jurisdiction (the same as if it were a Friendly Society) by which it might attack the defaulter. But this power will not, I fear, be practically of any use, for I think that no society will register. You will observe presently the labour, the penalties, the publicity, and other evils, to which you will be subjected *by registration.* It gives little or none of practical benefit; while it compels returns and information, which would destroy the secrecy—the self-management of such a society, and render their arrangements, objects, resources, and difficulties, open to the employers and to the Government.

Then we come to the *third* clause. Its entire blow is practically against the working men; as to them the law is entirely an exceptional one, for none but they will really come under its lash. And this exceptional operation is one reason, and perhaps the principal, why working men view the clause with such universal impatience and dislike. What is required by working men—such as regard labour with respect—is that there should be no distinction before the law between them and all others. A crime, great or small, should be neither more nor less than a crime with them than with others. The attempted distinction is offensive to honest men of all classes. It tends to maintain that separation of interests, sentiments, and habits which good men deplore. This third clause ought to be expunged altogether; or rather the crimes with which it seeks to prevent, should be removed from the summary jurisdiction of the magistrates, and remitted to the Common Law. I am, however, endeavouring to explain the Bill, and must therefore give a few more lines to this, its most objectionable branch.

This third clause enacts that certain (three) offences are to be punished with not more than three months' imprisonment, with or without hard labour. These three offences are using violence—threatening or intimidating—molesting or obstructing.

But the violence, threats, intimidation, molestation, or obstruction must be with a view to *coerce* any master to dismiss or to cease to employ any workman, or any workmen to quit any employment; and then this third clause proceeds to enumerate all the offences of the present Combination Act—and another, "hiding tools"—and to visit each of them with the same punishment, "not exceeding three months' imprisonment with or without hard labour." Now, so far the mere variation of words between the present law and the proposed Bill would be of no benefit whatever to the workman. The offences are the same, the punishments are the same.

6

One does not see why the Bill should, with reference to this branch of the subject—the description and punishment of offences—have been been brought into existence. There are, however, some attempts at making a difference which must be looked at.

First; there is an interpretation of the words "threat and intimidation." These, in order to be punishable under the proposed Bill, must be such "*as would justify a justice of the peace, on complaint made to him, to bind over the person so threatening or intimidating to keep the peace.*"

But this limitation is more apparent than real, and it is much to be feared that it would practically be of no help to the workman. The magistrate can construe, if he so please, any act, however vague or vapory, to be such as would call upon him or allow him to bind over. A sour look, a word of anger, a stray oath, with an "I go in fear of my life," would be sufficient for the purpose; especially if there were any "combined action" among the workmen going on in the neighbourhood.

There is another matter to be considered. This threat or intimidation must be "with a view to coerce." The present law says that the purpose must be to "force, or endeavour to force," a variation which makes the old law much more favourable to the working man than the proposed Bill. The case of "Walsby and Anley" gives a kind of judicial interpretation to the word "coerce," to which we may hereafter ask attention.

Then comes a definition of "molestation or obstruction"—of what those words are to mean when the proposed Bill is passed—which I must be permitted to hope it never will be.

A man is to be held guilty of "molestation or obstruction"—"if he persistently follow such person about from place to place; or if he hide any tools owned or used by such person, or deprive him of the use thereof; or if, with two or more persons, he watch or beset the house or other place where such person resides or works, or happens to be, or follow such person tumultuously through the streets."

Now, it is observable that the first and third of these definitions are—certainly the first—consistent with innocence. One man may persist in following another with a view of fair persuasion; and the same may be said of two or more watching the house. Bear in mind that it is only by working men getting together and reasoning on the considerations which belong to their trade, its locality and special circumstances, that they can have a chance of persuading each other to act together for their mutual protection. They have much to talk about to their fellow-workmen. And the employer frequently so manages as that the workpeople shall not have an opportunity, unless with considerable difficulty, of meeting each other. *He* has no one to talk with or persuade but himself. A clause which particularly operates against the free intercourse of workmen with each other admits of much oppression, and should be watched with extreme jealousy.

As to "hiding tools," it is one of the new offences to which a summary jurisdiction is given by the proposed Bill. But it is an offence which admits of calculation as to its amount of injury—the value of the tools and the loss of time—and it should be dealt with as other injuries are, by the ordinary civil or criminal courts. To decree a punishment by an irresponsible tribunal, perhaps themselves in some way interested, without a jury, so large, so severe, and so degrading as three months' imprisonment and hard labour for such an offence appears to me to be most unjust and unwise. Why should not the injury be considered by the ordinary courts as now, where the accused might give evidence on his own behalf, and have the benefit of a jury if he so wished. This coining a new offence to be tried and disposed of, and summarily, by a perhaps prejudiced and objectionable tribunal—for the magistrates who decide will, in all probability, be in the trade of or connected with the prosecution, is (so it seems to me) one of the most fatal blots to the Bill. The fact is that all these offences—violence, threats, intimidation, molestation, and obstruction, hiding tools, following—should be removed from the magistrate's jurisdiction and left to the operation of the ordinary law. Then you would have a judge, a man of superior mind, perhaps (for such things have been, the late Judge Talfourd was one) with a leaning in favour of those who live by work; perhaps, indeed, with a leaning the other way; yet still amenable to public opinion, who would have, in his *summing up*, to explain the law, to justify the prosecution, to give a commentary on the defence, to dissect and discuss the evidence. Then, too, you would have a jury, from whom you might remove an employer or his dinner companion, and many other advantages. You would have time, too, for your defence. Contrast this with what you painfully know to be the course now, and which may be and will be under the proposed Bill. A warrant is granted, the accused is taken into custody, and brought before the bench of magistrates, sometimes (for he is not entitled, as of right, to an hour's delay) with no time for his defence or witnesses. His case is soon disposed of, frequently not occupying half an hour. It happens sometimes that there is a difference of opinion with the magistrates, and they retire to a private room; what passes there is of course unknown to the accused and the public. There is no publicity, no explanation. And then comes the sentence, and the prisoner is removed. Everything has been against him. He has had none of what may be called the "protection of circumstances." What wonder, then, that so many miscarriages of justice occur. In London the magistrates are, by the influence of public opinion, supported in their desire of doing what is right according to their ability, but in the country there is little or no presence of expanded sentiment, no restraint against anger, self-esteem, or self-interest; the Press is ineffective, and it follows that the accused, for the most part, calculate on a conviction as a certainty; not, indeed, on the merits, but because the trial has been devoid of that forbearance and careful absence of hurry which to a poor man are absolutely necessary for his having a fair trial.

Why do I go into this branch of the subject so fully? and yet how feebly have I written and how much more there is to be said. It is because you are asked to continue this strange system, to give to it additional power, to add another offence to its jurisdiction, and thus to give your sanction to its endurance and increase. The Bill may pass, all you can do may not prevent its passing. From what I learn of an interview at the Home Office last week, I fear that it must pass. The attitude of the Trade Unions is regarded as an antagonism to be put down. But what I ask of you is to protest against its passing, to repudiate it altogether, *that it may not hereafter be thrown at you as your own work.* Insist upon it that workmen be not exceptionally treated; deny as loudly as you are permitted the assertion that working men require more than the ordinary restraints from crime and wrong. Denounce to your fullest power the anomalies and iniquities of the present law. But don't commit the folly of accepting the proposed Bill as a substitute.

The next paragraph of the proposed Bill says that besides all this exceptional crime and punishment you are also to be liable to any other or higher punishment than is provided by the Bill. Only you are not to "be punished twice for the same offence." This "once only" is, I suppose, a concession—you are not to suffer more than once for one offence. The effect of this provision is that the prosecutor is allowed to select such mode of proceeding against you, as he thinks most suitable to his purpose —either by Common or Statute Law, or by the proposed Bill.

The fourth clause protects you as to your civil rights, as the second clause protects you against the operation of the Criminal Law. No "restraint of trade" purpose is to render any agreement or trust unlawful. There is, of course, no objection to this; yet neither will it be found to have much beneficial operation. Its possible "good" indeed is very much qualified by the clause next mentioned.

Then follows an important clause—the fifth—one of the worst. It denies to your *Trade* agreements all legal power of enforcing them. It says these agreements are *not unlawful,* but that the law shall not give you the assistance to enforce them which it gives to other agreements. Its precise language is that it prevents your obtaining the enforcement by "any legal proceeding" of any agreement by which (1) the members of a Trade Union shall or shall not sell their goods, transact business, employ or be employed, or (2) for the payment by any person of any subscription or penalty to a Trade Union, or (3) any agreement for the application of the funds of a Trade Union (a) to provide benefits to members, or (b) to furnish contributions, &c., or to discharge any fine, &c. This enactment that certain things are "not unlawful," and yet specially legislating that the law shall not assist their enforcement is new, and worthy of thought and examination. It certainly has a look of inconsistency.

The peculiarity of these restraints of legal protection in the proposed Bill—the entire closing of the Courts against you—is that they do not exist in the present law. Many of such agreements might certainly be enforced under the present law. Practically, however, this fifth clause, though provocative of hostile remark, will have little or no operation; for in truth Trades Unions do not apply to "Courts of Law" to enforce their rules and agreements. For the most part they expel the dishonest member refusing to comply with his engagements. Still the clause, while certainly it is gratuitous, may by some who are sensitive be considered to be insulting. And the same remarks apply, in some degree, to the sixth clause, by which it is declared that the "Friendly Societies' Acts," "The Industrial and Provident Societies' Act," and "The Companies' Acts," shall not apply to any Trade Union, "and that the deposit of the rules" under the Friendly Societies' Acts (I have mentioned this before) shall have no effect. This gives permanency to "Hornby *v.* Close."

9

The next division of the Bill is headed "Registered Trade Unions." Its seventh clause gives power to a Trade Union to *register* such Union by subscribing their names, &c. I shall pass rapidly over this branch of the proposed Bill, for indeed very few, if any, trades will avail themselves of it. But still it requires some notice. The eighth clause gives power (and some special facility) to buy an acre of land and build upon it. The ninth and tenth clauses give power to Trade Trustees to sue and be sued, to prosecute and to be indicted; and that if the trustees die, the same privileges are to pass to their successors. Under the eleventh, twelfth, and thirteenth clauses, trustees are only to be liable to the extent of trade funds; and the treasurer of the union is to give security, and to account from time to time. And then comes an important clause—the fourteenth—which gives to the Society the same advantage—that of a summary process against a defaulter—as was supposed before the decision in *Hornby & Close,* to belong to every Society which deposited its rules with the Registrar of Friendly Societies. But you must bear in mind that this advantage, and the other clauses which follow the sixth, apply to such Societies only as are *registered.* You will observe presently that the "Registration" is indeed a very formidable affair, and all its forms (several of them most inconvenient in practice) must be complied with if any benefit is to be derived. The summary power against a defaulting treasurer, of bringing him to book, and of punishing him if he will not do what is right, is certainly a valuable one, but hardly worth all this trouble of form, accounts, and exposure which is made necessary as a preliminary, and which requires (and this is important) constant renewal; it must be gone through at the beginning, and also every year afterwards.

Then follow (15) the "Regulations for Registry," with regard to which there are six divisions—(1) as to the application to be registered, which, "*with a list of the titles and names of the officers, shall be sent to the Registrar under this Act,*" who, if satisfied that (2) the proper forms have been complied with, shall register; but (3) there are to be no two unions of the same name; then (4) if a Trade Union is more than a year old it must, before it can be registered, deliver to the Registrar "*a general statement of the receipts, funds, effects, and expenditure of such Trade Union in the same form, and showing the same particulars as if it were the annual general statement required, as hereinafter mentioned to be transmitted annually to the Registrar.* Then (5) after all this is done the Trade Union is to have a certificate which "*shall be conclusive evidence that the regulations of this Act with respect to registry have been complied with.*" Then (6) the *Board of Trade* may make regulations. After that there is a section compelling the registered Society to set forth, amongst other things, the following provision. I give it in full that you may the better enquire of yourselves whether any Trade Union in the kingdom is likely to avail itself of the benefits of registration—

"*The whole of the objects for which the Trade Union is to be established, the purposes for which the funds thereof shall be applicable, and the conditions under which any member may become entitled to any benefit assured thereby, and the fines and forfeitures to be imposed on any member of such Trade Union.*"

I do not think that any Trade Union will consent to give any Board of Trade the information which is thus required prior to and during registration. Trade Unions, and it is pleasant to think of this, may go on as they are now without registration, and consequently may avoid its attendant inconveniencies and penalties.

But there is still more form and ceremony to be incurred by the Trade Union which seeks the honour of Registration. By clause seventeen "*Every Trade Union registered under this Act shall have a registered office, to which all communications and notices may be addressed. If any Trade Union under this Act is in operation for seven days without having such an office, such Trade Union and every officer thereof shall each incur a penalty not exceeding Five Pounds for every day during which it is so in operation.*

After this we come to a very stringent clause—the eighteenth. *A general statement of the receipts, funds, effects, and expenditure of every Trade Union registered under this Act shall be transmitted to the Registrar before the 1st day of March in every year, and shall show fully the assets and liabilities at the date, and the receipts and expenditure during the year preceding the date to which it is made out, of the Trade Union; and shall show separately the expenditure in respect of the several objects of the Trade Union; and shall be prepared and made out up to such date and in such form, and shall comprise such particulars as the Registrar may from time to time require.*" A Trade Union making default as to these returns, is to be liable to a penalty not exceeding Five Pounds for each offence. And then there is another penalty for something else of "not exceeding £50 for each offence." All this will demand some considerable exertion and knowledge of arithmetic.

Then, as a kind of Central Authority to watch over the proper deposit of the rules and accounts of the registered Trade Unions, "the Board of *Trade shall be the Registrar under this Act; and the Board may appoint any person who is one of their secretaries or officers, or a Registrar or deputy Registrar of joint stock companies, to perform and exercise, subject to their directions, all or any of the duties, powers, and authorities of the office of Registrar.*" By subsequent clauses it is provided that the present Registrars of Friendly Societies are also to retain their present powers. (In going through the whole Bill I am compelled to touch upon many points which have little or nothing to do with the subject I am considering.) Then we come to a very important section of the nineteenth clause—

"The Registrar shall lay before Parliament an annual report with respect to the matters transacted by such Registrar in pursuance of this Act." So your debts, liabilities, assets, rules, objects, &c., are to be brought before Parliament at the beginning and every year afterwards, that is if you *register*. Is there a Club in London or in the kingdom that, without protest, would submit to such a supervision as this?

Now, I do not hesitate to express my strong belief that all this about *Registry* and *Registration* will, as it is read by every Trade Union in the kingdom, be received with derision, reduced by indifference. All the privileges (there may be an exception or two) so pompously conferred are not only of such a nature as not to be required, but there is no Trade

11

Union in the kingdom, not a Club, Lecture Class, or Reading-room, that would for a moment condescend to receive them at such a price, or at any price. All this exposure and interference would be entirely destructive of the purposes of a Trade Union—fatal to its existence.

Then comes the twentieth clause, which enacts in effect that in case of misrepresentation as to the Rules of a Registered Trade Union, the offender "shall be deemed guilty of a misdemeanour." On this clause —the smallness of the fault, the severity of the punishment—much might be said. But I have not time to go into it. The offence is a new one, not existing before.

(The fact is that these Registration Clauses are—when considered in connection with other parts of the Bill—of very little value. A Trade Society is not illegal under the present law—unless, indeed, it has some avowed illegal purpose. And no Trade Society has ever been attacked simply as a Trade Society. Registration does not propose to give any special legality to what is illegal now—with the exception of power to buy an acre of land, and to prosecute a defaulter by summary process.)

Now we approach another division of the Bill—"Legal Proceedings." By these it will be seen that the Bill would render the Law much more oppressive to you than that which now exists.

First (21), you are to be prosecuted, for any offence in connection with Trades Unions, by the Magistrates' Court (Petty Sessions) in the same way as for offences of the ordinary kind, in which summary jurisdiction is given. You may be taken by warrant, without any previous summons, without time for defence, and tried there and then, and sent off (subject to a right of appeal, as to which I shall say a word or two presently) to your two or three months with your hard labour, or without. Then comes a clause whose importance is not at once observed, but which may give practical impunity to much oppression. "*In Great Britain the description of any offence under this Act, in the words of such Act, shall be sufficient in law.*" These words in the latter part of the twenty-first clause of the Bill are fraught with danger. In London the magistrate has already (under a special Act applying to London only) the power of sheltering himself in his judgment and record of conviction by the precise words—"threats" for instance, or "intimidation," or "violence," or "molestation," &c. But as just intimated, this privilege does not extend to the country justice. There is some reason for the distinction. In the country the justice under the present law must declare what the molestation is; he must specify what the "threats" are, and if he were to give "black looks" as the substance of his judgment, the conviction would probably be quashed. But if this safeguard is removed, if the country magistrate, instead of stating the facts upon which his judgment is founded, be allowed to merely give his own conclusion, there will probably be fearful injustice. This is another of the several instances in which the law under the proposed Bill would often be likely to cover an unjust sentence. For the Justice, knowing that his conclusion was to be taken as a true conclusion, on his own mere statement of it, would probably be less careful than he ought to be. The very necessity for supporting the conclusion by a statement of the facts would frequently compel some amount of reflection, and so prevent injustice.

12

Then (22) there is given a power to appeal. This clause will go far to deprive the workman of the small chance of justice with which he is now—under the law as now existing—obliged to be contented. I have not time nor space to allow of my directing your attention to more than one or two points. The appeal is to the Quarter Sessions, that is to the Magistrates, no jury. *This will not do at all.* The appeal, if it is to have any moral force whatever, must be to a Judge of one of the Superior Courts, and who, as to questions of fact, must be assisted by a jury. A notice of the intention to appeal is to be given to the other side, and then comes what I cannot but regard as a most unfair deviation from the present appeal protection:—

"*The appellant shall, immediately after such notice, enter into a recognizance before a Justice of the Peace with two sufficient sureties, conditioned personally to try such appeal; and to abide the judgment of the Court thereon, and to pay such costs as may be awarded by the Court, or give such other security by deposit of money or otherwise as the justice may allow.*"

"*Where the appellant is in custody the justice may, if he think fit, on the appellant entering into such recognizance, or giving such other security as aforesaid, release him from custody.*"

So the justice is to decide on the sufficiency of the sureties (that is, the justice is to *fix the amount*—just what he pleases—two of £20 each and the accused in £40 is the usual sum) in which they are to be bound, and they are to be bound not only to try the appeal, but to pay the costs. And then the justice may, that is "if he think fit," release the appellant. He *may* release him or he may not. It is really very hard to have to write on such a Bill as is proposed, and which is put forward as a special boon to the working man, a great concession, and for which he ought to be thankful.

Now contrast the working man's position under the present law. The appeal clause of the present Combination Act (6 Geo. IV., c. 129, sec. 12) is this:—"*Provided that if any person convicted, &c., shall think himself aggrieved, &c., such person shall have liberty to appeal, &c.;" and that the execution of every judgment so appealed from shall be suspended, in case the person so convicted shall immediately enter into recognizances, himself in the penal sum of* £10, *with two sufficient sureties* in the penal sum of £10, *upon condition to prosecute such appeal, and to be forthcoming to abide the judgment of the Quarter Sessions, and to pay such costs as the Court shall award.*

So you perceive the difference. Under the present law the appeal liability is bounded by £20. The accused is bound in £10, and the two sureties in £10 together, not each in £10, but £10 for the two. And the release is to be immediate. No "think fit" of the magistrate is to detain his prisoner in custody for longer than is required for the filling up the recognizances. Under the proposed Bill the appeal liability will probably be £60 or more, and the discharge, pending the appeal, not immediate nor certain, but only "if the justice think fit." Now, is not the demand too much when you are asked to substitute the proposed Bill for the present law? What can your legislators think of you? A working man (and it must be remembered that there are many who do not belong to a

Trade Union, and are obliged to contrive with their own individual resources) can perhaps manage to appeal under the present law. £20 possibly might not crush him. But so large a sum as £60 might frighten many into the endurance of a judgment and sentence which they could prove to be unjust, but which they had not the pecuniary strength to resist.

The repealing (25) clause, and its schedule, repeal the present Combination Act, the 6 Geo. IV., c. 129, together with the short Act—22 Vic. c. 34—by which the first-named Act was explained, and to some extent mitigated, with regard to the offences " molestation and obstruction." It also repeals section 41 of 24 and 25 Vic., c. 100, and the valuable Act of a year or two ago, " to protect the funds of Trades Unions," the 32 & 33 Vic., c. 61; and which (it expired in August, 1870) it might be well to re-enact for a year or two longer. Now mark what follows, for it is most important—of more importance than anything else I have written in this long letter.

The repeal of the first-named Act, 6 Geo. IV., c. 129—the Combination Act—and for which the present Bill is proposed as a substitute, deprives you of the greatest safeguard and protection which you possess.

Throw your recollections back to the time when the Combination Act was passed, 6th July, 1825, forty-six years ago; then throw your minds still further back by five centuries to the 33rd year of the reign of King Edward the First, when an Act was passed against "conspirators who do confeder or bind themselves by oath," &c., and so on generally against "Combinations or conspiracies" of workmen to obtain higher wages or to lessen the hours of work. From that time to the 57 Geo. III. not a reign, and hardly more than five or six consecutive years at a time, passed by without some fresh law of restraint against combinations of working men. It would seem that as each law came into operation against the workman a fresh attempt was made by him to preserve his power to act in combination with those who, like himself, were helpless when acting alone. Hard words were used. Every combination by workmen was a "conspiracy." An Act of Geo. III. declares that "combinations in trade are public nuisances."

Notwithstanding, however, this unceasing contention of our Parliaments against the WORKING MAN, public opinion in his favour progressed and extended. It was kept alive principally by the efforts of that honest Legislator JOSEPH HUME. I pass by his many struggles. What was the result? It 1825 (6th July) Mr. Hume obtained an Act repealing at one blow all those Acts of hundreds of years, and substituting the present "Combination Act" in their stead. This last-named Act reduced and mitigated the offences which were the subject of the repealed Acts to the following:—*violence, threats, intimidation, molestation, and obstruction*—much the same as is done in the Bill now proposed. And then followed the famous clause 4—often called the SHEET ANCHOR OF LABOUR—a clause worth millions of money and centuries of exertion, and which clause the proposed Bill repeals, and possibly deprives you of for ever. Is it not "too bad?" But what character will you deserve if you suffer the repeal to pass without lifting up your voices in indignant remonstrance. I will give the clause in full (the fifth clause gives the same protection to employers that the fourth clause gives to the workman).

"IV. Provided always, and be it enacted, that this Act shall not extend to subject any persons to punishment who shall meet together for the sole purpose of consulting upon and determining the rate of wages or prices which the persons present at such meeting, or any of them, shall require or demand for his or their work, or the hours or time for which he or they shall work in any manufacture, trade, or business, or who shall enter into any agreement, verbal or written, among themselves for the purpose of fixing the rate of wages or prices which the parties entering into such agreement, or any of them, shall require or demand for his or their work, or the hours of time for which he or they will work, in any manufacture, trade, or business, and that persons so meeting for the purposes aforesaid, or entering into any such agreement, as aforesaid, shall not be liable to any prosecution or penalty for so doing; any law or statute to the contrary notwithstanding."

Now it happens that most of the combinations sanctioned by "Trade Unions" are mainly for the very purpose of raising the rate of wages, or diminishing the amount of work; and to such combinations a solemn legal sanction is given by this fourth clause. True, indeed, all other combinations are left untouched. And so the workman is left, as to them, to the mercies of the common law. The power of this fourth clause—its exceptional and large power in favour of combined efforts for better wages and less work—has been often alluded to by the superior Judges, and by many of them (for it is a mistake to suppose that they are all of one mind in their appreciation of the rights of labour) with approbation, and with a sort of dignified satisfaction. Whatsoever else is to happen, you must give your whole strength to the retention of this right to combine. *What have you done that you should be deprived of it?*

This fourth clause is incidentally mentioned in the case of "Walsby *v.* Anley," decided by the Queen's Bench on January 19, 1861. The prosecutor's case was that the two men who had been convicted had threatened him with "a strike." This "strike" was held to be a "combination among the workmen to *bring about an alteration in his mode of carrying on his business.*" What was wanted by the Society, from whom the two men were a deputation, was in fact that the prosecutor should discharge a particular workman who was working for him under the well-known "declaration." There was no charge of a combination as to wages or the hours of labour. If the prosecution had embraced either of these two (wages or labour) it would have failed.

In the "Reports of Cases in the Queen's Bench" the facts of "Walsby *v.* Anley" are fully set out, and then we have the judgments of the judges. Hill J. referred incidentally and apart from his judgment to the *fourth* clause of the Combination Act as "allowing certain combined movements as to the amount of wages to be given and demanded; does not that lead to the inference that any combination, as to the matters mentioned in the THIRD section, would be unlawful?" There were some other remarks by the judges, and then we come to the judgments. Of these I only give the one in which the *fourth clause of the Combination Act* is mentioned; and recognised as an exception from the ordinary Law of "Conspiracy."

CROMPTON J. "It is a matter of common learning that what one man "may lawfully do *alone* he may not do *in combination with others* when "the act tends to do harm to another. The present Act, by repealing "the older statutes, in effect *re-establishes* the common law as to *combina-* "*tions*, and sections 4 and 5 were therefore necessary, as it seems to me, "for the purpose of preventing it from being illegal for masters or men "to meet and to determine as to the amount of wages to be given or "demanded. In the present case the charge is that they endeavoured "by threats to force the prosecutor to limit the description of his work- "men. Section 4 does not render such a combination legal."

(It is observable that the repeal of this Act lets in the operation—as the Repeal Clause is worded—of all the prior Acts which that Act repealed; but this is a mere mistake, aud can be remedied by three or four lines.)

I have now, with but few omissions and those not important, gone through the whole of the proposed Bill, whose title appears as "The Trade Union Act, 1871," and I ask you to consider well, and to pause, before you confer on that Bill any portion of your approbation or sanction. Remember that the Bill is put forward as *your* Bill—as a compliance with your desire to have the protection of the law for the funds of Trades Unions, the same as is given to the funds of Friendly Societies, and which, I may observe, might be effected by an Act of a dozen lines, declaring (after the words in the second clause) that Trades Unions should be allowed to deposit their rules with the "Registrar of Friendly Societies," and to have the privileges of such deposit notwithstanding that their rules, or some of them, were in restraint of trade. Such an Act would give you that special means of recovering what had been stolen from you, which you required in the case of "Hornby *v.* Close." A re-enactment for a year or two of Mr. Bruce's Act, passed 9th August, 1869, would indeed be sufficient. The special protection is not perhaps so much demanded now as at the time when the "Hornby *v.* Close" was decided; for you have since found out another mode of successfully proceeding against defaulters by trial at the Assizes under the ordinary Common and Statute Law. Still there is no doubt but that the summary proceeding given by the *Friendly Societies' Act* would be useful; and you would also have, in addition, that "Assizes" remedy which I have just mentioned, and which would be useful in cases of importance and where you might wish to have a jury. But there is no reason why you should, in order to obtain the benefit of the "Friendly Societies' Act," give your sanction to a number of other clauses, most of which are directed against you, and will, if they become law, add to your present difficulties. Remember, too, that under the proposed Bill, this "Friendly Society," protection is only given to

Trades Unions *which are registered, and whose purposes, government, and mode of action, are therefore exposed to the common eye, and especially to the eye of the employer.* My own view is decidedly against the proposed Bill. But it may have virtues which I do not perceive. You will, however, I am sure, agree with me in protesting against the repeal of the fourth and fifth clauses of the present combination act—the law which now is—and which gives to you the right to combine for "more wages and less work." If this power, obtained with such mighty efforts, once goes from you, it may be long, indeed, before it is again restored. If it is to go, let its departure be the act *of the Government—not of you.*

It may perhaps be said that you ought to take advantage of the present disposition to consider the subject of Trades Unions, and to endeavour to temporise with the proposed Bill, so as to manipulate it from a weapon of annoyance into a shield of defence. There is something in that, certainly. And there are several suggestions which I should like to make to you in that direction. But I must restrain myself, for this letter has extended to far beyond what I had contemplated. I shall, however, be most happy at any and at all times to confer with you in discussion. The question is "What do the workmen want? What ought they to do?" For the present, I will content myself with the suggestion that Mr. Bruce be requested to renew for a year or two his Act which expired in August last, and to postpone for the present further proceeding with "The Trade Union Act, 1871."

Permit me to subscribe myself,

Gentlemen,

Your faithful servant,

W. P. ROBERTS.

H. W. FOSTER, PRINTER, 16, BEAR ALLEY, FARRINGDON STREET.

REPORT

OF THE

CONFERENCE OF TRADES DELEGATES,

HELD IN

THE TOWN HALL, LEEDS.

On December 2nd, 1871.

CALLED TO CONSIDER THE STATEMENTS MADE BY WM. NEWMARCH, ESQ., IN HIS ADDRESS TO THE SOCIAL SCIENCE CONGRESS LATELY HELD IN LEEDS.

Published by the Leeds and District Trades Council.

LEEDS:
FRED. R. SPARK AND CO., PRINTERS, "EXPRESS" OFFICE.

LEEDS AND DISTRICT TRADES COUNCIL,

PARKER'S TEMPERANCE HOTEL, BRIGGATE, LEEDS, DEC. 20th, 1871.

To the Officers and Members of the various Trade Societies of the United Kingdom and others interested in Trades Unions and Social Science.

GENTLEMEN AND FELLOW WORKMEN,

For the capitalist classes and their organs of the press to misrepresent the facts of the action of Trades Unions, and of high authorities to mistake the motives and objects of their representatives, has been so common for years past, that, the rule being so common and uncorrected, has led to opinions being held among those not experienced of a most erroneous and mischievous kind, in relation to the objects, purposes, and tendencies of the working classes Trades Unions.

The errors charged upon the unions, and the slanders charged upon their motives generally, has doubtless arisen from actual ignorance, and a partial and one-sided selfishness. In the one case, parties who should have known them better, and who ought to have ascertained the facts for themselves, have not liked to do so, for reasons little creditable to themselves. In other instances, want of experience, and the assuming of statements without proper evidence has led to what is now stated. And perhaps the unionist and representatives are not a little to blame in permitting errors and falsehoods to be related concerning them without contradiction. Often they have been charged so, that they did not know until too late, and published where they had no means of either correction or reply; often, if replied to, they have been injured, snubbed, and refused publicity, and often the men who could have corrected have refused to do so in disgust at the unfair and partial treatment they have received and seen. Nor should it be denied that their position as workers often unfitted them to be enabled to reply in a way satisfactory to themselves, though they know the truth—or to be acceptable to their opponents, who they know to be in error, and thus from various causes, as before said, unions have been misrepresented, and unionists, if active, slandered by and before the highest authorities. This was as usual the case at the Social Science meetings, held in Leeds in October last, and particularly in the address of William Newmarch, Esq.,

3

President of the Economy and Trade Section, October 10th, and which has been specially published and circulated by the Association for the promotion of Social Science.

The few representatives of the operative classes present felt grieved at what they knew to be false and unfounded charges upon the facts, purposes, and objects of Trades Unions, and so represented them to the Leeds and District Trades Council; and they felt so strongly upon the matter as to resolve upon a general council meeting of the trades, with a view to challenge Mr. Newmarch to prove his allegations, and to substantiate his charges.

The district approving of this course sent delegates and representatives to the Conference, held in the Civil Court of the Town Hall, Leeds, on Saturday, December 2nd, 1871, when they were prepared to hear and disprove the charges made, or they said, to submit to the censures thrown upon them. Yet, strange to say, William Newmarch, Esq., although respectfully requested, neither attended himself, neither forwarded deputy or evidence in proof, and even neglected the common courtesy to reply to the last communication sent to him by the Secretary. Working men will do well to contrast this course—one not uncommon to that economic school which classes labour with chattels to be bought, sold, and cheapened, in the best markets—with the conduct and treatment of Sir John S. Pakington, and who though only incidentally alluded to, as in error in our foreign trade relations, at once most courteously explained, and most candidly instanced what he thought to be the correct state of the case.

To Sir John S. Pakington the trades can have but one response, that of a thankful and profound regard, after all differences; and to Mr. Newmarch and all who have preceded him in that way, but one estimate of reprobation and condemnation, the due to both their conduct and teaching.

Conscious of having been slow to correct the errors published concerning them and unwilling to resent, the Leeds and District Trades Council believe it their duty now to publish and to make known as far as in them lies the facts of why they have thus acted, and to place before working men both unionists and non-unionists the result of the Conference; this they now submit with the request of a careful and impartial consideration of what has passed, and with the hope that the result, will to some extent, put a stop to the wholesale misrepresentations so often made, if they do not communicate some truths beneficial to the labouring classes in general.

Believing that individually working men can neither obtain the wages or conditions of work desirable for their bodily welfare or mental progess; believing that there exists in the increasing wealth and prosperity of the nation ample means for both; believing that unions properly acting for the above objects, are most likely to attain them, and so promote the well being of the great majority, and through them the whole nation, the Leeds and District Trades Council strongly advise the respectable and intelligent of the operative classes, to join Trades Unions, and to carry out such measures as may be proved beneficial alike to both capital and labour.

Signed on behalf of the Leeds and District Trades' Council.

E. C. DENTON, JOINER, *President.*
THOS. PLACKETT, COACHMAKER, *Secretary.*

4

The following Circular was issued to the Trades :—

LEEDS AND DISTRICT TRADES COUNCIL,
November, 9th, 1871.

To the Officers and Members of the Trade Societies of Leeds and District.

FELLOW WORKMEN,

The Leeds and District Trades Council beg to address their fellow workmen who had neither the privilege nor the opportunity of attending the Congress of the Social Science Association lately held at Leeds, in order to direct their attention to certain mis-statements of facts, and errors of imputation and inference made against them as a class; and especially charged against Trades Unions and those institutions promoted to defend the interests, and to improve the position of the workers in general.

Passing over the fallacies of the decline of our export trade from foreign competition, when every return proves its extension and growth* ; passing over the assertions that our present grossly unequal distribution of wealth is "just and reasonable," because we have the largest developments of gross produce,† we have to direct the special attention of Trades Unionists, and of the working men generally, to the pointed and strongly developed assertion of Mr. Newmarch that Trades Unions have the "most erroneous and mischievous *purpose* of seeking an uniform rate of wages *without regard* to the difference of skill, knowledge, industry, and character * * a doctrine which, if acted upon rigidly, would stop all progress. Put into plain language it *means* that there shall not be in the world, *any such thing* as superior attainment, that every art and handicraft shall be reduced to the level of the *commonest, most ignorant, and most stupid* of the persons belonging to it; that to secure *this* end every man to whom nature has given special ability, and who by effort and self-denial has brought that ability to more or *less* perfection (?) shall forego the superior rewards, the honours, and the advancement which he can easily and fairly earn."‡

This statement, so praised up and quoted by most of the press, the Trades Council knows to be false, and the inferences they feel to be so grossly unjust and injurious, that they indignantly urge all unionists to deny and repudiate this libel upon their character, unions, and objects.

And to this intent it is proposed to call a Conference of the Working Trades of the District, to be held in the BARRISTER'S ROOM, TOWN HALL, LEEDS, on SATURDAY, DEC. 2nd, *at Three o'clock, p.m.*, when delegates and representatives of the industrial interests are invited to attend. It is proposed to invite Mr. Newmarch and the Officials and Members of the Social Science Association, to charge these misrepresentations upon the authors, and especially to challenge Mr. Newmarch to prove his assertions, and to give just and reasonable evidence for his imputations against Trades Unions—their "purposes" and "meanings."

The Trades Council, along with tried and trusted friends, believe that it will be most inexpedient in maintaining the character and interests of the working classes, to permit these and other such errors and fallacies made against them by high authorities, who either do, or should know better, to go unchallenged and unrefuted.

The attention of the trades and working men being now directed to what has been going on, you are earnestly requested to join in the proposed Conference, and to send delegates and representatives well acquainted with trade rules, customs, and organizations, with a view of establishing the truth, and of repudiating the slanders and fallacies charged upon the them.

Signed on behalf of the Leeds and District Trades Council.

E. C. DENTON, JOINER, *President*.
THOS. PLACKETT, COACHMAKER, *Secretary*.

The Conference was accordingly held at the Town Hall, Leeds, on Dec. 2nd. There was a good representative attendance. Mr. E. C. DENTON (President of the Trades Council) was in the chair, and there were also on

* Made by Sir John Pakington in his opening address.
† Made by Mr. Newmarch in his address on Social Economics, Oct. 10th.
‡ Newmarch's Address, p.p. 11-12, published at 1, Adam Street. Adelphi, London. Price 6d.

the platform Mr. Ald. Carter, M.P., Mr. George Potter (London), Mr. John Holmes, Mr. W. Lishman, Mr. J. A. Greenwood (Castleford), Mr. J. Sanderson (York), Mr. J. Beagan, and Mr. J. Johnson. The trades represented were carpenters and joiners, coachmakers, glass bottle makers, engineers, masons, plasterers, clothdressers, brush makers, ironfounders, boiler makers, wheelwrights, miners, wood carvers, tailors, and shoemakers.

The Chairman said the statements to which they took exception had been asserted at many meetings of the Social Science Association, and when they were asserted at the Leeds Congress it was felt by the trades representatives in Leeds that they should take a public opportunity of refuting them. It was necessary they should be refuted. They had been made so often that the whole nation was beginning to believe them. (Hear, hear.) He called upon the Secretary (Mr. Thomas Plackett) to read the correspondence which had taken place in connection with the Conference. A letter was accordingly read from Mr. Newmarch in answer to an invitation to the Conference.

67, Lombard Street, London, E.C.,
Nov. 17th, 1871.

Sir,—I am favoured with your note of the 16th instant, and regret that I cannot be in Leeds on the 2nd proximo.

Nobody will be more gratified than myself, if it can be specifically proved that Trades Unions—not here and there—but generally are not open to the objections I have stated.

Yours obediently,
Mr. T. Plackett. WM. NEWMARCH,

To this letter the Secretary wrote:—

Wm. Newmarch, Esq. Leeds, Nov. 22nd, 1871.

Sir,—In acknowledgment of yours of the 17th, and on the part of the Leeds and District Trades Council, I have to say that it does not in the order of investigation, or logic, rest with us to produce the proofs which you say will be "gratifying" to you. Without notice, and of course without our knowledge, not being present, you made certain specific charges against unions and unionists, and made inferences therefrom very dissatisfactory to thousands who hold themselves wholly innocent of both the facts charged and the deductions made.

And you must know, Sir, it is for you when called upon to prove as publicly as you charged the statements made. We called the meeting noticed, to give you a fair opportunity; and when the evidence you have to bring of wrong against unions before us, we will, to the best of our ability, rebut the charge and disprove the evidence, or we will submit to be corrected accordingly. Until then, we cannot, and we feel we ought not to permit your charges to pass unchallenged. We think this ought not to be met by your requiring us to prove the negative to your satisfaction.

As a recognised authority, and as the official of an important public association, you made a public charge, and the onus of proof rests with you, with all its responsibility and results. Hoping yet to have the honour of your attendance, or of such representative or means as you may deem fitting,

I am, on the part of the Leeds and District Trades Council,
THOS. PLACKETT, Secretary.

No further reply had been received from Mr. Newmarch. A circular had been sent to Sir John Pakington, and the following correspondence ensued:—

Westwood Park, Droitwich, Nov. 19, 1871.

Sir,—I have this day received your printed circular convening a meeting of the "Leeds and District Trades Council" on 2nd Dec.

6

In that circular you impute to me in my opening address at the late Social Science Congress, "fallacies as to the decline of our export trade from foreign competition, when every return proves its extension and growth."

As I am sure you can have no wish to misrepresent me, I write to say that if you will refer to my address (see *Leeds Mercury*, Oct. 5th) you will find that I said nothing about "decline of our export trade." On the contrary, I referred to "the late satisfactory Board of Trade returns."

The subject to which I did refer was the necessity for a better system of technical education to protect the artizans of this country from the dangers of being passed by other European nations in the great race of manufacturing competition.

In proof of the existence of this danger, I quoted authorities which seem to me to be unimpeachable.

I beg to remain, sir, yours faithfully,

Mr. Thos. Plackett. JOHN S. PAKINGTON.

Leeds, Nov. 22, 1871.

Sir John,—It is with great respect that, on the part of the Trades Council of Leeds and the district, I beg to acknowledge and reply to yours of the 19th inst. The delay of reply has arisen partly to obtain the *Leeds Mercury* of October 5, and partly for consultation. You do us simple justice, Sir John, in supposing that we have no wish to misrepresent what you said. And, on the other hand, we believe that you had no intention to misrepresent the actual facts of the position of the trades of England in relation to foreign competition. But we still are of impression that you, in assuming the statements of your authorities to be true, taught by and through them, what all evidence proves to be incorrect.

Identifying yourself with your authorities, you clearly attempt to show—

1. That foreign Governments have trained their workpeople by a system of technical education into greater skill, better qualities, and a higher economic application than our own Government has; and, in consequence,

2. That our trade, both home and foreign, is not merely in danger, but that it has actually fallen behind in the race.

The point we objected to, Sir John, both as a fact and the misappropriation self-interested parties have put it to, is the fact and bearing of the last clause.

We are bound, as we do not wish to misrepresent ourselves, to show the grounds for our statement of what we believe to be the misrepresentation complained of. And in doing this we must distinctly hold that, as you identified yourself as believing with your authorities, that the statements fully bear the weight of your own belief and authority.

These are, "We must take blame to ourselves for having allowed many branches of trade, skill, and commerce, which yielded to England the greatest remuneration, brought her commerce the greatest reputation for themselves and their country, and afforded her people the highest wages and best work to pass away into the hands of better educated people in foreign lands guided by wiser heads."

In another part you state how the Paris Exhibition of 1867 showed "the rapid rate in which other parts of Europe were passing us by." Again, "I am of opinion the English workman is gradually losing the race." . . . "The enemy is at the gate, not in the shape of fleets or armies, but in the shape of a rivalry in manufactures and trade, in which, if we fail to maintain our position, our national prosperity will be undermined, and England's wealth, power and greatness will be a story of the past. (Hear, hear.) Let us then be wise in time. The train is behind time, and if the guard cannot recover his lost ground, there will be a collision," &c., &c.

Striking out the word "export" in the fallacies we objected to—(and that arose entirely from not having the *Leeds Mercury* at the moment before us)—we still think that in the common sense of the words above quoted by you, our charge still stands, viz., that you hold we are being beaten, and that our trade is declining, from the result of foreign competition. And if this was the fact, both export and home trade must necessarily show it. That it is not a fact, is proved by your own admission of the "satisfactory returns of the Board of Trade."

Of course the two positions are incompatible, and the onus of that rests with you, Sir John, to explain. Had you given us the figures and the ratio of our trade, both home and export, for any number of years; and had you stated its present position—either in amount, or in relation to population—you would have seen at once that a not unfrequent

7

phenomenon in humanity (in common with certain animals) had taken place, viz., a fright—or stampede—without either good ground or probable reality of danger—in the matter of foreign competition. We have the figures before us upon authorities, certainly not inferior to either Mr. Scott Russell, Dr. L. Playfair, or Mr. Mundella. But we assert that, in ratio to population, never at any time has our trade been in a more flourishing and paying condition, while it has for the last century grown beyond all precedent. We give one authority, viz., Mr. Gladstone, who stated at Wakefield, on the 12th Sept. last "That, while for the last twenty years the population had increased about 25 per cent., our export trade had increased at the rate of 400 per cent." And every last-quarterly return of imports and exports shows, as a rule, the increase of all departments of our manufactures and commerce.

We hold, therefore, that any statement of foreign competition having injuriously affected us is contrary to fact. And we may add that the recent experience of skilled artisans brought over to Newcastle from the highly trained technically educated countries of Belgium, Prussia, Denmark, and Sweden, does not bear out the opinions and fears given of their superiority. It has proved that both in skill, knowledge, and endurance, foreign workmen are decidedly inferior to our average; while in disorder, irregularity and drunkenness, they far exceed our worst.

Thus far we have replied in justification of our objections to the impressions we received upon reading your otherwise admirable address. And, as due to you, Sir John, we gladly acknowledge in many things our entire agreement with, and admiration and thanks for, the boldness of your statements of the evils and the remedies applying to the working classes. It is a fact that the growth of our manufacturing system has, in dwellings and other sanitary arrangements, told fearfully upon the health and lives of the working classes. That artisans live but 22 years against the wealthy classes 44, tells a tale for itself, and we thank you, Sir John, for pointing it out to be amended. And while the wealth of the large centres of production has increased in a way so amazing and unparalleled, it is a notorious fact that, as a rule, the portion the working classes receive does not enable them of themselves to provide the accommodation and means necessary for their own satisfactory existence, or for the country's economic progress. The idea of assisting the working classes by legislation may be stigmatised as "economically ridiculous," but in our simplicity as workers we hold that, legislation having in the Factories, Mining, and Workshops Acts proved highly beneficial to all parties concerned—to employers as well as the employed, and to capitalists as well as labourers—that, in the direction of helping the people further, when they cannot help themselves, or in doing better than isolated action can do for them, the Government of the country would not exceed their functions, public health and public good being the object and result.

In the want of a good technical education for all classes, and in the benefits likely to be derived from a better education, we also agree with you, but not, Sir John, for the mere advancement probable to our manufacturing and commercial interests. These might be of little or no advancement to us. But for the advancement of humanity itself we welcome the universal spread of education, and the development of knowledge, art, and science. With these we know all interests in every way will be advanced, whereas a mere increase of trade might tend to the increasing production by machinery; in which the wealth resulting might go, as it has before done, to the few, leaving the many no better, if not relatively worse, in the competition.

In these remarks, which we honestly and deeply feel, independent of sect or party, we trust that you will not think we are exceeding the bounds of our legitimate subject. If you do, attribute it to our perception that you do, and of our hope that you will sympathise with the working classes in another sense than that they are mere machines, beasts of burden, or commodities to be treated as such, and to be "bought and sold and cheapened in the best or lowest markets." This, a wretched, partial, and false so-called "Economic Science" has both constantly asserted and practised. And, because partial and false, however respectable the authorities of it are deemed, the working classes believe they are enemies not only to them and their prosperity, but to the public weal, and to the future progress and prosperity of the United Kingdom.

With the assurance of our highest esteem for your attention and reply to us, on the part of the Leeds Trades Council. I remain, most respectfully,

THOMAS PLACKETT.

To Sir John Pakington, Bart., Westwood Park, Droitwich.

8

Westwood Park, Droitwich, Nov. 29, 1871.

SIR,—I have no wish to enter into any argument or controversy with you, and I think our correspondence might well be left where it is were it not that I really desire to express the pleasure with which I have read your frank, fair, and very able letter.

You seem to think that an admission of the present prosperity of our export trade is inconsistent with a warning as to prospective danger to our manufacturing supremacy.

I cannot admit the inconsistency, or the construction which you seem to be still disposed to place on my words at Leeds.

What you have written refers chiefly to the present.

What I spoke referred chiefly to the future.

But you candidly admit that the imputation in your printed paper that I had spoken of "decline in our export trade" is not correct, and if there is difference remaining between us, it is not, I think, of sufficient magnitude to require further discussion.

I cannot, however, close this letter without acknowledging the courteous tone towards myself in which you have written, nor without expressing my satisfaction that you do not suspect me of sympathising in those unworthy and contemptible views with respect to our working classes, to which you refer with just indignation.

I remain, Sir, with sincere good-will,

Yours faithfully,

Mr. Thos. Plackett. JOHN S. PAKINGTON.

The CHAIRMAN followed up the correspondence by saying, in answer to Mr. Newmarch, that there was no such thing as a minimum wage among trades unionists in the sense in which many persons believed. Their rules did not admit it. He had worked for a firm where a minimum wage was the fixed rate, but the fault was not that of the workmen. (Hear, hear.) What they said was that after a man had learned a trade he should not work below a certain sum, but if there were men in any of their societies who were superior in skill, intelligence, speed, and general qualifications, they did not say that he must go down to this rate, but that he should go up and get as much as he possibly could.—Mr. JOHN HOLMES, for the guidance of the meeting, read the following extract from Mr. Newmarch's address, being that portion of which they had the greatest cause of complaint:—

"Not yet, but in course of time, as economic principles become popularly understood, we shall see trades unions purged of their most erroneous and mischievous purpose of seeking an uniform rate of wages, without regard to differences of skill, knowledge, industry, and character. There is no tenet of socialism more fatal in its consequences than this insidious and plausible doctrine—a doctrine which if acted upon rigidly for any length of time by large classes of men would stop all progress. Put into plain language, it means that there shall not be in the world any such thing as superior talent or attainment—that every art and handicraft shall be reduced to the level of the commonest, most ignorant, and most stupid of the persons who belong to it—that to secure this and every man to whom nature has given special ability, and who by effort and self-denial has brought that ability to more or less perfection, shall forego the superior rewards, the honours, and the advancement which he can easily and fairly earn."

Mr. LISHMAN (cloth dresser) pointed out that the practice of trades unions was to arrive at an average rate of wages amongst a given number of men, and to encourage rather than to prevent higher wages to superior men. He thought the constant efforts made by trades unions to elevate their members socially and pecuniarily were so far an answer to Mr. Newmarch. He moved the following resolution, which was seconded by Mr. MOORE (plasterer):—

9

Having heard the charges made against the purposes and objects of trades unions by Mr. Wm. Newmarch, as President of the Economy and Trade Department of the meetings of the Association for the Promotion of Social Science will be Leeds; and having heard of the course pursued by the Trades Council of this district to rebut the said charges and to challenge proof, this meeting is of opinion that the said charges are both false in fact and erroneous in conclusion; and they hereby thank the Council for their vindication of the truth and defence of the trades unions against the said false and erroneous stigmas and imputations.

Mr. POTTER supported the resolution. He thought it highly creditable to the Trades Council that they had organised that conference, and he had no doubt the steps they proposed to take would be supplemented by the action of the Trades Congress, which would shortly be held at Nottingham. (Hear, hear.) The trade to which he belonged had adopted a minimum rate of wages, and he believed that this principle had been followed out by the building trades generally. The rule was to fix an average wage accordingly to the fair value of the labour of the majority of the men, and it was the object of the unions to legislate not for the inferior men, but for the average skilled workmen. With the society with which he was connected the rule of admission was that a man should be proposed and seconded by men who had worked with him, and who could testify that he was at least an average man. If he was not an average man he was not admitted, because the society was intended for skilled workmen—for men who could give a fair equivalent in labour for a fair wage. (Hear, hear.) Mr. Newmarch's statement, therefore, that they legislated for the inferior men was contrary to fact; the inferior men did not belong to them. Their desire was to qualify every man to receive such an average wage as his skill and industry were entitled to, so that they lifted up and did not pull down. (Applause.) Their chief object, indeed, was to lift up, and he challenged Mr. Newmarch to give an instance in which a trades union had sought to prevent any man obtaining more than a minimum wage. They knew that wages graduated to the lowest. (Hear, hear.) They never uplifted themselves to the highest. What, then, they said was that an average skilled workman should not work under such a price, but that an employer should give to a superior workman as much as he pleased. (Hear, hear.) He pointed out that the same rule applied generally to capital and labour—that it applied notoriously to the bar, the press, and the medical profession—and contended that in principle there was nothing wrong in it. Mr. Newmarch had said, "There is no such thing in nature as an equal division of unequal earnings, and it is because socialism in essence means that and nothing else, that it is an impossible and pestilent delusion." If there was no such thing as an equal division of unequal earnings in nature, there was such a thing in society. (Hear.) In instituting an inquiry as to the distribution of wealth they must make a comparison between the present and a former time. One shilling formerly was equivalent to 2/6 now. More could then be bought for a shilling of what were deemed necessaries than could be bought now-a-days for 2s. 6d. of what was deemed essential. A working man might enjoy more comforts now than formerly; but if his position relatively to those above him was not improved, all else will go for nothing. He maintained that the increase of wages scarcely balanced the increase in the expenditure of the working classes. In 1688 England and Wales were

10

computed to contain about five and a half millions of people. In 1867 they contained 21 millions, or about four times as many. The income of the people in 1688 was estimated to equal £107,500,000. In 1867 it was reckoned at £662,000,000 and had therefore increased 50 per cent. more rapidly than the population. The figures he was using were Mr. Gregory King's for the earlier period, and Mr. Dudley Baxter's for the more recent one. The distribution of this wealth was as follows:—In 1867, 7,500 persons (or 1 in 2,800 of the 21 millions of people) possessed £5,000 a a year and upwards. In 1688, 160 persons (or 1 in 34,375 of the 5½ millions people) enjoyed an equal revenue. In 1867 42,000 persons (or one in 500) possessed from £1,000 to £5,000 a year. In 1688 only 6,426 persons (or 1 in 855) had an equal income. Coming to more moderate incomes, those varying from £300 to £1,000 a year, he found that in 1867, 150,000 persons (or one in 140) were so endowed. In 1688 40,000 persons (or 1 in 137) were equally fortunate. Of small incomes he found that in 1867 1,026,000 or 1 in 20 persons, were in receipt of from £100 to £300 a year; whilst in 1688 there were 405,000, or 1 in 13 persons, so circumstanced. These figures showed the following results:—Since 1688 the very rich, or those possessed of over £5,000 a year, had multiplied twelvefold. The moderately rich, or those with incomes ranging from £1,000 to £5,000, had also considerably added to their numbers; while the total of those possessed of between £300 and £1,000 a year in the two periods was about equal. But when we came to the incomes of between £100 and £300 the proportions were completely changed. One person in 13 was in this position in 1688, but only one in 20 in 1867. These figures were fearfully suggestive when we remembered that the annual increase of the capital of the country was estimated at £150,000,000. It would seem that this vast increase only added to the wealth of the few and did not increase the comforts of the many. But even this array of figures did not tell the whole story. The total income of the upper and middle classes was computed at the present day to be about £407,000,000, and Mr. Dudley Baxter had estimated the annual earnings of the manual labour classes at the present time at nearly £255,000,000, so that eight-thirteenths of the income of the country was the property of about one-fourth of the population, and five-thirteenths only fell to the remaining three-fourths. In 1688 the income of the country was equivalent to £107,500,000. Fifty-seven and a half of these millions were divided among the one-third of the population which then constituted the upper and middle classes, while fifty millions fell to those engaged in manual labour. The defence they made was this: A large capital had become a necessity; the master of old days with his two or three journeymen had been swallowed up in leviathan establishments, each large enough to find employment for a small town, and under these altered conditions there was an end of the old personal relations. The men were now merely part of the plant with which the master earned his profit; the master merely a necessity to the men in their efforts to earn bread. The relations between them were not based on their common humanity, did not recognise their common Christianity, and were simply a mercantile transaction in which gold represented the whole weight of obligation. The

11

chief material disadvantage in the present condition of the labouring classes he conceived to be a failure to improve, and not any positive deterioration. It was difficult to see how it could be otherwise. The working man lived from hand to mouth, required all he earned as soon as he had earned it, and showed the urgency of the necessities which beset him by his readiness to pay large interest for small loans and his inability to wait for any portion of his earnings. (Hear, hear.) As to Sir John Pakington's statement that they were driving trade from the country, he thought sufficient answer was to be found in the fact that during the last ten years the increase in the export trade of the country was as high as £173,000,000. In conclusion, he appealed to his hearers to place a stronger faith than ever in Trades Unions. Trades Unions were based upon instinct, sanctioned by reason, and vindicated by experience. The objects of Trades Unions were the protection and elevation of those who belonged to them, and the amelioration of the condition of the working classes in general. By the agency of Trades Unions the operatives of this country had been benefited socially and pecuniarily. Trades Unions had taught the working man to dispose of his labour to the best advantage, and helped him to repel the unjust aggressions of capitalists and employers. The proof of the value of Trades Unions was found in their existence, in their continual growth, and their popularity amongst the toiling millions. Under their action and influence the self-respect of the working man was cultivated, his capacity for enjoyment expanded, and his character morally and intellectually elevated. In times of slack trade, when intelligent artisans had walked miles in search of work, and when with wearied bodies and depressed spirits they had returned home unsuccessful, the out-of-work pay from Trades Unions had placed bread upon their table to comfort and sustain their wives and children. (Applause.) When prostrated by sickness, and prevented from following his employment, the trades unionist had found in the funds of his society the means of sustenance and relief; and when accidents befel him or death overtook him, his widow and children found consolation and assistance from the funds of these admirable organisations. (Applause.)

Mr. Carter, M.P., also supported the resolution. He commended the speech of Mr. Potter to the careful attention of the meeting and to workmen generally, and said he agreed with everything which had been stated in it with regard to Trades Unions. There could not be a better proof of the advantage of Trades Unions than the fact that in all directions the masters themselves were adopting the principle. The masters saw the advantage of Trades Unions in their own interest, and he did not see why they should complain of what the men did in this respect. He could not do better than repeat the advice of Mr. Potter—that as they valued their own interests, they should stand up for their unions. The tendency of these unions, as had been clearly stated, was to elevate workmen up to a certain minimum, and to then let men rise above it as far as they could. (Hear, hear.)

Mr. Joseph Pickles (engineer) and Mr. Turner Tetley (ironfounder) also spoke upon the resolution, and the former, amidst applause, quoted the testimony of Mr. Lampen, clerk to the Leeds Guardians, that he had never known a case of a trades unionist applying for relief.

12

The resolution was carried unanimously, as was the following one, which was proposed by Mr. BEAGAN (shoemaker), seconded by Mr. JOHNSON (shoemaker), and supported by Mr. TAYLOR (engineer) :—

That Mr. Newmarch, President of the Economy and Trade Department of the Social Science Association, having been respectfully challenged to prove the charge and statements he made against Trades Unions, delivered in the Mechanics' Hall, October 10th, 1871, and he having failed to do so, or to render to the Trades Council either evidence, proof, or grounds on the matter charged and denied by them, this meeting is of opinion that the said William Newmarch, Esq., has thereby lowered alike the office he held, the society he represented, and the doctrines he professes to teach.

The next resolution was moved by Mr. GREENWOOD. It stated :—

That a copy of the preceding resolutions be forwarded to the directors of the Society for the Promotion of Social Science, with the respectful request of this meeting that they will, as far as possible, prevent their association being made the medium of social ignorance and false allegations, both of which they believe, know, and feel to have been the case in the address of William Newmarch, Esq., on the 10th October, 1871. Other such and similar instances of ignorance and error they could point out, as in the address of Lord Brougham, at Bradford, and Sir William Armstrong, at Newcastle, and which they beg to represent, are exceedingly grievous and offensive to the intelligent and conscientious members of the working classes, who have not always means of knowing at the time, nor of having the opportunity to correct or to rebut. If, then, the promotion of social science be the wish and desire of the directors, they will do well to put it in such a shape, and on such a footing, as to be less open to the appearance of prejudice and passion in its observations, and to partiality and error in its teaching.

Mr. HOLMES seconded the resolution. He read extracts from the Brougham and Armstrong speeches to which the resolution drew attention, and expressed a hope that that meeting would have some influence with the authorities of the Social Science Association. This association was calculated to do a vast amount of good, but through prejudice and ignorance on the part of some of its supporters, it had the effect of spreading error, and it was very important that where possible it should be put right.

The resolution was agreed to.

On the motion of Mr. SANDERSON, seconded by Mr. ROYLANCE (mason), it was agreed to advertise the resolutions, and to forward copies of them to societies and the parties interested.

A vote of thanks to the Chairman and Mr. Potter concluded the conference. It should be stated that several of the delegates brought copies of the bye-laws, and in some cases the minute books, of their societies, and read extracts from them to disprove that statement of Mr. Newmarch which was being called in question.

After the conference the delegates held a social meeting, at which their thanks were given to Mr. George Potter, who had attended the conference by their special invitation, for his advocacy of Trades Unions, and his efforts to improve the social condition of the working classes. Mr. Potter, in acknowledging the resolution, said it was encouraging to him in his work to find that he enjoyed the confidence of the trades of Leeds, and expressed the hope that he would be more useful to them in the future than he had hitherto been.

Societies or others requiring a quantity of these Reports can be supplied at the rate of 5/- per hundred, up to Feb. 1st, on application to the Secretary, 22, Carr Street, Little London, Leeds, where all communications must be addressed.

THE

ADVANTAGES

OF

TRADES UNIONS,

BY

W. H. WOOD,

SECRETARY OF THE

MANCHESTER & SALFORD TRADES COUNCIL.

PRICE ONE PENNY

SALFORD:

PRINTED BY THE SALFORD STEAM PRINTING COMPANY, LIMITED,
217, CHAPEL STREET.

PREFACE.

IN publishing "THE ADVANTAGES OF TRADES UNIONS," the writer desires to correct many erroneous opinions entertained by those outside the pale of unionism, with regard to the operations of these much maligned institutions, and also to strengthen the faith of those who recognise such principles.

The employers of labour, throughout the country, have found out the advantages resulting from united action, and have formed themselves into a "National Federation of Employers," for the protection and advancement of their interests, which must be sufficiently conclusive that the position taken by the operatives has proved worthy of being copied by the employers, and that there are advantages to be obtained from combinations which isolated action fails to command.

Should the employers make use of their newly-acquired power to terrify isolated trades into accepting that which would violate their principles, the operatives will, no doubt, retaliate by the formation of a National Labour League, which would be empowered to create a National Strike Fund, to be used at the discretion of a Central Executive, who would be empowered to conduct all labour movements.

Arbitration and Conciliation are, however, very popular with the operatives, and those who have a just cause of complaint, need not fear it being thoroughly investigated before an impartial tribunal

The operatives do not regard the formation of Associations amongst the employers with any fear or jealousy, but rather rejoice, for employers by the adoption of the principles taught them by their workpeople, will be enabled to frame their scale of prices and work out their operations with a less suicidal policy, and that mistrust and uncertainty in the framing of estimates for work required to be tendered for may possibly be avoided if the employers will only further copy the principle of trades unionism.

Should the pamphlet, now issued, meet with the approval of those for whose benefit it has been written, the object of the writer will have been accomplished. There is no attempt made in its pages to apologise for outrages, intimidation, or coercion, as such acts are repudiated by all well-regulated trade societies, who ought not to be held accountable for the individual action of those, who, from a variety of circumstances, might be tempted to violate the law, the responsibility of which ought to attach to the offender Recent legislation has ignored the capacity of the ordinary law as applicable to all, for the punishment of offences, by singling out trade unionists for special legislation, and it is not improbable that the machinery of their vast organizations will be brought into operation to return to Parliament men from their own ranks to make an effort to redress their grievances, but that and kindred subjects are not treated upon for want of space, but may furnish arguments for a second treatise upon "The Advantages of Trades Unions."

THE

ADVANTAGES OF TRADES UNIONS.

In the race for wealth, which characterises this utilitarian age, gigantic undertakings have become the rule, and society seeks to resolve itself into combinations to accomplish everything by the formation of vast establishments, either in the shape of limited liability companies, or by the amalgamation of a few large capitalists who carry on business upon an extensive scale This state of things has destroyed the social character and relationship of employer and employed, which formerly existed, at a time when the workman laboured side by side with his employer, and when that mutual good understanding existed which bound them together by a bond of friendliness and sympathy with each other, and rendered the oversight of the labour of the workman, by an intermediate agency, unnecessary. Combinations of workmen were then little required, for the spirit of competition had not assumed its sway over the country such as is so apparent at the present day

The establishment of large mills and workshops where thousands are employed, under one firm, have called into existence a species of middle-men in the shape of managers, whose interest it is to cheapen production, by grinding down the wages of those who labour and accelerating the speed of the machinery under their charge The ties, that bound employer and employed together under the old system, and which enabled both to amicably adjust any differences that might arise, either as to the value of their labour, or the material upon which they were employed, no longer exist under the present system. The employer, as the responsible individual, could formerly settle the differences at once, for the margin of profit, if capable of allowing certain concessions, was known to him, and the friendly relationship remained undisturbed, but if on the other hand, the employer had undertaken a contract that would not afford an advance upon the current rate of wages, his workmen were not slow to recognise the unreasonableness of their claim, for the evidence was presentable where the employer and the employed worked in harmony with each other

The formation of mammoth establishments has destroyed all the social bonds that formerly existed, and the workmen find that not only Trade Societies are required for their protection in the towns in which they labour, but all trades of a national character have found it necessary to amalgamate their branches in one common central authority, whose seat of government is determinable by voting power, the suffrage being membership. But even these wide-spread organisations have not been considered sufficiently strong enough within themselves, thus the was not made for mere labour alone, and when we are told by such men as Professor LEONE LEVI, that if we reduce the working time by one hour per day, the loss on production is not less than £90,000,000, or three pounds per head over the whole population, or considerably more than the entire amount arising from Imperial taxation, we retort, by pointing to the increased wealth of the country, which has been doubled since the first movements for short time came into operation If the production of the country is £1,000,000,000 per annum, which is obtained by 3,000 hours labour, being at the rate of £300,000 per hour, the country is quite capable of sustaining the reduced hours, for the increase of machinery and its accelerated speed in all departments of industry, fully proves the elasticity of the influences at work to remedy the loss, if any there be, in reduced hours of labour.

But let us rely upon the evidence of practical men rather than the theories of a professor

Mr GEORGE ELLIOT, M,P, declares that the cost of getting coal in Aberdare, in South Wales, is 25 per cent more than in some parts of the North of England, where the Miners work five hours a day less

The late Mr. BRASSEY, when constructing the Paris and Rouen Railway, employed both French and English Navvies, the former worked from five in the morning till seven in the evening, whilst the latter only worked from six to half-past five, yet, the amount of work accomplished by the Englishmen, in the shorter hours, was much greater than that accomplished by the Frenchmen.

It has been proved, on the most reliable evidence, that the system of eight hour shifts have caused more work to be executed than in ten hours, and the system has been adopted where work has had to be executed within a stipulated period—notably so during the construction of the Trent Valley line

If the vast wealth sunk is liable to be placed in jeopardy by the reduction of the hours in which that machinery is engaged, why not let the machinery work sixteen hours per day and adopt the system of double shifts—the morning hands to commence work at six o'clock and work until two o'clock, and the evening hands go on until ten o'clock By this plan, the hours of the workpeople would be reduced to forty-eight hours per week of eight hours per day, the rent and taxes would be the same for the increased production, and the workpeople would have one-half the day for their own leisure and no doubt accomplish as great results in the forty-eight hours as they now do in the time they are now called upon to labour, and thus enable the capitalists to secure sixteen hours running of their machinery where they now only obtain nine or ten

Many failures in the Shipbuilding trade have been attributed to the influencing power of Trades Union wages. The knowledge of Mr SAMUDA controverts this statement, and in one portion of his evidence before the royal commission he says "that the deficiency of one firm

7

was equivalent to the whole amount of money they had paid for wages from the day they went into business to the day they stopped" So that it was not wages that caused that firm to collapse

Mr BRASSEY, in speaking of the cost of the South Devon Railway, stated, "that the amount of wages paid, afforded no adequate criterion of the actual cost of production The wages were advanced 50 per cent, and yet the work was executed at a cheaper rate at the higher scale of wages"

Mr BRASSEY further states that a similar remark applied to the works in Oxford Street, in connection with the drainage of the metropolis, which was constructed at a cheaper rate, per cubic yard, when the wages of the workmen were raised to 10s per day, than when they stood at 6s."

Mr HEWITT, the American Ironmaster, considered it "the most marvellous thing in Europe, that mechanics could be got at so cheap a rate"

The advantages of a Trade Society enables its members to strike for a redress of grievances, with a probability of success, which a few men could not hope to attain, and reduces the whole to a science by regulating the machinery with which it is guided, and by preventing undue influence

Contentions between employers and their workpeople are not of modern date. The earliest record that we have of any dispute between the hirer and the hired, occurred so far back as 3 500 years ago, when Jacob quarrelled with Laban over a question of wages, and we all know that the afflictions of the Israelites under the bondage of the Egyptians is as old as the history of the Bible, for, we are told that they had "to serve with rigour, and their lives were made bitter with hard bondage." So much so, that the Lord said, "I have surely seen the affliction of my people, Israel, which are in Egypt, and have heard their cry by reason of their taskmasters, for I know their sorrows and I have come down to deliver them," and the punishment inflicted before the glorious delivery of the people from their oppressors is sufficiently well known. Moses himself deprecated "the oppression of a hired servant," and the prophets strongly denounced those who "oppress the hireling in his wages, and grind the faces of the poor." Our blessed Saviour, himself the son of a carpenter, selected His apostles from the poor and lowly, with whom He associated, and His teachings are worthy of careful study St. Paul exhorted masters to give to their servants "that which is just and equal, to forbear threatening, and to remember that they too have a master in heaven, with whom there is no respect of persons" The Apostle James says, "Behold the hire of the labourers who have reaped your fields, which is of you kept back by fraud, crieth, and the cries of them which have reaped have entered into the ears of the Lord of Sabaoth"

No wonder then, that in this modern age of progress and education,

8

that the working classes should be found to step forth in defence of their labour, which the moral obligations of employers fail to recognise, either in the present day or in the past history of nations

Trade Union officials are often taunted with fomenting disputes but Strikes never take place without due deliberation. They are not entered into, as is too often alleged, at the instigation of officials, for they entail an amount of labour which that overworked class would rather shrink from than otherwise. They, in fact, receive as careful and thoughtful discussion as a cabinet of ministers would devote to the breaking up of all peaceful relationship with a country in alliance with their own, but if no other alternative is available they are carried in the affirmative, not by a select view, but with the full voice and concurrence of the whole of the members convened together to discuss the position of the case, and the relative merits of the same, and the consequences, however cloudy, are not shrunk from, for duty often points the way that angels would fear to tread, and the courage and forbearance evoked, call forth the substantial approval of those in other trades, who, finding the material of war, prove to the world at large how unrestrainable is the courage and perseverance that is implanted in every sympathising breast of the Anglo-Saxon race The Royal Commission on Trades Unions dispelled that illusion, which the outside world so long believed in, of the secrecy with which Strikes were conceived and undertaken, as well as the supposed secret organisation and operations of Trade Societies.

Strikes rarely take place unless trade is good, thus their action is not unjustifiable, but exactly in accordance with the pet theory of supply and demand. They never dare to strike with an overburdened labour market, unless to do so in self-defence At such times they are not aggressive, but repressive. With all their faults they are never found acting in opposition to the formation of employers' associations; and, whilst respecting the rights of others to combine, they claim the exercise of the same privilege themselves Such cannot, with truth be said of their opponents, for efforts have been made to not only destroy the workmen's Unions, but also to outlaw such as have rendered themselves obnoxious to the employing class through their prominent attitude in trades disputes by publishing their names and deterring other employers from engaging them.

If employers would, by the adoption of the principle of Industrial Partnership, share with the operative the fruits of his labour, then Strikes would cease Selfishness is inherent in human nature, for there always exists a strong and natural inclination on the part of most men to obtain the best side of a bargain in any business transaction, and wages, forming an important item in estimating for work, is the first to feel the keen edge of the pruning knife. The spirit of competition which pervades modern society has a tendency to induce employers to reduce wages, as well as the price of material, which is but the product of wages, also, to the lowest possible point, which naturally proves

9

productive of opposition, and the weakest must go to the wall: thus we find the necessity of a repellant power which shall form a barrier against reduction, and this power is the resistant attitude known as a "Strike" Harsh word though it be, it simply means that, just as the merchant, manufacturer, or shopkeeper, declines to part with their goods under a certain figure, so the operative elects to withdraw his labour until he obtains his own price for it. Employers rarely, if ever, raise wages voluntarily, or give more than they are absolutely compelled to do, and the only means left to obtain a just recognition of justifiable claims lies in the power to enforce them. Man, as a unit, is incapable of resisting or standing upon the dignity of his position, for his necessities often, thus situated, compel him to accept terms repugnant to his moral feelings. Withdraw all protecting agencies, and he becomes a prey to the power of those with whom, in the present state of society, rests the sole means of his obtaining a livelihood for himself and family Money is simply labour in a concentrated form, always capable of commanding service from those who require it, and the possessors of wealth wield a power equal to the collective strength of the largest number they are capable of employing A capitalist can also easier change the mode of investing his money than the tradesman can change his occupation, for the latter is wedded to his trade, and is often valueless, or comparatively so, in any other pursuit. The best wages paid are only equivalent to the actual necessities of a family. No large amount of money can be saved towards a maintenance in old age. In the event of the operative being called upon to vindicate his position in any dispute that may arise between himself and his employer, he has no means of standing out in defence of his principles. Isolation to him is certain defeat. The only alternative for him is to obtain the assistance of others by associating together in any contingency, and when the combination is perfect, and the material of assistance in abundance, then those discontented assume an offensive attitude, and strike for what they conceive to be their just rights Many societies, prior to entering on a contest increase subscriptions to raise a fund capable of supplying all anticipated wants, at the same time calculating the time when the exigencies of trade will most effectually serve their object. In this they simply adopt the same plan as their employers in watching the rise and fall of the market before buying or selling The employer, by the possession of capital and the adoption of the principles of combination, now so popular, is doubly armed; for his capital alone will suffice to be placed in the scale against the operative class in any suspension of labour, for capital consumes nothing, but must command interest wherever invested The workman ceasing to labour must starve or return to work thus the exigencies of his position compel him to accept what he conceives to be unjust terms, however much his conscience may recoil from it, sooner than the employer. A Strike in defence of his Trade Society, and in vindication of his liberty to belong to such, he claims as a just right, and is very tenacious of outward interference, for he knows that it is a never-failing source of strength in time of need. The employer obtains service at the hands of unprotected labour for a bare living, without affording any guarantee of making provision for the labourer in the future, for when such service fails to be remunerative, or sickness, old age, or accident overtakes the toiler, he is cast off, either to exist on the common rates, which is an injustice to the ratepayer, or he becomes a burden to his relatives and friends. The lodge-gates of the employer are closed upon him, and he becomes a waif on the waters of an unsympathising world—a human derelict, cast away when he has served all the purposes for which he was retained. Thus, by the force of circumstances, he is driven to seek a palliative remedy in a Trade Society, where superannuation funds exist for the aged, travelling allowance when necessitated to seek work away from home, out-of-work donation for the unwilling idler, and a burial fund for the departed brethren. No wonder, then, that he is prepared to suffer a Lock-out rather than cease his membership, and to strike when his claim to continue a unionist is challenged; for when his employer forsakes him, he can rely upon the provident habits of former years, which have secured a suitable provision by easy payments on work obtained, not the pittance doled out by workhouse officials, but an allowance that he can claim as his own; not as a privilege or a charity, but as a right, for he is simply recouping himself from money which he has invested in former years as a joint partner in the undertaking against a period of adversity, the administration of which funds hav[illegible]n so wisely economised as to produce a maximum cost of carrying out the scheme.

Whatever our other faults may be, ingratitude cannot, at all events, be imputed to us. In this we contrast favourably with our opponents, for it is a canon of their icy creed to care nothing for any one but their own individual interests, but the cardinal virtue of a trades unionist is to promote the happiness of the greatest number. Labour unprotected degenerates. The lowest-paid trades are those least protected. The richest city in the world contains within its precincts a pauper population that would fill half-a-dozen towns. One hundred and twenty thousand paupers within that city exist as a standing incontrovertible evidence that labour, unprotected, sinks to the lowest depths of the social scale, and that, without union, these ills that we see exist amid a plethora of wealth can only be avoided by capital and labour arranging their differences more in harmony with the principles taught by Christianity and sanctioned and approved by justice.

The numerous contentions between employers and employed in Strikes and Lock-outs have been set down as a great loss to the operative class, and statistics have frequently been given in support of arguments submitted. For instance, it has been urged that the loss of wages expended in a Strike does not secure a corresponding equivalent in advanced wages obtained. Granted, that in many cases the argument

11

holds good ; but cases can be adduced where the loss has not only been made good, but a large bonus secured within a very limited period. The painters record a Strike which cost them 1s per man, by which they realised 2s. 5d per week, and the joiners' last Strike in Manchester, although protracted, and costing £4 10s. per man interested in that locality, secured an advantage equal to 2s. per week. But if it is beneficial, as the employers assert, for the operative to give way, would it not be likewise equally as beneficial for the employer to do so ? Does he not make sacrifices in a contested struggle, as to which shall first yield ? What amount of capital in stock or machinery is lying idle ? what contracts unfulfilled ? what orders unexecuted, and others lost ? What sacrifice does he make in a change of hands unsuited for the work, and unaccustomed to it, not to mention rent and taxes going on without corresponding returns ? What about the loss on work that might have been executed ? The operative may fairly throw down the challenge as to which should yield the first; for it is the public, after all, that must pay for the enhanced cost of the work produced, if it is a question of time or wages ; and if it is simply a question of union or non-union labour, then, all that is lost, the objecting party should bear the brunt of the sacrifice made. But I maintain that the history of Strikes affords abundant proofs that the operative has, by striking, and being in a position to exercise that privilege, advanced his social position. What stronger proof need we than the fact of the building trades, every branch of which have not only increased their wages within the last thirty years from an average of 24s. per week to an average of 36s., as now paid, but have reduced their working hours from 60 to an average of 54 per week. There is scarcely a body of men that can be named, but who have, by wielding the same power, and using that mighty leverage, so reduced working hours and raised wages, that those who have attained to years of maturity, and remember the long hours and low wages they formerly worked, rejoice, and thank God that he has implanted within the human breast a desire to so use the gifts which have been granted to him, that they shall not only contribute to his own present good, but so influence the future that generations yet unborn will enjoy the fruits, the seeds of which their parents have planted, and result in deeds that will re-echo the aspirations of the present generation Strikes on behalf of reduced hours must conduce to the growth of intelligence, tend to make our countrymen excel all others in the rank and file of creative art, progress with machinery and other scientific aids, until life becomes a real pleasure rather than a heavy burden, and all will be enabled to enjoy the fruits of the earth and glorify the Giver of all good.

But it is to union aid, in directing and influencing the action at the proper time, that we must look for and attribute the beneficial results which have been secured and guaranteed , and this state of things has been brought about simply by united action and mutual knowledge of the best means of securing the desired end, which is a mathematical

12

demonstration that the power of union is the only means at the present time at the disposal of the workpeople to secure a fair and equitable settlement of the reciprocal advantages which should be secured to all through the demand for any work that requires to be executed That the principles of Trades Unions have proved acceptable to the operative classes is evidenced by their present numerical strength, while at the same time they have, through the medium of the apprenticeship system, afforded a high degree of skill in the excellency of workmanship, and, by the maintenance of a standard rate of wages, have perpetuated such an equilibrium of equality as enables employers to estimate for contracts with that certainty which, in the absence of a controlling power, would not exist, and, further, Trades Unions have improved the social position of their members, inculcated provident habits, and enabled them to obtain recognition of just claims which, in their isolated position, they could never hope to command. They have at the same time secured the advantages of imperial legislation in the rectification of many grievances, arising from pernicious laws under which they have laboured, and tended materially to the intellectual, moral, and physical improvement of their members

The workman, unable to accumulate sufficient capital to act on the defensive in reserving his labour until he can obtain for it a recognised value, invests a percentage of his earnings with his Trade Union, which thus becomes his banker, and when his employer disputes his right to claim a given rate of wages, the Union steps in, and with its collective power protects the individual, who singly would have to succumb, were no defensive power in existence to shield him from the power of capital The unprotected labour of the workman lying dormant, tends, through its inability to cope with capital, to become a prey to the latter, which can always hold out the longest. The cessation of labour increases the value of the work already produced, and thus aids the employing class in rivetting still firmer the fetters of the unwilling idler, who, without extraneous aid, perishes in the vindication of his rights, for it is a rare occurrance for the workman to strike except in self-defence, or in the belief that he has a moral right to the claim he makes Ground lying fallow increases in vitality, but the labour of the workman so unemployed swallows up the long savings of years of provident and industrious habits, withdraws his children from school, breaks up his humble home at a ruinous sacrifice, engenders hatred towards his employer, breeds discontent, and annihilates faith, sympathy, and that reciprocity of feeling which should exist in the relationship between employer and employed. The Trades Union here proves the only true friend in such a time of need, the axiom that many can help one when one cannot aid many, is practically proved The contributions of former years are called into requisition, and, if necessary, levies are made by the Union on those in work to defend its principles, sustain its reputation, and aid the victims of capital to assert their claim or maintain their rights, and kindred

13

Unions of other trades are roused up to battle for principles that they may at any time be called upon to defend, and thus in the expectation that they will reap just as they sow, furnish from their treasuries the material for a determined struggle.

Gigantic fortunes, princely residences, and vast landed estates obtained in an ordinary lifetime, must of necessity be the result of exorbitant profits And yet the possessors of these easily-acquired fortunes can be counted by their thousands, and the bulk of their owners started life with little or nothing. Who ever heard of a working man retiring upon the fruits of his industry as a journeyman? When sickness, loss of employment, or death ensue, the Trades Union is the only resource, and rarely, if ever, fails to afford aid in time of need. Thousands of pounds are so expended in travelling allowance, sick, out-of work, accident, funeral and superannuation benefits, and hence the degradation of receiving parochial relief is obviated, the dignity of the operative is sustained, principles of independence practically inculcated, provident habits encouraged, and the rates of the employing class saved at least to the extent of 50 per cent., besides the services of a multitude of officials dispensed with, for the duties are performed not with the severity and insolence of the parish authorities, but by fellow-tradesmen repaying that which has been committed to their charge in times of prosperity, against the hour of trouble and the day of trial.

We have been told that Trade Societies drive trade from the country, and interfere with the development of the resources of the nation, and that the principles of Free Trade should extend to labour. Well, if Free Trade is to become the rule, the floodgates thrown open, and the legitimate and natural channels of the stream no longer recognised, there must be no barriers to any The principle to be just must be general No trade, profession, or vested rights, must be any longer conserved. Land, Patent, and Copy-right, must cease to enjoy the protection hitherto extended to them Law, Medicine, and Divinity, must throw their portals open for all who claim to exercise such professions There must be no monopoly in our Colleges, Universities, or elsewhere The various services of our country—Civil, Naval, and Military—must accept the services of the lowest bidder From the Crown to the humblest office in the realm, freedom to attain and aspire to must be extended to all The humblest operative may intuitively possess the genius to command armies, or rule the senate, represent the authority of the country at foreign courts, negotiate treaties, or become a modern Lycurgus, defend the felon before the law, cure the sick, preach Christianity, or fulfil any of the duties now entrusted and monopolised by the employing class Free Trade is unlimited in its application It recognises no line of demarcation. Its standard may be unfurled in the palace as well as in the workshop, and if the humble artisan can obtain no recognition for a lifetime spent in the attainment of a knowledge of his trade, for an educational training and a servitude performed on the faith of securing honourable employment at its termination, at a commensurate rate for the sacrifice involved, then he may without compunction of conscience welcome revolution in the demolition of class distinction, and the abolition of all the safeguards for the protection of vested rights and honourably acquired privileges.

The social condition of the working classes of the present day, as contrasted with those prevailing half a century since, is no doubt greatly indebted to the existence of Trades Unions The aptitude for business displayed by the members in the management of their societies in all their various ramifications, redound to their credit, and prove the growth of intelligence called into existence and developed through their various operations

If Trade Societies should cease to exist, employers would still have workshops to contend with, and they might cause more annoyance and inconvenience to an employer than the evils now complained of, from the simple fact that they would each act on their own responsibility, being subservient to no general law or organised body, whose rules would naturally disclaim all measures not approved of by the members of the trade collectively

It is to the skilled labour of our artisans that we owe the majestic piles of masonry that embellish and adorn our cities the net-work of railways spread over the empire, whose tunnels pierce our hills, and are carried over the ravines and valleys of our land, the majestic ships, the value of whose freights would have exceeded the wildest imaginations of our forefathers, whose huge hulks are docked in every clime By their agency the mineral wealth of the world has been melted, moulded, and worked into a thousand shapes, from the humblest household utensil to the most majestic of bridges that span mighty rivers They have constructed machinery sufficient to clothe the world, as well as gather in the fruits of the earth And out of the wages received from their industrial pursuits they have founded benevolent aids in every department of labour Whilst nations have armed against nations, the operatives have formed friendly alliances with the operatives of other countries, and linked themselves together in chains that, instead of fettering each other, are golden links and symbols of universal brotherhood

---

The necessity for the establishment of some counteracting influence to the power exercised by the capitalist has been recognised by the ablest political economists as well as the best friends of the working classes, and the independent testimony of the following has been selected as a fair speciman of the general feeling of those who are favourably disposed towards Trade Unions:—

ADAM SMITH says:—"Workmen desire to get as much, the masters to give as little, as possible; the former are desirous to combine in order to raise, the latter in order to lower, the wages of labour"

15

"A liberal reward of labour tends to the increase of wealth, and to complain of it is to lament over the necessary effect and cause of the greatest public prosperity"

"We rarely hear, it has been said, of the combination of masters, though frequently those of workmen. But whoever imagines on this account that masters rarely combine is as ignorant of the world as of the subject Masters are always and everywhere in a sort of tacit, but constant and uniform combination not to raise the wages above the natural value Masters, too, sometimes enter into combination to sink the wages of labour."

Dr ANGUS SMITH, F R S, in a lecture delivered before the members of the Manchester Philosophical Society, said—"Supply and demand were terms used to skim over great difficulties They have been treated as merely physical, but when moral considerations enter, supply and demand alter their conditions, and THE PHYSICAL LAWS MUST GIVE WAY, because there are moral feelings in men stronger than their love of gain"

SIR A ALISON says —"I think Trades Unions are not only proper, but a necessary balance of the fabric of society. Without them capital would become far too powerful, and the workmen would be much beaten down, and I think the men have reason to believe that if they were deprived of the power of combination, their wages would be gradually brought down to such a point that they would be reduced to the former condition of the serfs in Russia"

Mr. J RUSKIN in a paper on "Wealth and its Distribution".—The idea that direction can be given for the gaining of wealth, irrespective of the consideration of its moral sources, or that any general or technical law of purchase or gain can be set down for national practice, is perhaps the most insolently futile of all that ever beguiled men through their vices So far as I know, there is not in history the record of anything so disgraceful to the human intellect as the modern idea that the commercial text 'buy in the cheapest market and sell in the dearest,' represents an available principle of national economy This supply and demand argument is especially the mercantile form of theft, consisting of taking advantage of a man's necessities in order to obtain his labour at a reduced price"

THOMAS HUGHES, Esq, M P, in speaking of the cold, unsympathising doctrine of the political-economic school, draws attention to the fact that capitalists treat the labour of their men on the same principles as those on which they treat a dead commodity They most vigorously applied to it the same law of supply and demand, thereby putting the living man and inanimate things on the same footing"

Mr. W. R. CALLENDER, in reviewing the labours of the Royal Commission, thus expresses his views —"If the Unions have not unnaturally augmented the price of labour, they must have prevented its undue depreciation, and promoted the object which they especially seek-to protect the workmen from any unfair attempt to reduce the level of wages, and to secure the equitable distribution of the sum which the combined labour of the trade ought to receive The cry against Unions confirms this It is no uncommon thing to hear it said that the necessity to pay a fixed rate of wages deters the employer from giving more to certain workmen, but this, if analysed, means, that he would like to pay more to a few and reduce the wages of a larger number, in other words, he would provide an article cheaper, by dealing with each workman separately, and gain an additional profit, while the public, so far from deriving any benefit, would pay the same price, and in times of depression would be burdened with additional poor rates to support those whose wages would be ground down by individual competition The public have no interest in this question as opposed to the Unionist In times of prosperity it is no injury to the employer, or to the public, that the work performed should be duly paid for, and the proportion of payment which each worker should receive is a matter for his own decision, and the benefits which the Unions furnish are invaluable when trade is depressed, and few will undervalue the habits of prudence and self denial which their support demands There are, unfortunately, a large number of persons who look to the laws of so-called political economy and the accumulation of profits, rather than the welfare of the people, they regard an increasing trade and cheap provisions as the best, if not the only test of our national welfare, and they have uniformly opposed any legislation or combination which might shorten the hours of labour, improve the condition of the workman, or reduce the possibility of accident to the lowest average"

Printed by the Salford Steam Printing Company Limited, 217, Chapel Street

18

THE PRESIDENT'S ADDRESS.

The PRESIDENT then said—Gentlemen, you will observe from the programme that the next business is "The President's Address." Before I enter upon this, however, let me say that I think we have to congratulate ourselves on the position of this Association. We have further to congratulate ourselves on the numbers that have joined us since our last Conference held in Glasgow. The Secretary informs me that the addition to our numbers is 14,500. (Hear, hear.) At that rate we shall very soon embrace all that remains in Scotland or in England out of the other Association, if we proceed in the way we have been doing; and I may just say, before I enter upon the remarks I intend to make, that I think this result is the more gratifying, as those 14,500 have joined without the slightest influence, and without the slightest exertion being made on our part. We have not spent a shilling, we have not spent a penny, and we have not spent a farthing to add those under our banner. (Hear, hear.) They are voluntary members, and, consequently, the result is the more to be admired on that account. Perhaps it may be your duty, before the close of this Conference, to consider whether or not we might assume a little more of the aggressive character, with a view of still further increasing our number. That will be for you to consider. Mr. M'Donald then delivered the following address:—

GENTLEMEN,—Meeting, as we do, in the town—if I may use the expression—where we first drew breath as an Association ten years ago, I have deemed it my duty to call to your recollection a few facts in the history of the miners, as a class that may interest some and instruct others. In 1844 a great effort was made to enrol the mining population into one vast organization. The effort for a time seemed to do well for success; but the want of knowledge of what would constitute a real association soon began to appear. Strikes took place at the moment when they ought not—when money was high in price and coals falling in price, and other things altogether wrong, till at last all collapsed, and union had almost died out. There were a few who held fast to the principles, and these men began to think that they ought to make an effort to remedy some of the evils by legislation. One cannot take time to ment on them all—there were the Judes, the Grays, the Swallows, the Tetlows, and others. They got the inquiry that gave the Act of 1850; they got the inquiry that led to the passing of the Act of 1855. The reason why the complete breaking up took place we have been unable to learn, yet the disintegration was completed with 1855. Mr Swallow, who was by far the strongest man of the party, entered upon other pursuits and settled down. Mr Jude, though always willing to help, was no more actively engaged in London or elsewhere out of his own county, urging for more legislation for the protection of the miner and his little ones. Mr. Tetlow, of the old party, alone entered upon the arena anew to aid in the work. All the time the fragments of union that existed got less by degrees and more minute, till, at the close of 1855, it might be said that union among miners in the whole country had almost died out.

It was in 1856 we crossed the Border first to advocate a better Mines Act, true weighing, the education of the young, the restriction of the age till twelve years, the reduction of the working hours till eight in every twenty-four, the training of managers, the payment of wages weekly in the current coin of the realm, no truck, and many other useful things too numerous to mention here. Shortly after that bone began to come to bone, union began to live, and by 1858 we were in full action for better laws. A Conference at Ashton gave the first

20

signs of life. Representatives from South Yorkshire, West Yorkshire, Staffordshire, Oldham, Scotland, and a few other places determined that things would not remain as they were. The year 1860 came, and we succeeded in getting a new Mines Bill; but the defects of that bill were so great that we made still greater efforts to have it amended.

HARTLEY CATASTROPHE.

The dire calamity that made the nation for the moment to stagger as a drunken man occurred in 1862. With the passing away of it came a strong desire that there should be a National Conference. At this very time there appeared on the boards—if we may use the expression—a new element outside of the miners advocating to some extent their claims. First we had a Mr. Colwell; next we had a Mr Maitland; then we had a Mr Towers; and later we had enlisted a Mr Whitehorn and a host of friends, all seemingly interested in the condition of the miner. It may be noted that some of the parties drew largely on the purses of several noblemen and gentlemen to carry out the works. With the aid of the paper that had been established, and the persistent communications of others, a Conference was convened, which assembled in the Co-operative Hall of this town on the 9th of November, 1863.

THE PARTIES COMPRISING THE CONFERENCE.

That a Conference so brought together should contain more than one element was to be expected—the habits of the miners were so diverse; their education or no education so different. The first day developed clearly the two great parties which exist among the miners now—1st. The Federal, which has always formed the National Association—those who believe in purely self-government, and those who think, and as we think rightly, that outsiders' interference on the part of any body, however learned, is wrong, to say nothing of the fact that faction might be called upon to rule, or ignorance sit in the seat of direction. 2nd. There was the party who thought that the government of the whole of the miners should rest in the hands of a few, though the leader and the whole of his followers might, from the nature of the case, be utterly ignorant of the many conditions under which the men were placed. Unfortunately, there was at the time a third party, who had been simply brought there to carry out the schemes, designs, and artifices of the adventurers that hovered around the miners. These cormorants whom they followed cared for nothing but the miners' pence and the miners' name to raise them to position.

THE CONFLICT.

Among those that attended the Conference that began on 9th November, 1863, were Messrs Crawford, Sheldon, and Normansell, who still are enrolled under the old flag of the Miners' National Association. The Conference closed in a state of open conflict. The Federal districts were so strong that they carried with them a few that really believed in the dependent principle, but who could not see their way to the reign of a Towers or a Whitehorn over the mining population—men who were utterly ignorant of their wants. The hostilities waxed stronger and stronger every day for several months, papers were distributed by thousands, and money was scattered freely to secure the destruction of all who opposed the objects of the outside party.

THE YEAR 1864.

The year 1864 was one of considerable commercial depression, which went into 1865. In trying to maintain good wages the union of Staffordshire nearly disappeared altogether, what remained being strongly opposed to the National Association. In Durham, where the men had joined the union in thousands, the organization was broken up by the doings at Willington and Sunny Brow. As it became apparent in 1866 that the efforts of the Towers party could not shake the hold which the leaders of the National Association had upon the members, they became every day more virulent in their attacks. Efforts were put forth in 1866 and 1867 to have union established in Derbyshire. Many thousands of pounds were spent on this enterprise, South and West Yorkshire bearing the greater weight of the struggle—districts that very notedly called for the principle of an enforced levy on all members in all the districts, were far behind in the payments. It may be stated that some bitterness of feeling against union seems still to exist among some of the Derbyshire employers. They fail to stop its growth, as they fail to see their own folly in trying to extinguish a principle that has its seat in the deepest recesses of the human mind—unity of purpose to secure great en's.

THE END OF THE TOWERS PARTY.

When they could not wrest the miners' funds from their hands, or the government of their associations from their power, after spending enormous sums, the Towers party totally disappeared, but not, however, without sowing the seed of bitterness, which may be found here and there to this very hour. According to their own statement, as published in the *Advocate*, a body of men were hired in various districts, whose duty it was to vilify by every possible means the leading men of the Association. This was so flagrantly carried out that personally we carried them before one of the superior courts and compelled them to apologise and pay all the costs of the action. Their efforts to form a separate union or association also turned out a most miserable failure. A full account of the doings of this party will form a curious episode in the history of the miners, that remains yet to be written—the energy they put forth and the perfect discomfiture of their designs; the end of many of them, and the position of some that do remain; the total disappearance of the chief; the paper which they had, and which would have been a power for good beyond measure if it had been rightly conducted, after some flickering efforts to live, also paled out its fires.

THE FALLING OF WAGES.

In all the districts, save that of Northumberland and South Yorkshire, the rate of wages was greatly reduced. In Lancashire, in fact, they had a struggle.

32

Though they were warned what would be the result, they hazarded a strike, which brought ruin on nearly all the unions. In 1869 Messrs. Pickard and Halliday almost began to despair of the continuance of union in the county of Lancaster. We strongly advised them to call a county meeting. That meeting we attended, and, as President of the National Miners' Association, I presided. We may state that we wrote the circulars calling the meeting.

FORMATION OF THE AMALGAMATED ASSOCIATION.

We have already noticed the desire of some to have the whole Association under one control. Lancashire, in particular, made this suggestion. In the fall of the year 1869 we went to the United States of America, and on our return we found that Wigan, Farnworth, Worsley, and what of union that existed anywhere in the whole county, had agreed to become the Amalgamated Association. After that they took to the left, and we held on our way to the right. Expressions were often heard that almost led us to believe ourselves to be members of a played-out institution; that some fine morning, or the reverse, some dull one, some of our really enthusiastic friends would be called upon to bury us, without a monumental slab to say we ever had lived.

THEY CROWD TOWARDS US.

The hopes and the fears of both parties have, from the hour of their departure, every day become more and more dispelled. Soon after Durham joined our ranks with its 10,000 strong, now swelled out to 33,000 or 40,000 strong—a mighty contingent force! South Yorkshire increased, the Northumberland ranks grew stronger. West Yorkshire grew like a gourd, and has yet the stability of a wide spreading oak. Cleveland joined our ranks. Fife, Clackmannan, Mid and East Lothian, Maryhill, Larkhall, Den, and Dalry, with Wishaw, made common cause with us, which make a force from Scotland alone of about 19,000 men. These were followed by Warwickshire, Leicestershire, Nottinghamshire, North Wales, North Lancashire, and part of West Cumberland, and the main body now numbers above 120,000 strong—living in peace and unity in all our districts. The little one has become a great nation, and we hold a shadow and protection for all as a great rock in a weary land.

THINGS WE HAVE DONE.

Now, gentlemen, we think it is time that we should say something of what we have done as an association, by ourselves and with the help of others. As we stated in the outset, all was in a state of chaos—we were without power in and around us.

1.—ALTERATION OF THE LAW OF MASTER AND SERVANT.

In this town we suggested that a National Conference be called to consider how to take joint action on the subject. The Glasgow Trades' Council formed an

23

executive and took the matter up. A Conference was called and inquiry by a select committee was got. An Act was passed, which though it did not afford all we desired, yet there was a great improvement on the former state of things. The dragging of men from their beds was ended, and the character of the law became partly civil.

2.—TRADES UNIONS.

We found, in dealing with matters through the members of the Legislature, that there was a strong suspicion among them that the conduct of trades unionists as a body could not stand the light of day; that it was right they should be under pains and penalties that the rest of the community should not be subject to. Our correspondence with Mr. Walpole and others could tell how far we succeeded in the matter. After a full and searching inquiry by a Royal Commission, composed of a few of the bitterest enemies of trades unions that the country had, they had to report they were not all evil. Nay, that under certain conditions they should be made legal. One thing he would make bold to say, and that is, that they failed entirely to connect one or any of the great associations of the kingdom with any outrage on life or property. We do not make any apology for the outrages that section of the Commission at Sheffield or Manchester disclosed. We beg to express our strong abhorrence of those crimes. But let us ask, has society itself not committed crime? Has it not often, by its rulers, outraged every sense of justice? Have not even the classes that so strongly condemn these men's acts, done many things that would vie with the worst of them? And has not even a portion of the press of England shouted with delight at the doings of the brutal despots of our time to gain power? True, I say nothing of our history, which has deeds recorded in its pages which sicken the heart of the most callous—all done in the name of law and order. The law was changed, though, as in the former case, not as we could fully desire, yet there was progress on the subject.

3.—TRUCK.

This abominable system had its existence, more or less, in every mining village; making its votaries the most dependent and the most wretched of beings. We clamoured, we petitioned year after year, though the whole working classes looked on with an utter disregard of the evil, till the attention of the Government was attracted, when a Royal Commission was appointed to investigate into the whole question. That Commission's proceedings were thorough and exhaustive. A Bill followed, and another, but still, though the promise was given, the law is not yet fairly altered. If the giant is to be destroyed, they will only do it by the united blow of the whole working classes. It is a source of gain, and the selfish will return to it the moment the excitement dies away, we fear.

4.—LAW OF COMPENSATION.

In our united capacity we tested more than once the full powers of the existing law. The case of Wilson *v.* Merry and Cuninghame will remain on the list of decisions so long as the English language lasts. It led us to see the folly of

24

further actions almost to obtain just and proper redress. Knowing well that this law is one that affects large masses of our fellow-countrymen, we have handed the matter over to the Trades' Union Committee. The change that we desired has not been effected. Bills have been brought before the House of Commons, and public opinion is being moulded on the subject. It only requires pertinent and persistent work to accomplish the object in view.

5.—THE MINES ACT.

On this matter we cannot fully enter; only a few sentences. We sought that (first) children should be educated; second, we asked that they should not be allowed to enter the mines till they were twelve years of age; third, we asked that their hours should be only 48 hours in the mine in the week. We got 54. We asked that there should be true and just weighing. All that we sought is fully attended to. Trained managers now are of necessity showing that the arguments we used had taken effect in a most wonderful manner. It is the distant future that will tell fully of the benefits of that Act when carried out in the letter of the law and its spirit as well.

6.—OUR SEAMEN.

We have not confined our attention to the wants of the miners alone. The seaman's friend, Mr Plimsoll, made an appeal to the nation. I say it without boasting in any way—the cry for help had a ready response from us. Our friends of South Yorkshire passed their vote for a thousand pounds. They passed their vote and they paid their money. They did not wait for some months and then vote again. The gift was noble, princely. It became the donors—the miners of South Yorkshire. The men of South Yorkshire are always great in their benefactions. Durham gives L.500; Cleveland its L.300; Northumberland, L.250; Fife, L 100; Ashton, L.50, and other districts followed the example; Stirling and Linlithgow, L.40. Even this day, the district of Aspatria, only joining us, gave the sum of L.11 10s; and Furness and Cumberland the sum of L.10—in all, L.2800. It is a thank offering for what has been done for us. It is a proof that poor Jack will not be neglected if we can help it. This will remain as a monument when all of us will have passed away. There are still a few of our local associations that have not done anything. Help in the good work we say to them. (Applause.)

7.—COUNCILS OF CONCILIATION.

We come to the last subject under this head, which I hail with as great pleasure as anything that has been—we mean the forming of Courts of Conciliation. There are men who tell us that the relationship of labour and capital gets hourly worse—that all society is suffering from these things—that there is a dark future for our country from the cause named. The writers of such statements show an utter blindness to facts—to historical facts—that put them down as utterly ignorant of the subject they undertake to write upon. In Scotland, the normal condition of the working miners was a state of strike, open conflicts almost constantly. In

England it was chronic strife. The moment the vanquished could recoup on either side the struggle was renewed. In Wales, the mine owners, who made their colossal fortunes out of the miners, had them shot down on the streets—out of love to them of course, and to make them good law abiding citizens. Their rights were heartlessly trampled upon—their privileges were set at nought; nay, they, in the estimation of many, had *none*. Your gigantic unions have changed all this. They have led employers to see that working men have interests as dear to them as they have. That if these are not respected they will be defended by all lawful and legitimate means. The unions have raised their members from a state of abject and miserable dependence that may be seemly in the eyes of those that profit by it, but a sorry spectacle to every lover of his species. According to the old regimen the workman could not consider all that was right for him—the employer alone could do that—hence the continual strife. In Durham we find employers and employed's representatives meeting periodically to settle differences, while the owner may be in distant lands, and the miner digging his coal far underground, getting his bread by honest toil. In Northumberland it is the same In West Yorkshire and South Yorkshire the storm has ceased to rage. In the Cleveland, that last birth among the mining districts of the north, they also settle disputes in this way; and now nearly 100,000 men are working under the principle of conciliation. Is that no improvement on the former barbarous state of things; is that no change to the advantage of this country? With them that are so blind as not to see the great and mighty benefits that flow from this we have no controversy. We say the progress is great. Let us invoke Him who holds the destinies of nations in His hands to inspire all to spread the principle. Let us all work to make it extend further, and grow deeper, till all the employers and the employed so settle their differences in every district in connection with this Association and over the entire world; as we ask it, let us work for it, and, when we arrange any disputes or differences by this means, implicitly and loyally carry them out, to show to all that our obligations, whatever they may be, we will honestly and faithfully fulfil.

Now, gentlemen, we come briefly to note a few things that we think we ought to hold together to promote:—

NECESSITY OF UNION.

Let us venture to say to the strongest union that we have in our connection that we feel satisfied that however strong they may be in all questions affecting the National interests—on all questions affecting the rights or privileges of individuals—they are stronger, more powerful from being associated; maintaining on internal and local questions their purely federal independence, as against centralisation in the hands of one individual, it may be, all the power. On this matter we wish to be very emphatic. If 20,000 men can create an impression, the complete organization of 120,000 or 130,000 on given principles must have a more powerful influence; for it is no mean army of adults, 130,000 men. Take all means, we would say, to more closely cohere than ever. Let there be a

26

perfect unity, with a perfect freedom to all districts in their own affairs; and one of the first great duties that we have to perform, we think, is to see that the

MINES ACT SHOULD BE CARRIED OUT.

I am aware that the impression of some is that the Mines Act, being passed, they have no more to do in the matter. Nothing can be more fallacious than this, as it would be sure to lead to the provisions of it being neglected. The Act is one that relates to a class. It is one that imposes restrictions upon another class, and that class is allied with the governing class of the country; and we will venture to say, let the matter alone, take away the knowledge of the vigilant eyes of this association and similar institutions being gone, and we would see in full swing all the pertinences and impertinences of officialism that flourished in our early days. The whole body of inspectors need attention. The Home Office did need remembering of its duty; what it may be under Mr Lowe we are not in a position yet to say. The inspectors that are we fully know; the mine owners we also well know, and the power we possess in capital and numbers is what they cannot pooh at with impunity, if only we maintain our existing unity and strengthen it in the bonds of peace.

THE CRIMINAL LAW AMENDMENT ACT.

We ought to join all others, and direct all our attention to have this law repealed; or, if not repealed, to have it so considered that all who perform an act that has the same effect shall have the same punishment. The chief offences under this act are the preventing of men of any class from working. The crime of rattening, which we detest, only means keeping a man or men from work. Picketing, in its very worst sense, is only stopping men from work. To follow through the street in a tumultuous manner, its result is only the preventing of men being employed. It is driving away the independent workman. Now, by common agreement, this is done among the employers, and yet no law touches them. On several occasions we called the attention of Mr Bruce to this. His answer was—'You do the same.' The thing cannot be done. By a letter—and we have ample evidence to prove it—a man may, and has been made to carry about the note to cause his own banishment from the district, or even the half of the nation. Could any cruelty be more refined than this? Yet no law is violated—no statute broken. Only this, we may do the same—an impossibility on the part of the miners. Only recently the coal and iron-masters met and agreed to establish the discharge note in South Wales. This was done by the Fothergills the Davises, the Lewises, and others. Now, the poor unfortunate individual who got the discharge note that makes him a marked man would be as fully barred from employment as if the entire owners of Wales were rattening, picketing, or tumultuously following him or them through the street. Starvation would be his or their lot. Then, we would ask, why should the iron lords and the coal kings do the same as the very worst unionist, and not be punished? If the ignorant miner, bricklayer, mason, or labourer is to be punished, the men being kept from

27

their work, the grand and well-fed, who live by the toil of the many, must be so treated. They must have their term of imprisonment as well as the workmen, or it must be removed from the Statute Book. There is another phase of the discharge note that leads the man to be starved—the refusal to grant it. Then, we should say the refusal represents all the evils of the Act—then the refusal should be the ground of three months' imprisonment, with hard labour. The solution to the whole question, however, is the total and instant repeal of the whole Act.

LAW OF MASTER AND SERVANT.

We are not insensible to the benefits of the modification that took place in the law affecting master and servant in 1867; nay, we hailed that change as a great boon, and we do so still. All honour to those who so nobly aided us to get the change which was then effected. The discussion then, and the facts in history since, have led us to see that the Act was only a half-way measure, and now we ought to demand the full change or modification, and make it just and equal. In the first place, by the 9th section of the Act, magistrates have the power to say what the damages may be, if the employer sues for it, in the way of compensation. In the case of a workman being wrongfully treated by the employer he takes him into court, but only for the sum that he really could have earned; in the case of the man being taken into court by the employer the magistrate can give what damages he may think fit. This is an engine of oppression, and should be removed. Next we come to the portion of the Act that may be criminal, and we hold that, as the whole commercial transactions of the country are carried on on a civil basis, there is no valid reason why workmen should be criminals for any part of a broken contract. There are some who say that criminal proceedings are necessary to enforce contracts among workmen. This we believe to be a false charge against them. They have no interest in being in a chronic state of contract breaking. The workman has nothing to gain by this state of things—on the other hand, all to lose. Labour is a commodity like cotton or corn—it should be sold without pain or penalties, such as are yet attached. In one word, though some of our friends may be opposed to it, the law, to be just, must be greatly altered. Another important subject to engage your attention is

THE LAW OF CONSPIRACY.

Now is not the time to enter into this subject at length. One thing stamps the recent decision in the case of the gas stokers of London as infamous is to have the law on this subject as it now stands. The men were convicted for conspiracy to break an Act that had only a three months' punishment attached to its violation.

LABOUR REPRESENTATION.

At several of our Conferences this subject has been very much debated. It has been said that the members of the National Association should endeavour to put some of their leading men into Parliament; by others it was regarded as akin to nonsense, and that the matter should not be interfered with by us as an association. From these conflicting opinions the last Council meeting had a strong desire to be delivered. They deemed it necessary that at this meeting it should be fully disposed of, and pass among the things that were. With this we very strongly agree. If anything is to be done it should be done soon; if nothing is to be done, then the dream should be dispelled—the bubble should at once be burst. As to the question of the representation of labour directly in the House of Commons, there can be no doubt as to its justness—as one of the fit things to be. We have the testimony of the highest in the land to the effect. Indeed, it seems really to be the fashionable cry of the M.P.'s to declare it. Some of the leading journals of the day advocate it. Common sense declares it to be the real representation. We would say to you decide on its merits. The question is one of great moment to the mining population— more so to them than any other class of the community. They are surrounded from the nature of things by a very large number of class laws. On the fully carrying out of those laws do their welfare and their safety greatly depend. A few members in the House of Commons that could intelligently deal with mining would be of the very highest importance to them. Take an example:—By the statute of 1872 the workmen have the power granted them to send some of the number through the mine or colliery to see if the same be safe. Suppose that a person was in the House that could ask for the returns as to the men visiting the mine and the state they found them. It would have more effect on the country than the inspectors and their assistants have to-day. Were the same twaddle to be repeated about the price of coal and the condition of the colliers, it could be met and put down in the way that would make those that slander repent. Let us again say, dispose of it Many outside are anxious about it, and many inside are the same. The subject itself cannot afford to be delayed, we therefore say— dispose of it. At our last meeting held the attention of the Conference was directed to co operation.

CO-OPERATIVE PRODUCTION.

Let me again call your attention to it. It is a subject that merits our attention in many ways. At the present moment the mass of the wealth of this country is passing into the hands of a few. This is the record of all history. Great wealth in the hands of a few is an evil to those who gain it. It is a curse to all society as well. Much capital on the one hand and much misery on the other is to be avoided. Co-operation in its lowest type tends to correct this state of things—Co-operative production would secure it. It would have all the profits going to the proper quarter. We want no revolution to secure this—we want no violent change to accomplish it. All we want is the army of provident, thoughtful, men. We would ask you earnestly to consider it, persistently to recommend it till the mass of the industry is carried on by that means and by that means alone. This dear old land of ours will become the better the greater the more humane—the more intelligent. It will by this means keep the lead of the nations of the earth, calling upon them to follow, till squalor and misery be unknown.

In conclusion, let me say that there are a number of other matters that should have been noticed. We may say something of them in detail. Our Conferences have ever tended to the welfare of our class, to the increase of our personal knowledge, and a general improvement in many respects. Let me earnestly hope this will be so on this occasion as well. To some of us the call may soon come to leave them and that for ever. Let this one be spent so profitably, so agreeably, that to the latest hour of our existence we may be able to look to it with pleasure, and not with regret. Let it be spent by all of us in cementing friendships, in making alliances that will prove beneficial in the highest degree to those we represent and to ourselves as individuals. One thing I feel satisfied of, as heretofore you will contribute to making the task a very light one for me. (Loud applause.)

VOTE OF THANKS TO THE PRESIDENT FOR HIS ADDRESS.

Mr. Crawford said he had often heard the President speak, but he had never heard him deliver a more instructive and interesting address than that he had delivered that morning, and he had great pleasure in proposing that the thanks of the Conference be accorded to him.

Mr. Greaves, in seconding the motion, endorsed the remarks of Mr. Crawford. He never heard the President deliver a clearer address or speak to more advantage. There was one point in the address with which he was particularly struck—the reference to the good feeling which existed between employer and employed, and he considered that it was the best policy to cultivate a friendly feeling between the two. (Cheers.)

The resolution was adopted.

THE CRIMINAL LAW AMENDMENT ACT.

A MEMORIAL TO THE RIGHT HONOURABLE HENRY AUSTIN BRUCE, M.P.,

Her Majesty's Principal Secretary of State for the Home Department.

Sir,—We, the undersigned, the officers of the Trades Union Congress Parliamentary Committee, appointed at the Fourth Annual Trades Union Congress held at Nottingham during the second week in January of this year, and representing, by delegation, T ades Unionists to the number of over 375,000, and indirectly nearly as many more, beg most respectfully to lay before you, as the representative of the Government for the Home Department, the feelings of our constituents with reference to the Criminal Law Amendment Act, together with some considerations as to its operation, so far as it has been already invoked by the employers as an instrument of oppression in the several cases, amongst others, hereinafter mentioned.

First.—Your Memorialists most respectfully submit that the new law called the Criminal Law Amendment Act is founded on the pre-supposition of criminal intentions on the part of that large section of her Majesty's subjects known as Trades Unionists; this, we humbly submit, is in itself an act of injustice, caused by an imperfect knowledge as to the aims, objects, and working of those societies called Trades Unions.

Secondly.—We submit to your consideration, also, that any act of violence against person or property, either by an individual or by combination, is fully met by the Common Law of Assault, and by the General Statute and Common Law, such as are applied to the other subjects of the realm; and that, therefore, the application of special legislation to acts committed by that numerous and influential section of our artizan classes which your Memorialists represent, cannot but be regarded by them as a special wrong, inasmuch as they are as incapable as any other portion of her Majesty's subjects, of openly or secretly, violating either the letter or the spirit of the laws of the land.

Thirdly.—Your Memorialists most respectfully call your attention to the cases which have been already tried under this new law, and we submit that great injustice and even cruelty have been inflicted by the way that the law has been interpreted and enforced. With regard to the cases of violence which have occurred, we have no word of apology to offer. These, we all admit, are and should be punishable. Reports of these cases, taken from the newspapers, will be found appended to this Memorial (and, so far as we can learn, the record is a complete one), and we would more especially draw your attention to the Bolton case, both as it was presented to the magistrates, and on the hearing of the appeal before the Recorder. All the other cases that have occurred in which violence was used, were, in point of law, assaults.

Fourthly.—We have further to submit that even were the law found to be necessary in order to reach and punish a class of offences not clearly within the pale of the Common Statute and Criminal Law, that, nevertheless, there are either no definitions where definitions ought to be, or that the definitions are so vague and uncertain that cases of great hardship have arisen, and are likely yet to arise from these defects, other than those mentioned as having already occurred under the working of this exceptional and class-made Act.

Your Memorialists, therefore, most earnestly pray that her Majesty's Government will bring in a short Bill to repeal this Act, or that they will support the repeal of this law if the repeal be moved by an independent Member of the House, which, if it could be proven to have ever been necessary for the purpose of restraining or punishing acts of exceptional violence, is certainly necessary no longer; and which so long as it shall remain on the Statute Book, will ever be regarded as obnoxious, and will, therefore, be a cause of discontent to all those who, though Unionists, and as such specially legislated against in the Act referred to, are, nevertheless, as law abiding as any other portion of her Majesty's subjects. So long as this Act remains on the Statute Book, the Trades Union Act, with all its advantages, will remain inoperative for good. But if the absolute repeal of the Act be deemed at present impracticable, we pray for such improvement to be made in its provisions as will prevent the recurrence of the abuses such as have been enumerated and pointed out as having arisen in connection with the operations of this Act.

And your Memorialists, &c.

Signed,
- Alexander McDonald, *Chairman.*
- William Allan, *Treasurer.*
- George Howell, *Secretary.*

Trades Union Congress Parliamentary Committee.

2

APPENDIX OF CASES

UNDER THE

CRIMINAL LAW AMENDMENT ACT.

In the Report of the Cases herewith annexed, we have deemed it best to reprint them as they appeared in the newspapers, without alteration or abridgement, as they appear to have been pretty accurately reported.

The first case which appeared before the Magistrates was at the Gateshead Borough Police Court, in the early part of August, 1871. A youth, 17 years of age named J F. A. Ward, was charged with throwing stones from the Tyne Bridge, and was punished with imprisonment for 14 days The lad was in no way mixed up with the strike, but appears to have caught some of the enthusiasm of those who took part in the nine hours movement. Whether he was, or was not, guilty of the conduct with which he was charged, and it seemed doubtful as to whether he did commit the act, yet, the charge arose out of violence, or an act which might have caused violence, and therefore we do not in any way quote this as a case of hardship. It is right to add that there was no prosecutor in the case, the youth denied the charge, and that he was convicted upon the evidence of policemen. He was also charged with unlawfully molesting, watching, obstructing, and persistently following two workmen, while returning from their employment, but this appears to have altogether broken down.

Another case was brought before the same court on August 8th, on which occasion William Taylor, a miner, was charged that he did unlawfully threaten and intimidate, with a view to coerce William Oliver to quit his employment. In this case it appears that the defendant had used some bad language, and that some pieces of coal were thrown at the prosecutor, but it does not appear that the defendant threw the coal. The evidence however, seems to show that there was actual intimidation used, and therefore we do not point to this as a case of hardship. Thomas Wharrier was charged at the same time with a similar offence, committed on the same day and on the same man, and in both cases they were sentenced to two months' imprisonment.

We have only to add that in both cases the punishment appears to us to be excessive, and could have been better dealt with under the law of assaults.

Another case was brought before the Newcastle Police Court on August 11th, when William Pearson, 18 years of age, was charged with causing annoyance, by following, in a disorderly manner, James Kennedy, for the purpose of intimidating him. No prosecutor appeared, and the case was withdrawn, but the youth was fined 2s 6d and costs for creating a disturbance. The youth denied the charge altogether.

On August 14th there were several cases before the Newcastle Police Court for alleged assaults on foreign workmen, when imprisonment was inflicted, some for six weeks and others for two months each. In these cases denials were given to the evidence, and it is doubtful whether, had they been cases of ordinary assault, the magistrates would have even inflicted a fine, but would have perhaps bound them over to keep the peace. We add this, not by way of apology for violence, or the semblance of violence, but to show that exceptional severity was used under this Act

On August 18th, a policeman was charged with assault, arising out of the case of George Rock, reported at length in another place, and although the evidence appears to have been quite as conclusive as in several of the cases above quoted, the defendant was dismissed.

On July 12th, three men were charged with intimidation, but the charges were withdrawn on the men promising not to repeat the offence. Under other circumstances the case would never have been brought before a police-court, but in the case of strikes an unusual amount of ill-feeling is always displayed on both sides.

CHARGE UNDER THE TRADES UNION ACT AT GATESHEAD.

(From the *Newcastle Chronicle*, Aug. 5, 1871)

At the Gateshead Borough Police-court, on Thursday morning, before the Mayor, Alderman Johnson, B. J. Prockter, and Henry Milvain, Esq, Thomas Dorrington was charged with unlawfully molesting and obstructing certain workmen, the same being employed at the works of Messrs Clarke, Watson, and Gurney, by watching the approach to the said works with a view to coerce such workmen to quit their said employment.

Mr. J. Joel was for the complainants, and Mr. J. A. Bush for the defendant.

Mr. Joel stated that the complainants carry on business as engineers in Gateshead, and the man at the bar was employed by them for some time in their factory, and, unfortunately for him and unfortunately for every one concerned in the struggle, he was at present out on strike with the rest of the men previously employed in the factory. If the prisoner and his companions had been satisfied with absenting themselves from their work and had left their masters alone, the Bench would not have been troubled with that case. With other men, however, the defendant had thought fit from time to time to go to the neighbourhood of complainant's works to hang about the gates, walking up and down, and pressing the men to come out of the factory. He would prove that from the commencement of the unfortunate dispute these men on strike had made a regular practice of frequenting the road day by day, and had never been absent from the place, two being relieved at intervals by two, and so on. After the Act of Parliament came into operation, which gave their worships power to put down this kind of thing, the men resorted to an ingenious method of upsetting its powers. They got a number of bills and distributed them in the neighbourhood of the works, and, when interfered with, they invariably replied that they were following employment as bill distributors. On Wednesday the prisoner was seen by Sergeant Brown and a police constable slowly walking past the complainant's factory from the direction of Hawks, Crawshay, and Co.'s works. The defendant was accompanied by a man named Turnbull, who was known to be the chairman of the branch league in Gateshead, but when the officer made up to them, Turnbull went away. The prisoner at the bar, having bills in his hand, thought he was sufficiently fortified to meet the policemen, and therefore waited until they came up Sergeant Brown asked what he was doing there, but the defendant only used an improper expression in reply. He refused to give any explanation as to his presence, and was therefore taken into custody. He should call Mr. Clarke, who would prove that the prisoner was one of their workmen, and that he was on strike, and he should also prove, by other witnesses, the facts he had called their attention to.

Mr Robson, the magistrates' clerk, thought Mr. Joel would have to alter the charge before he proceeded further.

Mr Bush thought it a very curious thing that the defendant had been arrested on a charge that was untenable. He must be discharged and dealt with on the second charge to be preferred by Mr. Joel.

Mr. Joel thought if he showed that the object of the prisoner was to coerce the masters or the men from following their employment, that was sufficient.

Mr. Robson replied that the defendant was entitled to know exactly what the charge was against him, in order that he might meet it. The charge was for molesting and obstructing workmen and not the masters, and therefore defendant must be discharged upon that charge and another preferred.

Mr. Joel, after some discussion, elected to prosecute the prisoner for molesting or obstructing the masters with the view of coercing such masters to dismiss or cease to employ some workmen, and the case then proceeded.

Mr. Wm Clarke, one of the firm of Clarke, Watson, and Gurney, stated that the defendant was formerly in their employment, but he was now on strike. He was a striker then and is a striker now. (Laughter) He would not say that he had seen Dorrington in the neighbourhood of the works, nor would he say he had ever recognised him there.

The Mayor : In fact you know nothing about him? Witness: No.

Sergt. Brown stated that when he spoke to the prisoner at the works on Wednesday afternoon he had a number of bills in his hand, and he saw him distribute them to two different parties. He observed defendant walk up and down near the works for about a quarter of an hour and stopped him. He told defendant he should like to have his name, and also what he was doing there. Dorrington replied, "What the h— have you to do with that." Witness asked him for a bill, and he refused to give him one, as he said they were his own property. The defendant refused to give his name or address, and he (the officer) therefore took him into custody.

Cross-examined · I took him into custody because he was walking up and down near the works, and because he refused either to give his name or to give any account of himself. The reason witness took the man into custody was because he was loitering about complainant's works. If he had given his name properly it would have been a different thing He would not say that he would take every man into custody who did not give his name.

Mr Bush—Then why did you take him more than anybody else?

Witness—Because he was walking up and down there.

Mr. Bush—Are these works under your special protection?

Witness—No.

Mr Bush—You asked him for a bill, and he would not give you one. Now, did you not want the whole lot of the bills?

Witness—No, I only wanted one.

Mr. Bush—Did you not take the whole lot from another man?

Witness—I took some, about one-half.

Mr Bush—Now, mind what you are saying; we have the man here

Witness—I took all from one man, half from another, and part from a third.

Mr. Bush—Do you know that he was actually distributing these bills?

Witness—I saw him stop two parties and give them bills

Re examined by Mr Joel—Sergeant Brown stated that the reason he took the prisoner into custody was that he refused to give his name or any explanation as to his presence near the works.

P C. Hunter having given corroborative evidence, Mr. Joel said this was his case, to which Mr Bush retaliated that it was a very shallow case indeed. Without calling upon Mr. Bush for the defence, the Bench discharged the prisoner.

Mr. Bush assured their worships that he had a complete answer to the charge. It was a very hard case for the prisoner, who had been arrested on a most frivolous charge.

Mr. Joel was afraid that the bills had brought about the decision of the magistrates.

Mr. Robson replied that the magistrates declined to say what had done it.

The defendant was then liberated.

THE ALLEGED PICKETING AT GATESHEAD.

TO THE EDITOR OF THE DAILY CHRONICLE.

SIR,—In the evidence given before the Gateshead borough magistrates in the case of Thomas Dorrington for alleged "picketing" at Messrs. Clarke Watson, and Gurney's factory, Sergt. Brown made a statement concerning me, which I take leave to say is a gross falsehood. Brown states that he saw me walking up and down past the gates of the factory for almost twenty minutes, and insinuates that I was engaged in watching the factory gates, and under the pretext of distributing bills I was endeavouring to keep the men from going to work.

The statement made by this pretentious policeman is an unmitigated falsehood, and in keeping with much more that the Gateshead officers of the law have said and sworn to concerning the engineers engaged in the strike for the nine hours.

I never was in the company of Mr. Dorrington in the streets, and have no more to do with the distribution of the bills referred to than Mr. Clarke himself. The men distributing the bills were legitimately engaged earning an "honest penny," and the gentlemen who engaged them can confirm this statement if required.—Yours truly,

JOHN TURNBULL.

16, Lawson Street, Byker Bar, Aug. 4, 1871.

4

ANOTHER CHARGE UNDER THE TRADE UNION ACT IN GATESHEAD.

At the Gateshead Borough Police Court, yesterday (Aug 8th) morning before Mr. Ald Johnson, Mr. Ald. Brown, and G Crawshay, Esq, John Kennedy was charged, under the Trades Union Act, with intimidating James McPherson, to prevent him following his ordinary employment as a blacksmith at Messrs. Black, Hawthorn, and Co.'s Gateshead.

Mr. Bush appeared for the defendant.

Mr. Chief-Constable Elliott stated that the prosecutor lately commenced work at Black, Hawthorn, and Co's, Gateshead, as a blacksmith, and the prisoner was a puddler at Messrs Hawks, Crawshay, and Co.'s. On Monday afternoon, Aug. 7th, Kennedy went into Martin's public-house, nearly opposite Black, Hawthorn, and Co.'s manufactory, and there met with what was called a strike picket One of the men of the picketing party induced the defendant to go over to the complainant to intimidate him, and prevent him going to his work.

In answer to Mr Ald Brown, Mr. Elliott said Kennedy went within the gates and intimidated McPherson.

Mr. Bush desired to know what the charge actually was.

Mr Ald. Johnson replied that defendant was charged with intimidating the man, and preventing him following his employment.

John McPherson, the complainant, on being sworn, said the defendant on Monday afternoon about half-past one o'clock, came up to Messrs. Black, Hawthorn, and Co's gate, and asked him if he was going to work. Witness, replied, "What is your business, what do you want?" Defendant then said if he (McPherson) went into work there, he would catch it.

Mr. Ald. Johnson—What did you understand by that observation?

Witness—I thought they were going to give me a beating or something of that sort, and I told him to walk off or I would get a policeman to him. He went away, and a policeman came up and asked me what he had been saying to me. I told the officer, and he went and took him into custody.

Mr. Bush—Did you give him in charge?

Witness—Yes, and that was the end of it, and I went to work again.

Mr. Shafto Robson—From what he said, "If you work there, you will catch it," are you afraid they will do you some harm?

Witness—No, they will do me no harm.

Mr. Bush—After that answer, I submit there is an end of the case, and I have no charge to answer.

The Chief-Constable—Stop a bit. There are two policemen to call yet.

P C. Brown stated that at half-past one o'clock, he saw the prisoner come out of the boiler yard to the gate, and hold up his hand in a threatening manner to the complainant. He did not hear what passed between them, but having heard from last witness, he went and took the defendant into custody, and charged him with intimidating McPherson.

By Mr. Robson—When Kennedy held up his hand in a threatening manner, both men were a yard or two within the gates

Mr. Chief Constable Elliott, sworn, said that when Brown brought the prisoner to the station and charged him with intimidation, he said he would tell witness all about it. He (Mr. Elliott) replied that he need not do so, but he had better keep his statement until he was brought before the magistrates. The prisoner, however, would tell him, and informed him that there was a picket at Martin's beerhouse, and one of the party gave him a gill of ale to go down to McPherson, and tell him that he was wanted at that house. Prisoner further said that he was very sorry he had been such a fool as to go to the man

Mr. Ald. Johnson said the magistrates all thought that the remarks made were all in favour of the prisoner, and they had no doubt that he would well consider the painful position in which he had been placed, and would not repeat the act again. Under these circumstances, therefore, they took great pleasure in dismissing the case

Mr. Bush said he was much obliged to their worships for their decision, but he had a very great ground of complaint against the police, because the defendant had done nothing that had rendered him liable to arrest. He was taken into custody, and had been locked up in durance vile since Monday last. Most respectable bail was offered and refused by the constable, the chief-constable, and some of the magistrates who were waited upon, and Kennedy, a respectable young man, had suffered this five days' imprisonment without the shadow of a case against him. Was the Act to be taken up in this spirit by anybody? He sincerely hoped not, and while saying this he would ask their worships not to lend themselves either to one side or the other in the great but unhappy trade dispute which at present existed in the neighbourhood. He stoutly maintained that the constable had no right to take the man into custody, because he saw no assault committed or heard anything to justify his apprehension of the man.

Mr. Elliott—After these uncalled-for remarks would the Bench allow me to explain?

Mr Ald Johnson said it was unnecessary. The law was framed to prevent a man who was a union man interfering with a non-union man, and there was every reason in that case to believe that the defendant was endeavouring to prevent him going on with his work. The magistrates considered the officer had acted perfectly right.

Mr Ald. Brown said if they believed that the defendant used the expression "you will catch it," he for one considered there was ample ground to justify the officer in taking the man and locking him up.

Mr. Bush was sorry to hear Mr. Ald. Brown say this, but he would call his attention to the following provision of the Act.—"A person who is a master, father, son, or brother of a master in the particular manufacture, trade or business in, or in connection with which any offence under this Act is charged to have been committed, shall not act as or as a member of a court of summary jurisdiction or appeal for the purposes of this Act" Mr Brown was therefore not allowed to sit upon the case.

Mr. Ald. Brown replied that he was not a manufacturer. He was only a shareholder in certain companies, and nothing more.

Mr Bush said that was quite enough.

Mr. Chief-Constable Elliott said one of the men who had been working was very violently assaulted on Saturday afternoon by three ruffianly fellows when returning home with his wages. He was completing the case, and it would be brought before the Bench in the course of a day or two.

Mr. Ald. Johnson assured Mr. Bush that he had no prejudice, and he should as soon protect a Union man as a non-Union man while in the execution of his duties as a magistrate. If, however, manufacturers were not to adjudicate upon

5

such cases, they would not get anybody to dispose of them but a stipendiary.

Mr. S Robson, the magistrates' clerk, said it was quite clear that those who were engaged in the same trade as that in which the dispute arose could not adjudicate.

Mr. Bush had thought it right to bring this out after the gratuitous remark of the alderman on the right of Mr. Johnson.

The defendant was then discharged.

CHARGE OF INTIMIDATING ONE OF SIR WILLIAM ARMSTRONG'S WORKMEN.

At the Newcastle Police-court, yesterday, August 15th, 1871, before the Mayor (Mr R. B. Sanderson) and Mr. Ald Nichol, John Lamb, 35 years of age, was brought up on a warrant charged with assaulting, threatening, and intimidating James Thompson, a workman in the employment of Sir William Armstrong and Co., on Saturday last, in Back Boundary-street, Scotswood-road. Mr Joel appeared for the prosecution, and Mr. Bush for the defendant

Mr Joel said he appeared for Sir William Armstrong and Co, who, for the sake of protecting their workmen, felt it to be their duty to proceed against the defendant under the 34th and 35th Victoria, an Act passed during the present session relating to violence, threats, and molestation of workmen. The facts he gathered were shortly these. The defendant, who was a man on strike, formerly worked at Sir William Armstrong and Co.'s works. The complainant, that was the person who had caused this information to be laid, formerly worked there, but when the general strike took place he thought it necessary for purposes of his own to go out with the general body of men. As the magistrates were aware, a proposition was made by the masters to the men to go in on certain terms When the proposition was made the complainant thought it right to accept Sir William Armstrong's terms and went in, and surely no one could complain of him for that. It appeared that he had six chickens, one of which he lost on Saturday, and he went to a neighbour's door for the purpose of inquiring after it. Whilst he was there the defendant came up to him He asked the defendant how he was, but he did not get a very civil answer, and it ended in the defendant putting his fist in the complainant's face and making use of language which he (Mr. Joel) would submit was of such a character as would justify them in coming to the conclusion that it was used for the purpose of intimidating the complainant. The defendant said "You are a low, lousy Scotchman for going in to work with such a mean, low crew." The complainant told him he need not trouble himself about him, when the defendant said, "You are a great deal worse than your own countrymen who came here and were ashamed and went back, but you have no shame. If there had been a man among you, you would have turned out when that low crew came to the place."

The Mayor—That's not what is alleged here.

Mr. Joel said the defendant used these words, or words to the same effect.

The Mayor—Let us see what you allege?

Mr. Joel replied he alleged, in the first place, that the assault was committed, that the defendant put his fist in complainant's face.

The Mayor—What are the words you say were used?

Mr Joel repeated the words given above, which the Mayor took down

The Mayor—And that is all?

Mr. Joel remarked that the defendant also threatened to smash the complainant's face, and took up an empty pail

Mr Bush thought that was leading the complainant.

Mr. Ald. Nichol said the man ought to state that.

The Mayor (to Mr. Joel)—The words you use are not in the information, and all depends upon the accuracy with which the words are reported here.

Mr. Joel said they might not be able to give the exact words, but the complainant was an intelligent man, and he would tell them what took place. He had only one observation to make, and that was, if this case was proved to their satisfaction, he was instructed to ask the magistrates to deal out such punishment as might be a caution to others. It might be unnecessary for him to add that, but, at the same time, his instructions were to ask for a severe punishment, provided the case was made out.

James Thompson, the complainant, was then sworn, and stated that at present he was a fitter in the employment of Sir William Armstrong and Co. He remembered when the strike took place, when he with other men went out. He went in again the first day the factory opened, and had been working there since He had chickens, and lived at No. 4, Back Boundary-street. On Saturday evening he lost a chicken, and immediately went to the yard of Sarah Pearson and made some inquiries. The defendant worked at Sir William Armstrong's factory, but was not working there now. He was one of the number who turned out. As he (complainant) entered the doorway he saw the defendant, and said to him, "How are you, Mr. Lamb?" The defendant drew up his sleeves, and came up to him, saying, "Be out of here, or I'll kick you out" He asked him to tell him what was the matter, and he would go. The defendant then said, "Out of here, you dirty lousy Scotchman." He (complainant) said, "Dear me, what's the matter," and the defendant immediately planted his fist in his face, and told him if he didn't go out he would smash his face, and that nothing would give him more satisfaction than to smash his face. Defendant put both his hands upon his (complainant's) chest, and shoved him back a little bit. He staggered back three or four feet, and said he would go out. He did get out, and the defendant shut the yard door with all his strength. When he went out he did not speak to the defendant. He went into a common passage leading to Scotswood-road, he could see the defendant, who was still in his own yard. When he got into the passage he stood leaning against the palisading, but did not say anything to him. Defendant came up, took up an empty pail, and said, "If you don't be out of this I will smash you down with this. I'll drown you with water." He placed the pail under the water tap, but took it up again. There was a corner where there was a long pole, which the defendant got hold of, and he said, "If you don't go out I'll smash you out with this," or words to that effect. He was in the passage, and the defendant was in his own yard. The words used were either "I will smash you down" or "Knock you down." He made answer, "Use your liberty." The passage belonged to another property, and he was leaning over the partition when he said "Use your liberty." The defendant, who still kept in his own yard, came to him again, put his fist on his nose, and told him he would smash his face if he did not leave.

The Mayor—That is not leave your employment, but leave the place

Complainant—Leave the place where I was work-

6

ing. He then proceeded with his evidence The defendant came to him, and stood inside the palisading which divided them. The defendant walked up his steps, sat down, and immediately said, "You are a low, lousy Scotchman for going in there to work among that low crew" He said, "You need not be the least surprised at my going in, I have told you all along that I would go in when the factory opened." He alluded to Sir William Armstrong's factory. The defendant said, "Your own countrymen thought shame when they came, and went back; but as for you, you have got no shame in you. If there had been a man amongst you, you would have left when that low crew came to the place." He understood the defendant meant either the Belgian workmen or the men from Woolwich.

Cross-examined by Mr. Bush, complainant said it was defendant's yard where he went to look for the hen. Defendant told him to be out, and he didn't go. He told him again, and he didn't go, he still stood there. He had satisfied himself about the hen, but was in conversation with a neighbour when the defendant came into the yard. —[Mr Ald Nichol—It is not the private property of the tenant.]—Defendant had at last to put him out partly. As soon as he got out he went round to the next passage, and leaned over the palings He looked at the defendant, but did not jeer or grin at him. He went because he wanted to see the man. While this was going on he was in the borough and the defendant was in the county; the defendant's yard being in the county, and the passage where he (complainant) was standing in the borough.

The Mayor—Is it a fact that this occurred in the county and not in the borough?

Mr. Bulman, clerk to the magistrates, asked the complainant if he had spoken to the defendant while he was in the county.

Complainant replied in the affirmative.

The Mayor—What could you come here for then?

Mr. Ald. Nichol said the whole occurrence appeared to have taken place in the county, and the case would have to be dismissed.

Mr. Joel said they would have to go to the county magistrates. The case had not broken down, but only the borough magistrates had no jurisdiction. The complainant would have to lay another information.

The defendant was then discharged.

MORE INTIMIDATION CASES IN NEWCASTLE.

At the Newcastle Police Court, yesterday (August 17th, 1871), before Mr Alderman Nichol and Mr. Alderman Wilson, Daniel Weallans, 20 years of age, was charged as follows:—That he did, on the 16th of August, in company with other men, unlawfully watch and beset the engine works of Messrs. R W. Hawthorn and Co., Forth Banks, with a view to coerce workmen to quit their employment."

Mr Joel appeared for the prosecution, and Mr. Bush for the prisoner.

Mr. Joel said the prisoner was in custody for committing an offence under the third division of the Criminal Law Amendment Act, 34th and 35th Victoria. The facts were shortly these. The complainant in the present case, a young man named Stephen Carter, worked at Messrs. Hawthorn and Co.'s, he being an apprentice there. On the 15th of August, the prisoner was seen by a policeman near to Messrs. Hawthorn's factory. The officer told him to go away, for he had no business there. On the day in question the prisoner was again standing near the factory in company with other two men, and as the complainant was going to the factory, one of them entered into conversation with him, which conversation would leave little doubt that he had committed an offence under this Act of Parliament. The prisoner was not the man who made use of the words which were addressed to the complainant, but he was standing by hearing the conversation, and he was besetting the works, therefore he (Mr. Joel) thought the magistrates would be of opinion that the prisoner was a party concerned in these illegal proceedings, and was amenable to the law just as if he had made use of the words.

P. C. Robert Service was then sworn, and stated that he knew the prisoner, who was a forgeman, and worked at Hawthorn's before the strike. He knew that the prisoner was now on strike, and that he lived in the Forth Banks. On the 15th of August he saw him near Hawthorn's foundry door. He saw him a few minutes there, and requested him to go away. He saw him twice on the 15th, and requested him twice to go away. The prisoner was standing against the company's door. There were five or six others with him. The first time he went away, but the last time he was rather cross, and said he thought he might stand there out of the sun. On the 16th he saw prisoner opposite the engine works, a little past seven in the morning. There was no person with him at that time. He never lost sight of him until about a quarter-past eight. He was near the entrance to the engine works, about seventy yards higher up the Forth Banks. Other two men came up to him, and they were together about five minutes. The complainant Carter was proceeding down the Forth Banks to the works, and when he got up to them he stopped. The prisoner and the other two men remained about two minutes with him, but he (witness) heard nothing. When they saw him coming they walked away.

M. Joel—They walked away sharply.

Mr. Bush—The witness did not say so.

Mr. Ald. Nichol—Don't put words into his mouth.

Mr. Joel asked witness whether they walked or ran

Witness replied that they walked away sharply, and after he had said something to Carter, he proceeded after them and apprehended Weallans. When he apprehended him he said he suspected him for intimidating Carter, and the prisoner replied, "It was not me that spoke." He afterwards charged him at the police station. The inspector wrote out the charge as it now stood, and he (witness) read it over to the prisoner, who made no reply.

Cross-examined by Mr. Bush—Witness said he did not see the prisoner, or either of the other two men speak to Carter. He didn't ask the prisoner to give him his name. He would swear that the prisoner did not refuse to give his name.

Mr. Bush—Didn't a publican come out and say to the prisoner, "Don't be dragged off for doing nothing, give him your name, or you will have to go to the station."

Witness replied he did not ask for his name, for he knew it, and where he lived. He took him to the superintendent to see what he thought of the case. That he considered to be his duty. The prisoner was kept in the room a quarter of an hour before the charge was made out. The prisoner was about 250 yards from where he lived when Carter stopped.

Stephen Carter, the complainant, stated that he was an apprentice in the employment of Messrs. Hawthorn and Co. On the morning of the 16th

7

inst., about eight o'clock, he was going to his work As he was proceeding from the Cattle Market down the Forth Banks he saw three men standing near the bridge. They were about a yard from each other. The prisoner was one of the men. They stopped him, and one of them, an old man, spoke to him He did not know whether the prisoner was near enough to hear what took place. The prisoner had worked at Hawthorn's factory, but was not working there now. The prisoner was on strike He did not know the old man's name, nor did he know the other man. The old man asked him whether he was going into the factory to work, and he said "Yes." He also told him he was an apprentice at the works. The old man said, "It did not matter to me, all the men would be turned out when the men came in" He (complainant) made no answer. While this was going on he saw the policeman, and after that he left them. The prisoner did not speak to him.

Cross-examined by Mr. Bush—Complainant said he knew the prisoner, and they had spoken to one another in the factory. The prisoner did not speak to him.

This being all the evidence, Mr Ald. Nichol said he thought the case was not made out, and the prisoner was discharged.

Richard Larner, 45 year of age, was brought up on a warrant, charged with unlawfully threatening and intimidating Thomas James Waters.

Mr. Joel, who appeared for the complainant, said this was a case under the same Act of Parliament. He stated the facts of the case, and then called the complainant.

Thomas James Waters, who resides at No. 45, Noble-street, where the defendant also lives, the complainant having two rooms on one side of the same stairhead and the defendant two rooms on the other side He stated that he was in the employment of Sir William Armstrong and Company. On the evening of the 9th of the present month, he was getting his tea in his own house between six aad seven o'clock, when the defendant came in and said, "You blackleg ——, you are not the man for me" The defendant also threatened him. After calling him a blackleg ——, he said if he could not break his (complainant's) head there were plenty of others down the street for to do it, and if he did not leave the factory, he would help to drag him out He had business at the factory, and on going down the street after getting his tea the defendant put his head out of the window and said, "There goes the blackleg ——, I could whip him any time." When defendant came across the step of his door he told him to get away out of his house.

Cross-examined by Mr. Garbutt, who appeared for the defendant—Complainant said he never had any disturbance with the defendant whilst they worked together. He took out the warrant on Tuesday. He might have taken a razor up when the defendant came into his house on Saturday evening. His house was not on fire on Saturday evening, but his blind was. He was at the house of Mr Pest at the time. He did not know that the defendant went into his house to try and save his children. He was sure he did not take a razor up. There was a razor in his house, but it was not in his hands. He did not threaten to cut the defendant's throat.

Mrs. Waters corroborated the evidence of her husband; and a neighbour, named Ann Hurst, stated that she heard defendant call Waters a "blackleg."

Mr. Garbutt addressed the magistrate on behalf of his client, contending that the latter had not committed any offence under the 2nd sub-section of section one, which stated that any person using language of a threatening and intimidating character as to justify the justices in binding him over to keep the peace had committed an offence, and they must not stretch the old law of binding over to keep the peace

The bench retired, and after a short absence returned, when Mr. Ald Nichol said the defendant would have to go to prison for two months.

Alexander Barnes, 50 years of age, was charged on August 17, 1871, with being drunk and disorderly, and also with using intimidating language to workmen in the employment of Sir William Armstrong and Co

Mr. Joel said the prisoner was in custody on a charge preferred against him under the last part of the third division of the third sub-section of the Act. It would appear that the prisoner followed a number of men going to Sir Wm Armstrong's Ordnance Works on Wednesday evening. They followed them a considerable distance, and hooted and called them "blacklegs" The prisoner was the one who followed and hooted. If his instructions were correct, he thought the magistrates, after hearing the evidence, would be of opinion that the prisoner did follow the workmen with a view to intimidate them.

P C. Beattie said he was stationed at Sir Wm. Armstrong's works. He knew the defendant Alexander Barnes, but did not know him previously to Wednesday night, when he saw him about a quarter to eight o'clock Some workmen were coming from Armstrong-street in the direction of the factory gates, and they were followed by hundreds of persons. The prisoner was in front of them and next to the workmen. He heard the prisoner shout at the top of his voice "You blackleg ———." They followed the workmen about a hundred yards. There was great disorder and much disturbance. They shouted, and he heard the prisoner make use of the word "blackleg."

P C John Wilson, said he was on duty on Scotswood-road on Wednesday evening, and saw the prisoner in front of the crowd shouting "You blackleg —— "

Mr Ald. Nichol. Did any other person make use of those words.

Witness I cannot say, but I heard the prisoner distinctly.

Witness further stated that the prisoner formerly worked in the factory, but was now out on strike.

Prisoner: How am I on strike when I was locked out?

George Makum said he was at present working at the Ordnance Works. They left work at a quarter to six on Wednesday evening, and about a quarter to eight they were going down to the factory, but not to work. The crowd followed them about two hundred and fifty yards. The prisoner followed them into a public-house, and tried to urge the other men to attack them. He heard the prisoner say, "Down among the dead men. We are all Englishmen. Down with the blackleg ——" They followed them down from from the public-house. The prisoner said · "Here they go, I'll set them on."

The prisoner, in reply to the charge, said he had been at Gloucester-street, where he had his tea and a glass of whiskey. As he was on his way home he saw a lot of men going along the street He followed up, but did not do anything.

8

Mr. Ald. Nichol said they thought the case clearly proved, and the prisoner would have to go to the House of Correction for two months.

A lad named George Ramage, 15 years of age, was charged with causing a disturbance opposite Messrs. Thompson and Boyd's works, Barrack-road, on Tuesday night A policeman stated that the prisoner was shouting through a window, but he could not tell what he said.

The prisoner, who promised to conduct himself properly in future, was discharged.

Two young men, named John Elliott and William Morgan, were charged with being concerned along with others, in violently assaulting Thomas Tranery, interpreter at Messrs. Hawthorn's works

On Wednesday night, Mr. Tranery was taking four or five foreign workmen to their lodgings, when he was attacked by a number of men, who knocked him down and gave him a black eye. The prisoners, who are apprentices, working for Messrs. Hawthorn, were amongst the crowd and were taken into custody.

Mr Joel, who appeared for the prosecution, said he had gone into the case, and could find no evidence to show that they took part in the assault, and, in accordance with his advice, the charge would be withdrawn.

The prisoners were then discharged.

ADDITIONAL INTIMIDATION CASES IN NEWCASTLE

At the Newcastle Police Court yesterday morning (August 19, 1871), before the Mayor (Mr. B. B. Sanderson) and Mr. Ald Nichol, Andrew Watson, 45 years of age, was brought up on a warrant, charged with threatening and intimidating George Marmen

Mr Joel, who appeared for the prosecutor, stated that the defendant was charged under the second subdivision of the first section of the Criminal Law Amendment Act, relating to violence, threats, and molestation of workmen, with using threatening and intimidating language to the complainant, who worked as a labourer at Sir William Armstrong and Co.'s Works, and lived at No. 8, Back George-street. After stating the facts of the case, Mr. Joel called the complainant, George Marmen.

The latter, on being sworn, said—On Monday night, about twenty minutes past six, he observed his wife at the window, and heard some one from the outside speak to her. He afterwards went to the window and saw the defendant outside When he went to the window the defendant said to him, "You blackleg —— come out, it is a long time since I have been working" Defendant, who was feeling his arms, also said, "I would like to try them" (He complainant) made no answer. The defendant then said, "Before he would be a blackleg he would eat the stones on the street," to which he replied, "It would not fill his belly." Defendant also said, "A young man like you should be ashamed to take the bread out of other people's mouths" He told the defendant there was plenty of work for him as well as for him (complainant) if he liked to go to it Defendant then turned to go down the street, but then turned back to him and said, "Come and have a glass of beer." He made no answer. The defendant then said, "You know better not to come. The dark nights are coming on, we will drive you into the Tyne, the whole lot of you." He was afraid the defendant would do him some injury.

Sarah Marmen, the wife of the complainant, gave corroborative evidence, and in answer to the defendant, said he was drunk.

The Mayor—Was he very drunk?

Witness—Yes, he had had a good sup.

P C. Wm. Davidson said he apprehended the prisoner about twelve o'clock on Thursday night, and charged him with threatening and intimidating the complainant, to which he replied that he knew nothing about it. The defendant said he was on strike.

In answer to the charge, defendant said he had previously been on good terms with the complainant He was drunk, and knew nothing about it till the policeman came to his house on Thursday night. He had no intention of doing the complainant any harm

The Bench sentenced him to be imprisoned for one month, the Mayor remarking that if he had been sober probably he would not have used such language, and on that account they had inflicted a much less punishment than they would otherwise have done.

James Wise, 30 years of age, who had also been apprehended on a warrant, was then placed in the dock charged with threatening and intimidating Charles Millar.

Mr. Garbutt appeared for the prosecution, and briefly stated the case.

Charles Millar, the complainant, was then sworn, and said he resided at the Ouseburn. He was a rivetter, and had worked for some time at Messrs. Mitchell and Co's shipbuilding yard, Low Walker. When the strike took place he turned out with the rest of the workmen, but last month he re-engaged and went back to his work. The defendant, whom he had known for some years, lived about 300 yards from him, and worked at Abbot's up to the time of the strike, but was on strike now. About nine o'clock on Wednesday night he was in the Ouseburn not far from his own door when the defendant came and asked him why he wrought for Messrs. Mitchell, to which he replied he believed it to be his duty to do so. The defendant then called him one of Mitchell's "blackleg ——," and said as long as he lived he would insult him, and he would have him out. He was in fear in consequence of this threat.

James Millar, complainant's son, corroborated the statement of his father.

P.C. John Thompson, who apprehended the defendant, stated that the latter had had a drop of drink, and said he was very sorry for what had occurred.

The Mayor told defendant he had no business to interfere with other people working.

Mr. Ald. Nichol said he could not understand how men on strike were continually drunk.

The defendant was sent to gaol for a month.

ALLEGED PICKETING AT GATESHEAD.

Yesterday morning, August 18th, 1871, at the Gateshead Borough Police-court, before Ald. Bigger, B J. Prockter, Esq., and John Sowerby, Esq, three engineers, named Henry Gunston, George Rock, and John Burns, were summoned for that they did, on the 2nd day of August last, unlawfully loiter and remain on a certain footway called South Shore, so as to obstruct the free passage along the same, without reasonable cause.

Mr Bush, who appeared for the defendants, said these summonses involved a principle which was contained in an application for a summons by himself on behalf of Mr. Sinclair against Brown, the police-officer, and no doubt they were issued in consequence of that application, and

9

such being the case, it was only fair that the case against the policeman should be taken first.

Mr. Elliott, Chief-constable, said that Mr. Bush's remarks were entirely uncalled for. The cases were brought before the Bench independent of the application made by him, and entered in the books before they had any knowledge that he was going to apply for a summons

Police-sergt. John Brown was then called, and stated that on the 2nd inst, about three o'clock in the afternoon, he was on duty on the road between Hawks and Crawshay's and Clark, Watson, and Gurney's Works, and saw Gunston and Rock delivering bills and stopping persons on the footway. He went to them and asked them to give him a bill, and they gave him some of them, on which he told them that they had already been cautioned about standing on that footpath, and that they would probably be summoned. About half an hour on the same afternoon he saw Burns delivering bills, and stopping persons. He went to him and told him it was likely that he would be summoned as he had previously been cautioned, after which he went into the Ship Inn, and subsequently went away. There was a woman complained to witness, the same afternoon, about the men standing about.

In cross-examination by Mr. Bush, witness said the woman was not in court, and he did not know who she was Gunston was not walking till he saw witness, but he began to walk when he saw him. Mr. Clark, of the firm of Clark, Watson, and Gurney, was beside him when he spoke to the men He did not know that he was exactly watching his proceedings, but he was there. Mr. Clark was not obstructing the footway, he was on the carriageway. He would swear that he did not lay hold of either of the men by the coat and wrench the bills out of their hands.

P C Hunter corroborated Sergeant Brown's evidence, and, in answer to Mr Bush, said that all he saw the defendant do was to shove bills into hands of people passing by.

Wm. Clark said that on the day named he was going down the South Shore with the officer Brown, when he saw the two men standing about, and as soon as they saw himself and the officer they moved on. He had seen them so often before, and had spoken to them so often, that he hardly remembered all that took place on this particular occasion.

Mr. Bush—Were you obstructed at all?

Witness—Yes, I had to go off the footpath.

Mr. Bush—Well, that was a great hardship, wasn't it? But I suppose it was a matter of choice with you?

Witness—If I had not gone off they would have had to do so.

Mr. Bush—Very likely, and that is the offence against these men. Now will you venture to swear that on that particular day you saw either of the defendants there?

Witness—Yes, I will.

Mr Bush, in defence, quoted several cases from legal authorities to show that the delivering of bills was not an unlawful obstruction, and that it had been so held in all their courts from all time. These men, he said, were not vagrants, but were following a perfectly legitimate calling.

The Bench retired, and after a brief absence returned into court, when the Clerk to the Justices inquired of the witnesses whether they saw any passenger having to leave the footway in consequence of the defendants distributing their bills.

Sergeant Brown said he saw the woman who said there must be something amiss, as there were so many men about.

Mr. Bush—Is that woman here?

Witness—No, she is not.

The Clerk—If the case were adjourned do you think you could produce her?

Witness—No, I am afraid not, as I don't know her.

Mr. Prockter—The Bench are not satisfied that there is sufficient evidence to convict the defendants, and they are dismissed. And now, my men, take care and don't do it again.

Robert Finch and George Rock were charged with a similar offence, which was alleged to have been committed on the 8th of August, between eleven and twelve o'clock in the morning. The witnesses in this case were P.C. Brown and P.C. Thompson, and their evidence was almost the same as that tendered in the last case, except that they both swore that they saw the defendants repeatedly stop and obstruct the foot passengers, and cause them to leave the footway for the carriageway.

Mr. Bush, in defence, said the remarks made in the last case by him would also apply in the present one. He then called one of the defendants to give evidence.

Robert Finch said he and Rock were distributing bills on Tuesday, the 8th of August, when the constable came up to them and said, "Now, my lads, you will have to be away from here." They walked down as far as Newall's wire works, and returned near to Hawks's works, when Brown came up to them again and collared Rock, and demanded his name, where he lived, what he worked at, and where he worked last. Rock gave him an answer to each of these questions, after which they went to have a glass of ale at the Ship Inn. When they came out Brown again went to them and said, "Now, Rock, are you not going to go away?" He took Rock by the collar and said to him, "Give me them handbills," and after a few words between them he took hold of them and wrenched them out of his hand. Rock threatened to prosecute him for what he had done, and stated that he had got his number, on which Brown offered the bills back to witness.

The defendant Rock gave similar evidence, after which Mr. Prockter said the Bench were of opinion that they did cause an obstruction, and they were each fined half-a-crown and costs, 11s. 6d. in all, or fourteen days' imprisonment in default.

Andrew Vollan, William Batey, Joseph Spence, Thomas Walker, Matthew Batey, Richard Brass, John Potts, and Thomas Stobart were summoned for a similar offence.

Mr Elliott said that as the cases against the whole of these men were the same as that proved against the last defendants who were convicted he would, with the permission of the Bench, withdraw the charge against them. He had no wish to deal harshly with them, but he merely wanted them to understand that they could not proceed in the way they had been doing with impunity.

Mr. Bush said that as he appeared for the men he had no objection to take advantage of Mr. Elliott's great generosity, though the men were unconscious of having committed any offence.

Mr Sowerby asked if the men would all promise not to do it again.

Mr. Bush said that as they had not committed any offence, it was somewhat odd to ask them not to do it again. That was the advice given by the Bench in the first case heard, although the offence was not proved against the men.

Mr. Prockter thought they had better go on

with the cases if the men were disposed to offend again.

Mr Bush said he would undertake that his clients would not repeat the offence they were charged with, and they were therefore all dismissed.

SINGULAR TRADES UNION CASE.

From the *Newcastle Chronicle*, August 15th, 1871.

At Leeds Assizes yesterday, the case of Purchon *v.* Hartley and others, involving a singular phase of trades unionism, was tried before Mr. Justice Hannan and a jury

Mr. Price, Q.C., Serjeant O'Brien, Mr. Kemplay, and Mr. Beasley were for the plaintiff, and Mr. Seymour, Q.C., Mr. Field, Q.C., Mr. Waddy, and Wilberforce were for the defendants.

Serjeant O'Brien in opening the case of the plintiff, said that he (Purchon) came before the jury to assert the right of individual action and of liberty and labour. He was a glass bottlemaker, and had been in the employment of Messrs. Beetson, of the Rotherham and Masbro' Company for nine or ten years, being the foreman in the establishment since the year 1862. Plaintiff was a member of the Glass Bottlemakers of Yorkshire Trade Protection Society. He was also a member of the branch of that body, whose meetings were held at Rotherham. Among his fellow-workmen were two of the defendants and a number of other men, all belonging to the same society. In March last year the defendants Hartlev and Henderson were fellow-workmen with the plaintiff. Another of the defendants named Brown, was a member of another branch of the society. For some reason they had taken it into their heads to form a settled resolution with the object of rooting the plaintiff out of Beetson's employment On the 28th November they made a report to somebody, and upon that report being brought to a gentleman who acted as secretary, they called a meeting. That meeting was held on the 29th March, at the Effingham Arms Inn. Plaintiff was summoned there to answer for his "high crimes and misdemeanours." They said they had the blackest case that had been investigated by any such bodies as theirs. Plaintiff attended that meeting, where, after sundry threats, he was informed that he was about reducing Beetson's shop into a black-sheep shop. They threatened to root him out, and told him they would do so. They passed a resolution that would show the spirit in which they sought to overawe and oppress his liberty of action. That resolution was to fine every man of their body 5s. if they were seen with, or heard speaking to, the plaintiff, whose salary was 52s. per week. They said that nothing could save the establishment from becoming a non-union society but the rooting out of the plaintiff, and they then and there adopted a resolution that a deputation should wait upon the men employed—the other members of the society—by Mr Beetson, the object of the deputation being to induce his fellow workmen belonging to the same society to go to Mr. Beetson in a body, and announce to him that unless he discharged this man (plaintiff) they would send in their tickets, which, in the language of the body, meant that they would strike, and engage in a conflict with Mr Beetson, which might have the effect of ruining his business Mr. Beetson heard what the men had to say, on the 30th of April; and he on the 9th May, gave notice to the plaintiff that he was discharged from his employment. For five months plaintiff was out of work, which, at the rate of 52s. per week, amounted to about £52. Subsequently he got employment at a remuneration of only 40s per week.

Plaintiff was examined and cross-examined at considerable length, in corroboration of the facts stated in the above opening.

There was quite a "cloud of witnesses" on the side of the defendants, but many of them were not called, owing to their evidence being of necessarily similar character with respect to the transactions at the meetings held in the matter of Purchon, the plaintiff. Their evidence was to the effect that the plaintiff had exceeded the work prescribed by the union to from 15 to 20 gross per week plus.

The jury returned the following verdict:—"We find that great credit is due to the plaintiff for exposing the unwarrantable conduct of the defendants; and we censure the defendants very strongly for their heartless and malicious conduct towards the plaintiff, and we find that unwarrantable interference and undue influence were used by the defendants with the object, without the slightest foundation, of removing the plaintiff from his late employers, and we find for the plaintiff, damages £300."

A WORKMAN CONVICTED OF PICKETING.

At the Hammersmith Police-court yesterday, Jan. 11th, 1872, John Turk appeared to answer a summons which had been taken out under the Criminal Law Amendment Act, 34th and 35th Vic., cap. 32, sec. 1, relating to violence, threats, and molestation, by Mr. Henry Gwynne, a general engineer, carrying on business with his brother at the Chancellors, Hammersmith, for molesting him by watching his premises to coerce him.

Mr. Francis, instructed by Mr. Cole, appeared on behalf of Messrs Gwynne, and Mr. Croome for the defendant, who pleaded not guilty

From the opening address of Mr. Francis, it appeared that at the end of last year the men in the employ of Messrs. Gwynne sent in a petition, the defendant's name being attached to it, to them, asking for the adoption of the nine hours' movement on the 1st of January, for their recreation and culture. The request was declined on account of the contracts which Messrs. Gwynne had to complete, but they agreed to adopt the nine hours' movement on the 1st of April next. The men were understood to be satisfied with that arrangement, but on the morning of the 16th of December they threw pieces of paper into the office giving a week's notice to leave. The men, to the number of 103, were then paid and discharged, according to the rules of the establishment, that they should not belong to any combination The defendant was one of the men, and on Monday last he distributed in front of the works handbills, of which the following is a copy:

"Nine Hours' Movement.—Jan. 6, 1872.—Fellow Workmen,—We, the late workmen of the Hammersmith Iron Works, having endeavoured to come to some arrangement with our employers, J. and H. Gwynne, with respect to the Nine Hours' Movement, to commence on the 1st of January, 1872, beg to inform you that we have failed in this matter, and were locked out on the 13th and 14th December, 1871. We, therefore, appeal to you as fellow workmen, not to enter into any engagement with our late employers unless on the condition that the nine hours per day commence at once. By so refusing you will forward our cause as well as your own as working men. It is only by uni-

NOTE—We have since learned that the young man John Turk had to suffer all the indignities of an ordinary criminal for three days previous to being bailed out. His hair was cropped after the true felon type, and for three days he worked on the Treadmill as a common felon And this is the law which Masters' Associations wish to perpetuate and enforce—GEORGE HOWELL, Secretary.

11

ing that we, the working classes, can rise to an elevation never yet attained. We remain, fellow workmen, yours truly, the locked-out men of the Hammersmith Iron Works"

Mr, Henry Gwynne said he had seen the defendant walking up and down in front of his premises, and speaking to his men, and others whom he believed were seeking for work.

Cross-examined The defendant lived in the street opposite to his premises He did not think it was usual for a man who was out of work to be walking about the street in which he lived. On Monday last he saw the defendant following his men and distributing the handbills. He did not lock out the men.

Alfred Pike said he started work for Messrs. Gwynne on Saturday last. He was not a willing witness. A handbill, like the one produced, was given to him on Monday. He could not swear to the defendant. Several men outside of the premises asked him to give up his work, and said if he did not he would not be able to get work at any other place The defendant was present He also said that the defendant must have been present when the bill was given to him, as he passed him half a minute before. Witness said he should not leave.

After a number of witnesses had been heard, Mr Croome submitted that the evidence had failed to bring the defendant within the terms of the section. He said it had not been shown that he molested Mr. Gwynne in any way, and argued that there must be some actual molestation to justify a conviction.

Mr. Ingham said he had no doubt that coercion was used to the men, but the question really was whether, in coercing the men, the master was coerced. He should much rather submit that question to another court, but he was afraid that the Act of Parliament threw the responsibility upon him.

Mr Francis argued that the action of the defendant was a coercion upon the master, for unless he adopted the nine hours' movement no men should work for him. He again referred to the handbill, quoting the concluding words, "By so refusing you will forward our cause," in support of the argument that there was coercion.

Mr. Ingham said he believed there was. As a threat had been held out to Pike, it was a very serious thing, and to mark his sense of the impropriety of it he should order the defendant to be imprisoned for two months

Mr Croome asked the magistrate to state a case with reference to the point of law, but Mr. Ingham suggested that he should appeal, and he agreed to do so.

The proceedings excited considerable interest.

EFFECT OF THE NEW TRADES UNION ACT.

A STONEMASON CHARGED WITH INTIMIDATION.

Thomas Wearden, stonemason, of 32, Hennionstreet, was charged on remand with "molesting one Thomas Cooper, of 75, Craddock-street, Little Bolton, stonemason, by bessetting the place where he, the said Thomas Cooper, worked, with a view to coerce the said Thomas Cooper, being a workman within the meaning of the act 34 and 35 Victoria, chap. 32, to pay to a certain permanent association called the 'Friendly Society of Operative Masons,' a fine of 40s. on the 2nd inst."

Mr. Richardson appeared for the complainant, and Mr Hall defended.

Mr. Richardson having opened the case at some length, called the following evidence —

Thomas Cooper, stonemason, the complainant, said that on the 2nd of August he was working for Mr. James Martin, contractor, Bolton, who was building a new Wesleyan Methodist Chapel in Grecian-street. At half-past seven in the morning the defendant brought a card to him. Witness was a member of the Friendly Society of Stonemasons, and became a member about May, 1871. Both card and book of rules of the society were given to him by the defendant. He knew that he had been fined on July 27th, 40s. Defendant said he should want 5s., to which he replied that he had not got it. He said, "You would most likely get it from the foreman, Mr Clague," but he replied "I shall not ask him" He then said "Will you bring it in the morning?" and he said "No, I shan't." He (defendant) then went to his work. Clague and the defendant were both members of the society, defendant being shop-steward. He saw him talking to the foreman at noon the same day, and to Mr. Martin. They should have resumed work at one o'clock. The foreman did not give the usual signal to resume work, and all the men packed up their tools and went away, defendant amongst the rest The same evening he saw the foreman at his (witness's) house, and he brought him his wages. The workmen resumed work next day. He had not since been to the society.

James Martin, builder and contractor, said that on the 2nd August he had a contract for the building of a Wesleyan Methodist Chapel in Grecian-street. He employed Clague, Cooper, and Wearden. About noon he remembered the defendant speaking in reference to Thomas Cooper. He said, "There is a man on the job who is objectionable to us." Witness said, "Why is he objectionable?" Defendant said, "He is objectionable because he has not paid a fine." Witness said "Have you taken means to make him pay?" He replied, "We have, and he refuses," Witness said, "You had better let the matter rest, and I will discharge him at the week's end." The men did not resume work that afternoon, but did so next morning, except Cooper.

James Clague, the foreman, who refused to give evidence "before receiving his wages," was awarded 2s. 6d. He was a member of the society referred to, and was employed on the chapel in Grecian-street. On the 2nd of August the men left at one o'clock, but it was by an order of the witness. Wearden told him that Cooper would not pay the 5s asked of him. When the men went away he told them that it was foolish to go away. He went in the afternoon to Cooper's house to pay him his wages. He did not know that it was the rule that if Cooper refused to pay a fine the men would leave in a body. He did not know a single rule of the society although he was a member. [Major Hesketh—The witness is the most unsatisfactory one I have ever known. Witness said he had not a retentive memory.]

This concluded the complainant's case.

Mr. Hall then addressed the court for the defence He hardly knew how to approach such a paltry and contemptible case. He contended that there was no case against the defendant. No evidence of molestation had been proved, that he beset the place, or that what he did was done with a view so coerce the complainant, but amounted merely to a bare request to pay the fine. The magistrates were asked to send the defendant to gaol with or without hard labour, merely for asking a payment of a sum of money. The facts of the case were of the most simple and innocuous character, and it was a pity that the first case under the new Act should be of such a paltry and contemptible character.

The magistrates then retired to consider their decision, and after an absence of half an hour, they returned into court, and the Mayor said they had patiently considered the matter, and their decision was that the defendant would be committed for one month. They hoped that as it was a really important case, to test the operation of a new Act, that the matter would be fully sifted by another tribunal on appeal.

Notice of appeal was then given, the defendant entering into sureties in the meantime.

THE APPEAL BEFORE THE RECORDER OF BOLTON.

IMPORTANT TO TRADES UNIONISTS.

THE CONVICTION OF A STONEMASON FOR MOLESTATION QUASHED.

At the conclusion of the criminal business, a case of considerable importance to trades unionists came on for hearing. It was an appeal by a stonemason named James Wearden against the decision of the magistrates of this borough, under the Criminal Law Amendment (Violence, Threats, &c.) Act, ordering him to be imprisoned for one month for having molested another stonemason named Thomas Cooper, by besetting the place where he worked, with a view to coerce him, he being a workman within the meaning of the Act, 34 and 35 Victoria, chap 32, to pay to the Friendly Society of Operative Stonemasons a fine of 5s. Mr. Leresche and Mr. Hopwood appeared for the appellant, and Mr. Edge for the respondents.

The Recorder presumed that all formal notices would be admitted?

Mr. Edge said he was instructed not to take any objection, but to have the case heard upon its merits. The case had attracted a vast amount of attention out of doors, but with that they had nothing to do. Whether it was policy or not that there should be such a clause in the Act as that under which the appellant had been convicted, it was not for them to discuss, but they were there to decide whether the man Wearden had contravened an Act of Parliament which was passed last session. It appeared that both Cooper and Wearden were members of the Operative Stonemasons' Society, and on the 2nd of August last, when this offence was alleged to have been committed, they were working together under Mr. James Martin, builder, who had contracted for the erection of a chapel. Wearden was shop steward of the society, and Cooper having infringed some rule of the Association, he was fined 5s., which it became the duty of Wearden to collect. Accordingly, on the date named, he went to Cooper, and tendered to him a card upon which appeared this fine. Cooper said, "All right," but Wearden remarked, "I want 5s. from you." Cooper replied, "I can't pay you." Wearden told him he might get the money from the foreman, to which Cooper answered, "I shall not get it, for I don't intend to pay the fine." Wearden then went away, and he (Mr. Edge) believed returned to his work. Had the matter ended there, possibly there might have been no offence committed. He did not think it necessary to argue that there was any offence up to that time, but the whole case turned upon what was done subsequently. In the course of the forenoon, Wearden went to the foreman, James Clague, also a member of the Stonemasons' Society, and represented to him that Cooper would not pay this fine, and that the other men had an objection to continue to work on the premises with him. He believed Clague was asked to discharge the man, but that he refused to do so, and referred Wearden to the master, who was not at that time present. Wearden went back to his work, and shortly afterwards, seeing Mr. Martin with the foreman, he went up to him, and made a similar representation to that which he had already made to Clague. He said to Mr. Martin, "There is a man upon the works who is objectionable to us." Mr. Martin asked who it was, and was told it was Cooper. He then asked what was the matter, and he (Mr Edge) thought he was informed that Cooper refused to pay a fine. Mr. Martin said, "Will you let the matter wait till Saturday, and then I will discharge the man?" Wearden made no reply, but walked away. The men on the job, who were also members of the society, worked until noon, when they left for dinner. What took place during the dinner hour he (Mr. Edge) was not in a position to say, but instead of resuming work at one o'clock as they ought to have done, they packed up their tools, and went away, Cooper leaving the place at the same time. Some time after that, Clague went to Cooper's house, paid him his wages, and said something which might or might not be evidence against Wearden, and on the following morning, the whole of the other men resumed work. These were the facts of the case.

The Recorder said he was anxious to understand what Mr Edge's contention was. Supposing there had been no demand for the fine, but that Wearden had gone to the master and said, "I wish that man (Cooper) to be discharged," would that, without any communication with the man at all, have been an offence under the Act?

Mr Edge submitted that it would have been.

The Recorder—Under what section? How could that be a besetting of the man if Wearden had never communicated with him?

Mr. Edge—It is not a besetting of the man, but a besetting of the place, with which Wearden is charged.

The Recorder—But he has been convicted of besetting the man.

Mr. Cottingham—No, of "molesting" the man, by besetting the place where he worked.

Mr. Edge submitted that taking the whole of Wearden's conduct together, he was guilty of molesting this man Cooper (as described in the third sub-section of the Act), with a view to coerce him to pay a fine of 5s., imposed by a permanent association.

The Recorder—And you say the act of besetting was in going to the master, and that the evidence of intent was the demand for payment of the fine?

Mr. Edge—Yes. They had not to give evidence of actual molestation, because the mere act of besetting was molestation as defined by the Act. It was not necessary to do more than give evidence of the fact that Wearden beset the place where the respondent worked. He should submit that in considering the case, the Court must look at the conduct of Wearden in two aspects: they must first look at his conduct as a workman, employed on the job where Cooper was engaged, then they must look at his actions as an officer of an association, in demanding payment of a fine. By virtne of his character as a workman on those premises, Wearden had certain rights and privileges, and if he merely went about his ordinary work, they could find no fault with him. But immediately, although he was there in the capacity of workman, he diverged from that character and, as an official of a society, went to Cooper and demanded a fine from him, then it was precisely as if he was not a workman, but as if he had come there for that express

13

purpose. Assuming this to be so, he (Mr. Edge) contended that the Court must judge of the conduct of Wearden precisely as if he had been a stranger. Had a stranger gone to that place, demanded from Cooper payment of the fine, and, on being refused, acted as Wearden had done, and waited to see the master, with a view to get the man discharged, he (Mr Edge) submitted it would have been a besetting of the place within the meaning of the statute.

The Recorder—Although his object in waiting was not to interfere with Cooper, but to see the master, who did not happen to be there at the time?

Mr. Edge—Yes, because the words were "if he watch or beset" It was perfectly clear that besetting was something different to watching.

The Recorder—Not necessarily so.

Mr Edge—The word "beset" had a wider meaning than the word "watch." It had a Saxon derivation, and meant "to set by."

Mr. Cottingham—"To sit down before"

Mr Edge—But Webster gave it a wider meaning, and always used the word "to besiege, to harrass, to annoy."

The Recorder—But that would not be applicable to a place

Mr. Edge—Certainly not, and therefore I contend the word "beset" is more to place oneself by a place. I say if a stranger were watching a place, to make a complaint to the master, he would be besetting that place within the meaning of the Act Besetting implies a degree of continuity, and I submit there was a continuty in this man's actions

The Recorder—Does not that land you in this dilemma? If besetting implies continuity, it must imply not merely continuity of action, but continuity of intent. And if the besetting was not for the purpose of annoying the man, but for the simple purpose of seeing the master, is that such a continuity of intent as to bring it within the meaning of the Act?

Mr. Edge should say it was, because if the man did not pay the fine, Wearden intended to appeal to the foreman or the master to get him discharged With regard to the point as to whether it was with a view to coerce the man to pay a fine, he thought he need not trouble the Court at any great length. Whether Wearden molested the man within the terms of the Act of Parliament, it was perfectly clear that his intent was to coerce him to pay the fine. That was abundantly proved by what took place between Wearden and the master.

It was then agreed that instead of taking the evidence again, the depositions as taken before the magistrates should be read over to the witnesses Accordingly, the depositions of Thomas Cooper and Mr Martin were read, Clague, the foreman, although he had been subpœnaed, did not appear, and therefore his depositions could not be received.

The Recorder said it did not matter much, as the whole contention was raised on Mr. Martin's evidence.

Mr. Leresche, in reference to a statement in the depositions of Cooper, that he did not get his wages until pay day, explained that the men in this trade might leave at any time, no notice being required. He then proceeded to address the court for the appellant, observing that this act had created a great sensation throughout the country. It was vital, and would supersede the legislation of previous sessions. There had been a great struggle on the part of trades unionists to get their claims recognised as lawful. At one time, combinations were considered, *per se*, to be illegal, because they were combinations to effect unlawful objects. But the law was ameliorated so that workmen might be allowed to combine for certain objects. But while these associations were made lawful in the sense that their members were not outlaws, still they did not possess certain facilities which were given by the Friendly Societies' Act and others of the same kind. The reason of that was because it was supposed they combined against certain matters of public polity, as the restraint of trade, and things of that kind. But the Legislature last session, which passed this particular act, absolutely legalised all these societies, and declared that they should not be unlawful by reason of these particular matters, and it gave power to any body of men to associate as a trades union for the carrying out of objects which might even be considered a restraint of trade, and to register themselves as a friendly society.

The Recorder—You are referring, I suppose, to the Trades Union Act, of which this is a supplement.

Mr. Leresche—This was chapter 31, and the one which my friend is seeking to convict under is chapter 32—the penal statute.

The Recorder—The Trades Union Act not only legalises trades unions for civil purposes, but it expressly provides that an agreement or contract for the payment of this very fine or penalty shall be a legal agreement

Mr Leresche—Precisely. But the effect of this decision was this: By putting this interpretation on the word "besetting," the justices of Bolton, by a side wind, had repealed the whole of the provisions of chapter 31, because it would be impossible for the officer of any society to seek for the payment of any penalty imposed by its rules from any member without his being liable, if this construction be right, to be charged with molesting that man by besetting him. According to Mr Edge's argument, besetting might be the besetting of the house where the man resided, the place where he worked, in fact, any place where he happened to be—the street or otherwise; and the state of things would be this. while the previous Act had decided that it should be lawful for a society (a registered society) to adopt rules whereby certain fines and penalties might be inflicted on members, yet if any person went and simply said, "Will you pay such and such a fine or penalty?" that could be held to be a besetting.

The Recorder did not think it could be contended that the demand by Wearden, the shop steward of the society, for payment of the fine, could possibly be a besetting contemplated by the Act It would be obviously absurd to say that there might be a lawful agreement for the payment of a fine, but that the shop steward should not ask for it, and therefore Mr. Edge said he did not think he could contend that this was a besetting within the meaning of the Act of Parliament. The learned counsel for the appellant must address himself to the second point in Mr. Edge's statement.

Mr Leresche—My friend says you are to consider the proceedings of Wearden, not by the light of that demand, which is the only matter in connection with anything which goes to enforcing the penalty or fine, but he says you may look at a number of things which went to get Cooper discharged——

The Recorder—With a view to coerce him to pay the fine.

Mr Leresche—No.

The Recorder—What I understood Mr. Edge to say was this: "I grant that the asking for the fine was no besetting, but if you find a man loitering about a place, for the purpose of falling in with the master or foreman, in order to compel a particular member, by fear of losing his employment to pay the fine, I say that is a besetting with a view to coerce him to pay the fine; it is an illegal way of endeavouring to obtain a legal demand."

Mr. Leresche—But both men were there lawfully about their business.

The Recorder—You may take it that on the facts as proved I shall, without difficulty, say that Wearden, as the shop steward, was the proper person to collect the fines due to the society, and that in asking even at his employment for payment of this fine, did nothing which by possibility could come within the third section of the Act.

Mr. Leresche said with regard to the expression "loiter" that fell from the Recorder, it appeared to him that there was no loitering in the matter On the contrary, it was clearly shown that the two persons were working the whole of the time up to noon. The offence with which Wearden was charged was that he molested the respondent by besetting the place where he was. The only way in which that could be brought home to Wearden was that he was there unlawfully, because the idea of besetting was that of some person going to a particular spot where he had no business, and seeking to obtain his object by force. It was the idea of an invader coming to a place where he had no business to acquire possession of it by harassing the persons who persons who occupied that place. Wearden was charged with molesting the respondent by besetting the place where he worked.

The Recorder said there could be no doubt that the evil intended to be provided against by this third sub-section was that of picketing. The only difficulty in the matter was whether, that being the intention of the Legislature, these facts could be held to constitute, under the special circumstances of this case, anything like a picket Because although picketing would ordinarily be carried on by two persons, the amendment introduced into the act, after it left the House of Commons, made it an offence for one person to do what hitherto had been done in combination—

Mr. Leresche—But the man was perfectly lawfully there.

The Recorder—Yes, but supposing he had been at work on a different job, and he had come from a job on Manchester-road to some job on Chorley-road, at 10 o'clock in the morning, in his capacity of shop steward to collect this fine, and supposing that that had happened which did happen in this case, that the master was not in, and that he had gone to his work on Manchester-road again, and after the lapse of a certain period had come back when the master was there, would he have been besetting Chorley-road then?

Mr. Leresche—I apprehend not.

The Recorder—But, Mr. Edge contended that it was the waiting for the master which constituted the besetting.

Mr. Leresche—What is the waiting for the master? It exists only in the imagination of my friend. Where is there any evidence of it? When the work was over at dinner time, Wearden spoke to the master, but how could that be molesting when the man had gone away to dinner. I could have understoood it if the offence had been that of coercing the master to discharge the man, but Wearden is not charged with that.

The Recorder—But, as contended by Mr. Edge, his object in going to the master was to compel Cooper to pay the fine by fear of discharge.

Mr Leresche said it had never resulted in that The fine had never been paid. If Wearden had said to the man, "Unless you pay that fine, I will coerce your master to discharge you," it would have been a different thing, but the man was never troubled from the moment he refused. Mere molestation or obstruction was not an offence unless it was done with a view to coerce the person so molested or obstructed to do a particular act. Assuming that Wearden tried to coerce the master to dismiss this man, that was not a molestation or obstruction of Cooper with a view to coerce him to pay the fine. But he (Mr. Leresche) submitted that was not what Wearden went for. He never went to coerce, but he went to persuade the master to discharge Cooper, and whilst it was an offence against the act to attempt to obstruct a master, with a view to compel him to discharge a particular workman, it was no offence whatever on the part of any workman to go to a master, and persuade him to dismiss a workman. On the evidence there was nothing to show any act on the part of Weardon, which went a single step beyond trying to persuade Mr Martin to discharge a person who was objectionable. The amplest liberty was given to a workman to endeavour to bring his master or his mates to the same view as himself, and as long as he was content to deal with discussion, without resorting to any force or coercion or intimidation, he did not violate the law. Mr. Leresche admitted that the men left work at dinner time, but what was the evidence with reference to that? It was simply that they did not choose to come back at one o'clock, when they might have done; but there was no objection on the part of the men to give notice.

The Recorder said he would not trouble Mr. Leresche about the strike, because there was no evidence to connect Wearden with it The sole point was not that there was a strike, because that might be legal or justifiable, but Mr. Edge's case was this: "I say there is evidence from Wearden seeking the fine, and going to the master so immediately subsequent, that his intention was, by some act, to coerce the man to pay the fine; and I say that his being on the premises continuously from the time he demanded the fine until the master came, may be held to be, and I contend ought to be held to be a besetting within the meaning of the third section."

Mr. Leresche said Wearden was upon the premises the whole time for a lawful purpose. If he had gone to a place where he had no business to be—if his presence there was unexplained by a lawful reason, something might be said about the matter. But the man being employed the whole time negatived the possibility of such an interpretation being put upon it as that he was there for the purpose of besetting the place. He hoped the result of this appeal would be that the conviction would be quashed, and that people woule have the assurance that this was a sensible and reasonable interpretation to be put upon the act, quite consistent with the principles of trades unions and with the interests of employers all over the country.

Mr. Hopwood followed on the same side, and urged that this was a case of the thinnest description, and the magistrates below thought it necessary to go to the fullest verge it was possible for them to do. Now let him consider the Act of Parliament. The Act related to violence, threats, and molestation. By the previous law, it was held to be a threat to menace a man with personal injury to his property, or to his injury in any way, but this Act said "No, that is not to be so for the future; the only threats you can complain of for the future are such as intimidating any person in a manner that a justice of the peace could bind him over for." If that was remembered, it would throw a light on this question of besetting, because to make a man guilty of besetting for the reason that he had spoken to another about a subscription might bring upon trade unions all the mischief which the Legislature intended to relieve them from by passing this Act of Parliament. It must be something like a threat of personal injury or violence before they could make a threat penal. The apellant might have gone to the respondent and said, "If you don't pray, I will get you discharged," and that was not within the meaning of the Act. As to besetting, he contended that, bearing in mind the abolition of threats as a ground of penal legislation, it must by personal action cause some terror or fear in the man's mind The words of the Act implied the presence of somebody who was objectionable to another in order to molest or obstruct, the word "molest," or "obstruct," actually having a physical force meaning by this Act of Parliament. It was argued by the other side that every inference to be drawn by the man's acts of a doubtful kind were to be argued against him. Witness the argument about the man being in the place in two capacities But watching or besetting meant a continuity, a repeated observation, a continuance there for a space of time. He (Mr. Hopwood) came to this conclusion, that it was no offence for Wearden to ask Cooper for the money, even if he had threatened him when he did so; and his remaining at work and going up to tell the master was not a watching or besetting. Taking each act separately, not one of them would support the charge, taken continuously also they failed; and it was only by wresting the meaning of evidence, every inference being construed against the defendant, contrary to the usage of the criminal law, that anything like a justification could be obtained for this conviction.

Mr. Edge rose to make an explanation, and observed that there was no doubt whatever that under the 31st chap., relating to trade unions, men might agree to pay subscriptions and fines, and, as the Recorder properly observed, it would be no offence whatever to ask Cooper for the payment of a fine, even though it might be done in working hours.

Mr Edge was proceeding to make some further remarks, when Mr. Leresche took exception to his doing so, and Mr. Edge gave way.

Several dictionaries were then consulted as to the meaning of the word beset, and the definition to "harass persistently," taken from Webster, was the one which seemed to find most favour with the Court.

The Recorder then said—This case has been argued with great ability and ingenuity by Mr Edge, but I am clearly of opinion that the conviction cannot be supported. In so far as it is necessary, construing an Act of Parliament, and especially an Act of Parliament imposing penalties for a violation of a criminal or quasi-criminal enactment, I take it that we must look at the object which the Legislature had in view in passing the Act and the evil which the statute was intended to remedy, and that we must not allow either our personal opinion of the severity of the enactment or of its leniency to induce us to strain its terms either on one side or the other for the purpose of bringing within it or excluding any particular class of cases. Now, looking at the phraseology and the terms of the Act itself, it appears to me clear that the intention of the Legislature was to deal with acts having relation either to violence, or threats, or intimidation directed against some person, and that the fundamental principle upon which one should construe the Act of Parliament would be to look and see whether the act complained of was directed hostilely against some person who was sought to be coerced or dealt with by the party who was complained against. The very title of the act itself is "An Act to amend the criminal law relating to violence, threats, and molestation," and then, I am afraid with but indifferent success, the Legislature has proceeded in the first section of the Act to define what violence, threats, and intimidation usually mean. It is worthy of observation that under the definition of the threats which are intended to be made criminal under the Act, they must be threats of personal violence—such threats as would justify a justice of the peace, on complaint made to him, binding over the persons so threatening or intimidating to keep the peace. Well now it is perfectly obvious that no justice would be justified in binding over a man to keep the peace who simply said to another, "If you don't do this which I desire you should do, I will endeavour to subject you to some temporal disadvantage." That is not the sort of threat or act which is contemplated by a suretiship to keep the peace. Therefore it seems to be clear, as has been contended, that under this Act of Parliament it would be no offence even were such threat to be held out as a means of compelling the payment of that which the previous Act has decided shall be a legal fine inflicted under a legal contract, or agreement in connection with a trade union. Then when we come to that portion of the section to which attention has been more especially directed, we find the words to be: "Every person shall be liable to the penalties who shall do any one or more of the following acts, that is to say, molest or obstruct any person in manner defined by this section with a view to coerce such person to do certain things," and then it goes on: "A person shall, for the purposes of this Act, be deemed to molest or obstruct another person in any of the following cases, that is, if he shall persistently follow such person, if he hide the tools or clothes of such person, or if he watch or beset the house or other place where such person resides or works" It seems, to me, therefore, to be clear that the intention of the Legislature was to use words which should meet some mischief which was directed to the annoyance or to the intimidation of the person sought to be affected and sought to be coerced into doing a certain act. Now I am not prepared to say that a fellow-workman in the same employment and in the same place of business may not be guilty of an offence under this third sub-section. It is quite possible that if one workman employed in the same employment were persistently to "harass" (to use the dictionary interpretation of the word "beset") a fellow-workman by going to that portion of the building or that portion of the work upon which he is employed, for the purpose, by that harassing, annoying, and besetting, of compelling him to pay a fine or to do any acts which are defined in the section, that might be a besetting within the third sub-section; but I am quite clear that the mere application for the payment of a fine, even though that were immediately followed by an application to the master to discharge the workman, could not be said to be a besetting of the place within that sub-section. It must be, I think, something analogous, something just and generous with watching,—it must be a continual, a persistent, an habitual act. It is not sufficient that a man should simply call for the enforcement of a legal contract.

It must be such an annoying or habitual harassing, attendance, loitering about, hanging about, or some act of that character, which is directed against the person, who is sought to be influenced to do the particular act. Well, now, in this case, I am unable to see any evidence which would justify me in saying Wearden beset the place where Cooper, the complainant, worked. The facts appear to be that he made application by presenting the card for a fine which had been incurred under the provisions of the society to which both belonged, that he was the proper person to make that application, that he made that application in a proper way, that it was a legal application to make, because it was expressly enacted by the previous statute, by chapter 31 of 34 and 35 Vic, that such an agreement as he sought to obtain the advantage of should be a legal agreement. It is expressly enacted by sec. 4, cap. 31, of the 34th and 35th Vic., that while the trade union contracts, including among them agreement for a payment of any description or penalty to a trade union, should not be by that statute made of a character enabling any court to entertain any legal proceeding with the object of obtaining, enforcing, or recovering damage for the breach of them, still nothing in that section should be deemed to constitute any of these agreements unlawful.

Mr. Lereschc—And in the schedule there is provision made for fines and forfeitures.

The Recorder—Quite so; and there is no doubt there are other provisions in the section which justify what has been argued upon that ground. It is, therefore, I think perfectly plain that in asking for this fine, the appellant Wearden was guilty of no offence. I think it is equally clear that under this statute he was guilty of no offence even if he went to the master and said to him "Because this man has not paid the fine, I wish you to discharge him." That is not an offence within the statute. That has been felt strongly by Mr Edge in the course of his able argument, and he has therefore contended that while, so to say, the two extremities of the case must necessarily fail him, he can connect the two by reason of the man remaining upon the premises where Cooper was working after the time that he demanded the payment and up to the time when he spoke to the master, and that that remaining is a besetting within this section. Well now for what purpose did he remain? I am not to draw an inference unfavourable to him unless there be something from which I am compelled to draw it. He remained on the place where he had been working before Cooper came there to work at all, and on which he continued to work during the whole time that Cooper was there. He went to the same work, in the same way, and in the same place, and I can see no difference between the mere circumstance that he happened to be employed upon the same job with Cooper and his being employed on another job in another street of the town for the same master, going back to his work after he had demanded payment, and then coming at losing time for the purpose of making a complaint when he thought the master would be there. It could not be said under those circumstances that he beset the place. Can it be said that in attending to his work, because it happened to be at the same place, it was therefore a besetting? I think not; and I think, therefore, the conviction cannot be sustained. I only desire to make this further observation. The question of the strike has very properly not been brought into this argument. There was no evidence to connect Wearden with the strike personally, and the sole question seems to me to turn in this case upon the question whether his remaining at his work in the same place can fairly be said to be a loitering or an habitual harassing of the complainant by besetting the place where he worked. I think that would be a straining of the section—I think that would be an inference too violent to be drawn, an inference which could not be drawn impartially from the evidence, and therefore one which in sitting here I ought not to draw. I do wish, however, to say this further. It is a matter for regret, I think, that a case of some importance should have to be decided upon grounds so very slender as these. I can't conceive readily a case upon which a conviction could arise which sails further away from the real mischief which was contemplated to be dealt with by the Act of Parliament. What was intended undoubtedly was to render illegal the system of picketing, and I can see no resemblance between what Wearden did and the evil which this subsection was intended to remedy. No doubt this case has excited much attention. I think it extremely inconvenient that through the papers, either by leading articles or legal arguments in other places, questions should be discussed which still stand for judicial decision It has been perhaps fortunate that this argument has not in the least taken the course which has been taken in the discussions to which I refer, and which have proceeded upon an imperfect knowledge of the facts. The facts lie in a very narrow compass, and I have the less difficulty, therefore, in at once deciding that this conviction ought to be quashed.

Mr. Lereschc—With costs?

The Recorder—I think you should not ask for costs.

Mr. Lereschc—I won't trouble you then, Sir.

This concluded the business of the Sessions.—*Reprinted from the Bolton "Daily Chronicle," January 12th*, 1872.

THE CRIMINAL LAW AMENDMENT ACT IN SCOTLAND.

IMPORTANT TRIAL.

On Tuesday, February 27th, 1872, a case, involving a point of much importance to trades unions, came up at the Small Debt Court, at Airdrie, before Sheriff Logie. An action was brought by Edward Burns, a working mason, to recover pecuniary damages from four fellow-workmen, named respectively Crosby, Lockhart, M'Lay, and Gardner, as compensation for loss sustained by him through his being deprived of employment by an unlawful combination on their part, they having told his employers that if he was not discharged from the employment they would quit work, in consequence of which the pursuer was dismissed.

Mr. T A. Macfarlane appeared as agent for the defenders, and stated the defence to be a denial that the men had unlawfully combined for the purpose alleged. Proof was accordingly led at some length, Mr. Rose conducting case for pursuer.

Pursuer, on being examined, deponed that he had been in the employment of Messrs Gibson and Goldie, builders, Coatbridge, since May last. The whole of the defenders were in the same employment. On the 22nd January last Crosby told pursuer that he must pay down 7s 6d. for the society. Pursuer did not know what kind of society it was—it was not a sick society. He never saw the rules of it, though he had paid for them. He did not know the name of the society, but he had joined it about the time he went to that employment. The four defenders and pursuer were all the workmen Messrs. Gibson and Goldie, employed. On the 23rd January the defenders came out to work, but did not commence They went outside of the gate, and Mr. Gibson followed and spoke to them On coming back, Mr Gibson told pursuer that "they had all struck against him," and though he was sorry to part with him, still he would have to dismiss him. Pursuer was dismissed on the spot—it being the rule in the work that no notice is required to break the engagement. Pursuer asked Lockhart, one of the defenders, what was wrong, and he answered that they were not going to work unless he (pursuer) left, and joined the society.

Mr. Macfarlane, for defenders, said in opening he thought it necessary to state that it was doubtless the opinion of many of them, as it was of himself, that the present law in regard to combinations among workmen should be very much altered. What Mr Rose had said about the rights of working men was rather what they would wish to see than what actually did exist In reference to what had been said in regard to the common law of the land, he might be allowed to state what he understood such to be. At common law, a party was entitled to do anything which was consistent with his legal rights, no matter what might be the consequence to another. Mr. Macfarlane quoted authorities in support of this position, and argued that the act of the defenders was quite legal, even supposing that all that was alleged against them were true. The Trades' Union Act, he said, had made it a matter of perfect legality for working men to combine for the purpose of giving, withholding, or regulating labour, repealing the old law which rendered illegal any combination for restricting trade. It was stated that these men in an unlawful combination went to their masters and said that unless Burns was paid off they would not work. This was a simple withholding of their labour, which was quite a legal act; and he contended that there was not a distinct threat contained in what these men were alleged to have said.

Mr. Gibson stated that the whole of the parties to the case were in the employment of his firm, and that the four defenders had on the day stated struck work because the pursuer would not join the society. He asked the men why they were not starting work, and one of them answered that they did not want to run the risk of being fined for working beside Burns. It was either Crosby or Lockhart who spoke. He dismissed Burns that day. His only reason for doing so was that the other men would not work beside him. Burns was a good workman, and it was against the will of the employers that he was dismissed, for they had a contract in hand and were requiring his services

Mr. Goldie, the other partner of the firm gave similar evidence.

Proof was then led for the defence.

Richard M'Lay, one of the defenders, said he never said anything to his employers about leaving if Burns was kept on He never heard of any law of the society imposing a fine on parties who worked beside a non-union man. He did not go to work on the 23rd Jan.

Mr. Rose—Can you give any explanation how you did not work that morning?

I can work when I like.

On the question being pressed, he said—

We were going to look for another job.

Why did you leave the employment?

I had no reason in particular

You returned to work afterwards?

Yes, Mr. Gibson sent for us, and we went back to work.

That was after Burns had been dismissed?

Burns was not there when we went back.

Wm. Gardner, another of the defenders, said that he never threatened pursuer in any way. He never told his masters himself, nor authorised others to tell them, that he would not work unless Burns was dismissed He also did not work on the 23rd January last

The Sheriff asked why he did not work that morning, after going out

Witness—I did not go out to work.

Why did you not work?

I don't know what reason I had, I might be unwell.

Why did you begin work after Burns went away?

I cannot say.

Did you commence work because Burns was not there?

I cannot say.

Did you get any intimation from the society that you would be fined if you worked beside this man?

No.

Did you get any intimation at all in connection with Burns?

I was told that if we did not all join the society it would be made a "black job." He denied hearing Mr. Gibson ask why they had left their work, and said he never heard anyone say why they were leaving.

This closed the evidence

Mr. Rose, in addressing the Court for pursuer, said he thought he was entitled to say that it was the common law of the country that a working man was entitled to protection from any interference on the part of his fellow-workmen, and if any of his fellow workmen combined together unlawfully to get him expelled from his employment, he was entitled to sue for damages. The defence set up was a denial that this had been done in the present case; but he submitted that the evidence clearly showed that such had been done—that the pursuer had been very unfairly and unjustly dealt with by these four defenders It was distinctly shown that the dismissal of pursuer was occasioned by what these men said to their employers. Had this Society been approved of by law, there might have been a defence set up; but this was not the case. The sole question for his Lordship to decide was—were these parties entitled to act as they did? If they were, then he submitted that no working man nor employer in the country was safe. Had the employers in this instance not complied with the threat of the defenders, they would have had their contract stopped, and the pursuer, by not complying with their threats, had lost his situation.

THE DECISION.

On Tuesday, Feb. 27th, 1872, Sheriff Logie gave his decision in the action of damages brought in the Small Debt Court two weeks ago by a mason, named Edward Burns, against four fellow-workmen, named Crosby, Gardner, M'Lay, and Lockhart, and which was reported in our columns. His Lordship's decision is as follows:—

This is a case of very great importance, particularly to the working classes; and, as an issue such as the present is fortunately one of rare occurrence, I am fully sensible of the responsibility that devolves upon me in endeavouring to arrive at a right decision. This action has been brought by the pursuer for a sum of £12, as the damages due to him by the four defenders, his fellow-workmen, from their having unlawfully combined to cause his masters to dismiss him from their employment under a threat that unless they did so they would all leave their work, the consequence of which was that he was dismissed, and lost his employment. The facts, as brought out in the evidence, were that pursuer and defenders were all engaged as fellow-workmen at a building in course of erection under Messrs. Gibson and Goldie, builders. The pursuer appears at one time to have belonged to the United Operative Masons' Association, and had fallen into arrear with his subscription. To entitled him to be re-instated as a member of the Association, he required to pay a sum of 7s. 6d. of entrance-money, and one of the defenders having reminded him of this, he declined to pay, stating that others were as much in arrear as himself. In this state of matters, the defenders came to the building as usual at 6 a.m. on the 23rd January, 1872, but did not begin to work. When their master came and inquired why they were not working, he was informed that pursuer owed 7s. 6d. to the Association, and that they themselves would not resume work until he either paid that sum or was dismissed from his employment. This was confirmed by the evidence of Gibson and Goldie, the masters, both of whom were examined, and explained that, as they were under contract to finish the job in a certain time, and as workmen were difficult to be got, they were compelled to dismiss the pursuer rather than loose the services of the other four. They had no objection to him as a workman, and had no intention of dismissing him, but to the loss and inconvenience to which they would have been put if they had refused to dismiss him, and if the threat of the defenders had been carried out. They accordingly dismissed the pursuer, whereupon the defenders resumed work. For this proceeding on the part of the defenders, I am of opinion that they are liable in reparation to the pursuer. The old Combination laws having now been repealed, it is unnecessary to advert to them; but the Act of last session has an important bearing on the present question. By that Act, 34 and 35 Vic., chap. 32, sec. 1, it is enacted that every person who shall "threaten or intimidate any person in such manner as would justify a Justice of the Peace on complaint made to him to bind over the person so threatening or intimidating to keep the peace—or shall molest or obstruct any person in manner defined by this section with a view to coerce such person, being a master, to dismiss or cease to employ any workman" shall be liable to imprisonment with or without hard labour for a term not exceeding three months. No doubt the present action is not brought under the statute, and had criminal proceedings been taken against the defenders, these would probably have failed, because the defenders took no steps against the masters by threats or intimidation which would justify their being bound over to keep the peace, nor did they molest him in the manner defined in the Act by following him about from place to place, watching or besetting his house, or by following him in a disorderly manner through the streets. But it being established by the Act, that to coerce a master into dismissing one of his workmen is illegal, it matters nothing, as regards the injured party who suffers by this illegal proceeding, whether the *modus operandi* be by threats of violence and personal annoyance, so as to subject themselves to punishment criminally, or by a combination to refuse working along with him, so as to force his dismissal by a fear of the work being brought to a stand still. As a general rule, whenever an unlawful act has been committed, by which an innocent party has suffered pecuniary loss, the sufferer is entitled to reparation at the hands of the aggressor. "*Altevum non lædere* is one of the three general precepts laid down by Justinian, which it has been the chief purpose of all civil enactments to enforce. In consequence of this rule, every one who has the exercise of reason, and so can distinguish between right and wrong, is naturally obliged to make up the damage befalling his neighbour from a wrong committed by himself. Wherefore every fraudulent contrivance or unwarrantable act, by which another suffers damage or runs the hazard of it, subjects the delinquent to reparation."—Erskine, b 3, p. 1, s. 13. It is, therefore, the illegality of the Act which gives the claim for reparation. Any criminal procedure which may or may not follow against the aggressor has no effect whatever as regards the civil rights of the injured party to reparation. If a man assaults another he may be tried criminally for the assault, but he is also liable to the assaulted for the injury he has sustained. If he breaks a neighbour's windows he must make up for the loss of the broken windows, as well as stand his trial for malicious mischief. In short, as concisely stated by Guthrie Smith in his work on Reparation —p. 5, "The punishment inflicted in the interest of society is no bar to the private remedy." It can hardly be disputed that if the defenders had coerced the masters into the dismissal of the pursuer by proceedings which rendered them amenable to the criminal law, they would have rendered themselves liable to indemnify the pursuer, but, if so, the same principle applies to the present case. They have been guilty of an unwarrantable act, by which the pursuer has suffered damage. It may, no doubt, be said that the defenders were at perfect liberty to leave their employment if they saw fit, and that it was optional to the masters either to part with them or with the pursuer. But the illegality consists in this, that they combined together for the illegal purpose of coercing their masters to dismiss one of their fellow-workmen, not from any personal objections they had as to the man, but for the purpose of concussing him into the payment of a certain fine demanded of him by an Association of which they were members. As was said by Justice Crompton in the case of Walsby, 19th January, 1861—30 *Law Journal*, p. 121:—"I think that several workmen have no right to combine to procure the discharge of persons obnoxious to them by threatening to leave the employment at once in a body unless those persons are forthwith discharged." "What a man may do singly he may not combine with others to do to the prejudice of another."

I have carefully perused the recent case of Weardon to which I was referred by the agent for the defenders, in which the conviction of a stonemason at Bolton for molestation was quashed, but it appears to me to have no bearing on the present case, because in that case the prosecution was a criminal one against Weardon under the same Act for conduct similar to that followed by the present defenders, and the Recorder held, and I think rightly held, that he had not been guilty of such violence or intimidation as rendered him amenable to the criminal law, and he therefore quashed the decision of the magistrates. In the circumstances of this case, as proved in evidence, *I am of opinion that the pursuer is entitled to damages from the defenders for having illegally combined to coerce their masters to dismiss him in the manner already detailed;* and that the pursuer having not only lost his employment, but having been unable to find it elsewhere, as appeared from the evidence when the case was heard on 6th February current, that he is entitled to £3 3s of damages, for which sum I decern against the defenders jointly and severally, with expenses.

Mr. Macfarlane gave notice of appeal against his Lordship's decision.

WOMEN SUMMONED FOR MOLESTING THE STRANGERS.

At the Merthyr police-court on Monday, August 14th, 1871, a number of women were summoned, for that they "unlawfully did, with divers others, molest John Howells, by watching and besetting a certain place, to wit, the Navigation colliery yard, Mountain Ash, where he then happened to be, with a view to coerce the said John Howells, he being then a workman, to wit, a collier of Mr. John Nixon and others, his partners, to quit the said employment" The defendants' names were Catherine Rowlands, Elizabeth Carpenter, Mary Jane Summers, Jane Callaghan, Hannah Byewater, Ann Taylor, and Catherine Sullivan. The latter did not put in an appearance, and the summons as against her was therefore adjourned for a fornight.

The offence with which the above-named were charged was committed on the 10th of August, whilst one woman, who also did not put in an appearance, named Emma Beecham, was summoned for a similar offence, which had been committed on the 9th of August.

Mr. H. P. Linton prosecuted on behalf of Messrs Nixon, Taylor, and Cory. Defendants were asked if they had employed an attorney, but replied in the negative.

The first witness called was one of the strangers a man named John Howells, who, on being sworn, said—I am a collier in the employ of Messrs. Nixon and Co. I came to work at their colliery last Tuesday week. I remember Thursday last, the 10th instant. On that day I left off work at half-past three, and when I came up the pit, and went to wash at the washing place, and whilst I was there, and during the time I was walking from the pit to the place, I heard shouting, but I cannot say where it came from. I have heard shouting on other occasions when I was leaving my work. On that day there was a train of coal trucks between me and the turnpike road, and therefore I could not see anyone who might have been there.

Defendant Carpenter—Have you seen me shouting?

Witness—No, I have not.

Mr. Fowler—No, he says he saw neither of you shouting.

P.S. Thorney was the next witness. He said—I remember the men coming out of the pit on Thursday last, and Inspector Hodgson was at the time on the turnpike road; it was about half-past three o'clock, and the women, numbering several hundred, immediately they saw the men commenced shouting very lustily. I noticed defendants among the women in particular. Inspector Hodgson told me their names. I wrote them in my note-book.

Each of the defendants asked the officer whether he had seen them shout, and he replied that he had. Catherine Carpenter and Jane Callaghan denied positively that they had done so, though the other defendants admitted shouting "hurrah! like all the other women."

Mr. Linton intimated, after the defendants had made their defence, that that completed the case for the prosecution.

Mr. Fowler—I will give my decision at the Aberdare police-court to-morrow.

At Aberdare on Tuesday, August 15th, the defendants again put in an appearance before the magistrates with the exception of one, whose case was adjourned for a fortnight.

Mr Fowler, after taking some time to consider his decision, said the defendants had been summoned under the new Criminal Law Amendment Act. The summonses charged them with molesting one John Howells with a view to coerce him to leave his employment. He thought the charge against the defendants had been proved. Every man had a right to dispose of his time and labour as he thought right and fit, however and wherever it was wanted. It was a very serious thing to interfere with a man's liberty in those respects. They were now convicted of the offence, and would be sent to Swansea gaol for one week.

THE CRIMIMAL LAW AMENDMENT ACT.

DEPUTATION TO THE HOME SECRETARY.

On Thursday afternoon, March 21st, a deputation waited upon Mr Bruce at the Home Office to present a Memorial praying for the repeal of the Criminal Law Amendment Act, or for such alteration or amendment as would remove its most glaring injustice

Mr Alex McDonald, the chairman, and Mr Geo Howell, the Secretary of the Trades' Union Congress Parliamentary Committee, were present, as were also Mr Geo Potter, Mr W Leigh, Mr R Applegarth, Mr J. Normansell, Mr J. Whalley, and several other representatives of Trades' Unions Delegates from the Miners and the Factory Operatives were also present

The Deputation was introduced by Mr Mundella, M P for Sheffield, who said that the gentlemen present represented the Parliamentary Committee, which had been appointed by the Trades' Union Congress held at Nottingham in January last, to draw up a Memorial for the consideration of the Right Hon Gentleman, the Home Secretary. In introducing them he (Mr Mundella) was not called upon to express any feelings of his own in the matter, although they might be well-known, as the gentlemen present would give every necessary explanation, and furnish documents to support their Memorial. Mr. Bruce had expressed his willingness to endeavour to repeal any law which operated unfairly on any class, as it was not the wish of the Government for a single clause in this Act to work prejudicially, and he had also stated in another place the undesirability of oppressive punishment being inflicted on any class, whether belonging to Trades Unions or otherwise He would now ask Mr Howell to read the Memorial, and to supply any necessary information

Mr Howell, after thanking the Right Hon Gentleman for receiving the deputation, read the Memorial as printed on the first page He said —

Sir,—In presenting this Memorial to you, as the representative of the Government for the Home Department, we beg to accompany therewith, for your consideration, a complete record of the cases that have been heard since the passing of the Act—so far, at least, as the Committee have been able to ascertain We present them just as they were reported in the newspapers in order that it might not be thought that we have tampered with them, or toned them down while condensing the reports.

The only cases which have been condensed are those in which violence appear to have been used, or threatened; these we condemn in common with all right thinking people

The deputation represents a very large section of the Trades' Unions, who demand the repeal of the Act as a

20

piece of exceptional and unjust legislation, but we know that the Government would have great difficulties to overcome in attempting to effect the repeal of an Act passed so recently as the last session We do not want to embarrass the Government, knowing, as we do, that all the weight of the opposition was brought to bear in support of the Lords' Amendments But we do feel justified in asking the Government to restore the Act to the condition of the Bill when it left the House of Commons by repealing the Amendments of the House of Lords, and to modify certain other sections and sub-sections, so as to bring it more into harmony with the best spirit of the age in which we live

Had the Bill passed as it left the House of Commons, no doubt it would have been tried for some time before any serious attempt would have been made to alter it But we are worse off now than under the old law, and thesefore a most earnest effort will be made to modify it

Recognising as we do the difficulty in the way of repeal we shall aid the Government in any attempt to amend the Act in the direction pointed out in the Memorial, and found to be necessary on a careful consideration of the cases It will be found that some of the cases are very cruel and unjust, in others the punishment is exceptionally severe, in consequence of the animus displayed towards those on strike In the cases at Bolton and Hammersmith which have been appealed against, the cost of appeal is too great to be attempted except by such large societies as the Masons and Engineers (Mr Bruce —The original decision by the Bolton magistrates was absurd, and quite contrary to the spirit of the law) In answer to Mr Bruce, Mr Howell said—The Committee have furnished reports of nineteen cases in the Appendix to the Memorial, but these did not include any acts of violence, as they were summarised at the beginning of the Appendix, as it was not deemed necessary to reprint them in full Some of the cases occurred at Gateshead, and some at Newcastle during the high feeling engendered by the nine hours' strike.

Then there was the case at Bolton, which was reversed by the Recorder, one at Hammersmith not yet decided, one in Scotland, and one in Wales In all these cases the great difficulty was clearly to define the meaning of certain terms The word "besetting" was not clearly defined, and although we are not able to suggest another word, we do not think the one in question an appropriate one The word "besetting" suggested criminal intentions on the part of the workmen Another objection was to the word "coerce," which was so liable to be misconstrued by those who had the decision of these cases, as at Bolton and Hammersmith The memorialists therefore hope that the Government will take this matter into their early consideration with a view to immediate action, for there is no redress but for those who could appeal against the magistrates' decisions.

Appealing is most expensive, and can only be instituted by societies with large funds; but where appeals have been made they have been, so far, successful Still it is most desirable that the law shall be so clearly defined, that these appeals shall be no longer necessary for working men.

In conclusion, sir, we trust that the Government will take this matter into their early and serious consideration, with a view to the removal of the grievances complained of

Mr. McDonald said he had no wish to repeat what had been so clearly laid down by Mr Howell He would wish it to be understood, however, that the Association with which he was connected had no desire to harass the Government, and if the Act in question could not be entirely repealed, he hoped that the Government would take away certain objectionable words, and then the Trades would be able to go on with the Act as it stood, which might then be a good and useful act, especially in cases of rattening, if not met by the ordinary course of law

Meanwhile, although it had repealed a most obnoxious act, the present one did not work either harmoniously or properly. With regard to the Scotch case, mentioned by by Mr. Howell he (Mr McDonald) knew the whole particulars of that case, it was, in fact, a reversion of the decision of the Recorder of Bolton, except that the one was a civil action, and the other a criminal one In the case of Wearden it was imprisonment, in the Scotch case it was the infliction of a fine

Mr Potter said the workmen hoped that after all the trouble that had been taken by them, some consideration would be given to their case They were quite content to act upon the law when it worked harmoniously, but they were worse off now than before Formerly it was difficult to get a conviction against the workmen, but under the new law they could do so easily This Act was only injurious as it affected the men, and therefore the masters were not likely to make any effort to get it altered, as it worked to their interest. Many of the Members of the Government and several Members of Parliament had promised their support to the Bill as it stood, and it was thought by working men that if the Government had done its duty the Lords' Amendments could not have passed

Mr Bruce reminded Mr Potter that all the members of the Government voted for the original Bill

Mr Potter said he was aware of that, but if the Government had used the same influence on this, as on som other questions, so many Liberals would not have voted against the Government on this occasion If the Act was restored to what it was when it first left the Commons, they could work under it As it now stood it was most unsatisfactory Working men wished to be guided by the law but they did not wish to have exceptional legislation, they only desired that labour should be equally cared for, and that the power to combine should be unrestrained

Mr. Normansell said Mr Howell had some suggested alterations to suggest as agreed upon

Mr Howell said the Committee was most desirous to give all the aid in their power for the purpose of amending the Act, they had therefore gone into the matter carefully with a view to suggest some alterations which they thought would be effective They did not bind themselves to the precise wording of the suggested alterations, as that must be left to a Parliamentary draughtsman, but they desired to submit the following —

Clause 1 —For the last sub-section (3) of the last paragraph of Section 1, substitute provisions to the following effect —

1 If two or more persons tumultuously watch or beset the house or other place where such person resides or works, or carries on business, or happens to be, or the approach to such house or place
2 Or if with two other persons together, they follow in a tumultuous and disorderly manner in or through any street or road
3. If with one or more other persons they beset or watch with threats of harm to person or property, &c

This the Committee thought would meet the case for both parties The Act as it now stood seemed to be on the supposition that there were always two or more Union pickets to one non-Union workman who had gone in on strike The very reverse was the fact Often the two pickets had to meet ten or twenty persons on the other side, and surely no possible intimidation could take place under such circumstances Moreover, two pickets were a better guarantee for not breaking the law, as one man frequently lost his temper, and molestation was the result.

Then as to the better definition of the terms, to prevent any possible mistakes, "besetting" must be defined as meaning something more than merely being in a certain place The other term "coerce" appears to mean *induce*, and not what it is ordinarily supposed to mean, *compel*. Surely this was never meant by the Government when draughting the Bill (Mr. Bruce nodded assent) The other terms, "violence, threats, and intimidation," in the first section of the Act seem also to be misunderstood by the magistracy, especially by the unpaid Magistrates in many parts of the country where the Employers sit to try such cases

Mr Bruce said, in reply, that the Bill in question had certainly been brought in at a time when it did not stand a good chance of being fairly and calmly deliberated upon He thanked the deputation for their information, and for the documents which accompanied the Memorial, and would undertake, on the part of the Government, to give them his careful consideration After alludsion to a rattening case, which, it was explained, had not been included in the list, Mr Bruce said the Government were very desirous to make the law act harmoniously to all classes of working men, but he thought it would not consent to a repeal of this law, inasmuch as it had not been found yet to work unsuccessfully There were two points to be considered, first, they should see in what this Act had fallen short of what was desired, and, secondly, if it was found to have fallen short to endeavour to place it in that position as when it first left the House of Commons, but they must remember that the amendments that came from the Lords had been adopted by the vote of the House of Commons Of course it might be thought by some that the amendments of the House of Lords were hardly necessary, and they were entitled to their opinion In conclusion, however, he said he would look carefully over the cases, and see how the Act had been interpreted by the different justices. Although the decision in some instances had been at variance, in time, when they were fully known, they would become more uniform

Mr Mundella thanked Mr Bruce for his attention to the deputation, which then withdrew.

LONDON: PRINTED BY H. W. FOSTER 16, BEAR ALLEY, FARRINGDON STREET.

TRACTS FOR TRADES UNIONISTS.—No. IV.

THE

NATIONAL FEDERATION OF ASSOCIATED EMPLOYERS OF LABOUR.

REPORT UPON THE MEMORIAL PRESENTED TO THE HOME SECRETARY BY THE NATIONAL FEDERATION OF ASSOCIATED EMPLOYERS OF LABOUR, DECEMBER 13TH, 1873.

BY

HENRY CROMPTON, ESQ.,

BARRISTER-AT-LAW.

TOGETHER WITH THE

CORRESPONDENCE BETWEEN MR. HENRY CROMPTON AND MR. T. R. JACKSON.

PRINTED BY ORDER OF THE SHEFFIELD TRADES UNION CONGRESS, BY THE PARLIAMENTARY COMMITTEE.

GEORGE HOWELL, SECRETARY,
27, VILLIERS STREET, STRAND,
LONDON, W.C.

PREFACE.

The following Report upon the Memorial presented to the Home Secretary by the National Federation of the Associated Employers of Labour was prepared by Mr. H. Crompton, at the request of the Parliamentary Committee, and read by him to the Trades Union Congress at Sheffield. A resolution was passed by the Congress that it should be printed as a pamphlet together with the whole correspondence between Mr. H. Crompton and Mr. Jackson.

GEORGE HOWELL,

Secretary of the Parliamentary Committee.

THE NATIONAL FEDERATION OF ASSOCIATED EMPLOYERS OF LABOUR.

A Report upon the Memorial presented to the Home Secretary by the National Federation of the Associated Employers of Labour.

This memorial, though mentioning incidentally the Trade Union Act and certain Statutes in restraint of employers, is limited to considerations upon the Criminal Law Amendment Act, the Master and Servants' Act, and the Law of Conspiracy. It sets out with a very fair statement of the law as it was before the appointment of the Trades Union Commission, and of the various alterations which had been made from time to time, omitting, however, some of the most important judicial decisions, like that of Vice-Chancellor Malins, which brought matters to a climax, or that of Mr. Justice Lush, which declared picketing to be perfectly legal. The memorial goes on to quote largely from the well known report of the majority of the Trades Union Commission, but makes no allusion to the fact that the wholesale charges which had been brought against Trades Unions, and out of which the Commission arose, had not been substantiated, or that the whole Commission had acknowledged the beneficial side of the Trades Unions.

The memorial then enters upon a history of the passing of the Criminal Law Amendment Act. It quotes the opinion of the National Association of Factory Occupiers, that "watching," "besetting," or "picketing," might be as well effected by one person as by three, but no mention is made of the legal and constitutional objections which were raised to the making such actions penal. No mention is made of the fact that an eminent and notoriously just judge, the Recorder of London, thought it right to leave his party and vote against such a law being passed. The employers say nothing of the important and often cited declaration of Mr. Bruce, that "the effect of the Lords' Amendment would be that any one man who stood at the door of any works for the purpose of ascertaining who were going to work there, might, without any great stretch of the law, be held to be there for the purpose of coercing other people."

The quotations made, chiefly parliamentary, not judicial, are so one-sided as entirely to negative the claim of the employers to an impartial, or to use their own word, to an "imperial" treatment of these questions. There is not a single allusion from first to last to the wrong and unjust convictions that have occurred. They do not say whether they approve or disapprove of such convictions as that of

Turk at Hammersmith. They preserve a complete silence. They do not even attempt to deal with the facts which have been proved to exist.

The memorialists "submit that it is the duty of the State to take care that no alteration in the law shall be permitted which may result in giving power to associations or individuals to oppress other people, either as a matter of right or upon such easy terms, as to be worth while buying at the cost of the prescribed penalty." The Trades Union Congress can heartily accept this somewhat mercantile conception of penal law, and further insist that wherever the inadequacy of law allows such things to be, that the law should be made more stringent, as in the case of truck, of women and children in factories, and that where a law is used as a means of oppression, it should be abolished, as in the case of seamen being compelled by penal law to go to sea in rotten ships; as in the case of the Criminal Law Amendment Act; the Master and Servants Act; and the Law of Conspiracy.

The memorial then proceeds to answer the various objections which have been raised against the Criminal Law Amendment Act.

1. That it is a penal Act, specially directed against trade offences and working men. The only answer is the old one, that the Act is verbally applicable to both employers and employed. The Act, which makes picketing a crime, is said to be applicable to the employers. Our objection, however, is not stated with sufficient force. It should be thus:—That the Criminal Law Amendment Act is a special law against a part of the population in respect of crimes of violence, which all classes of citizens can commit and do commit.
2. That these offences might well be dealt with under the general law. As answer to this, the memorial says that by general law we mean the common law. We mean nothing of the sort—nothing so stupid. By general law we mean the general criminal law applicable to all citizens, whether by statute or by virtue of the old common law, and as distinguished from a statute directed against a class, and exempting the rest of the community from punishment. Any defect or insufficiency in the laws dealing with violence ought to be remedied by general enactments.

The memorial then proceeds with the assertion that if the Criminal Law Amendment Act be repealed, no legislative provisions will remain to meet the offences now dealt with by the act. Our answer must be twofold, first by reference to the Criminal Law, the extent and severity of which is constantly misrepresented; and, secondly, to the fact that a very large number of these convictions might have taken place under other laws. It is necessary to repeat that the only mode in which the Criminal Law Amendment Act has had a beneficial effect has been in cases of threats of violence. The memorialists do not answer the question so often put, of why is a workman who uses a threat about his employment to be guilty of a crime, and a man who threatens a woman exempt from punishment? We say that the law should deal out equal justice to both. It does not, and therefore even in that part of the Criminal Law Amendment Act which refers to threats, there is injustice and class legislation.

3. The third objection refers to picketing. Here the memorial states so erroneously the views we have constantly advanced, that it is necessary to reproduce the reasons why picketing ought not to be a criminal offence.

Picketing only involves standing still in the street. A law which says that any person standing in the street shall be punished is obviously unjust. So is any law which says that if a man stands still in the street he shall be held to be there intending to coerce. Such an inference cannot be drawn; it is illogical and unjust. Yet this is what the memorialists propose. They distinctly say that picketing ought to be criminal. It can only be criminal by means of such an inference. And many of the wrong convictions that have taken place have occurred because the view to coerce has been inferred either from the watching or from the fact of a strike. It would be exactly the same error if a man were punished for having a knife in his pocket, because he might stab with it. The absurdity of this extraordinary conception of criminal law reaches a climax, when the memorialists give as a reason for picketing being criminal, that it "facilitates physical and moral coercion, that it enables those on guard to whisper threats which they would not dare to print, and that it is a most potent means of interfering with that individual liberty which it is the chief duty of the law to protect." Not merely in the opinion of the employers are threats and intimidation to be punished, but everything which facilitates moral coercion or enables people to whisper threats, quite irrespective of whether there has been coercion used or threats uttered. And yet a threat to murder is no offence by our law! Such extravagance is its own answer. The memorial says that this act "rests upon sound principles of social economy, and upon the same basis as all criminal law." We say that it violates these principles, and does not stand upon the proper basis of penal law, because it employs neutral words without limitation or definition; because it includes acts which are perfectly legitimate, and some that are morally right; because it mixes up highly criminal proceedings with those which have nothing criminal in them, and thus confuses all sense of right and wrong, all distinction between that which is criminal and that which is not; lastly, it punishes offences in the case of workmen with three months' hard labour, which are generally punished with a 40s. fine as a maximum, a gross and glaring inconsistency which no man can deny.

It is somewhat curious that the memorial should quote a passage from the report of the majority of the Trades Union Conmmission which entirely contradicts their own assertions and substantiates the statements we have put forward, and which we now maintain. It is as follows:—"We are not prepared to propose new legislation on the subject of picketing, believing that the existing law is, at all events, equal to repressing flagrant abuses." That is to say, the Commission reported that the law, as it stood before the passing of the Criminal Law Amendment Act, when picketing was legal and when these cases were usually brought before the highest tribunals, was strong enough to put down the abuses which might arise out of picketing. Thus the employers' quotation proves our case exactly.

4. The 4th objection of the Unionists to the Act, that of the summary jurisdiction, is misrepresented by the memorial. They say that the Trades Unions "asked a Secretary of State to tamper with the administration of justice," and "that working men should be exempted from the jurisdiction of the magistracy, to which all other citizens are amenable." Both statements are devoid of truth. The exact opposite is true. We have always protested and we now protest against class legislation and class administration.

Such are the arguments upon which the memorial asks for the retention of the Criminal Law Amendment Act. There is, indeed, one other page of foolish inflammatory writing, about Welsh Colliers being ready to murder Irishmen, and other matters tending to excite prejudice, which I pass over as containing nothing requiring a reply, and as unworthy of the attention of an assembly like this.

The memorial deals very shortly with the Master and Servants Act. It reprints the report of the Committee of 1865, and then proceeds to enunciate two propositions, both of which are untenable. The first is, that under the 9th Section, imprisonment can only be given for disobedience to the order of the Court. They miss the point of our objection to that Section altogether; namely:—That a workman can be directly fined irrespective of actual damage, and in default, sent to gaol; and that seemingly without the protection afforded by the 11th Section. But the position of this question has now changed. Some comments had been made in the *Times*, by an anonymous correspondent, upon the views held by Trades Unions on the Master and Servants Act; and I wrote two letters in answer, in which I insisted that our objections to the Master and Servants Act were to its criminal clauses, to the magistrate having any direct power to fine or imprison; and I further insisted that whenever imprisonment was used to enforce payment it should only be after other means have failed, and as a last resort. In addition I made the following proposal:—

Repeal of the 14th section.

Retention of the rest of the Act in its present form, with the following exceptions:—

1. That the power to fine be struck out of the 9th section, which would leave the magistrate power to award damages up to the amount of £20, such damages to be enforced by the provisions of the 11th section, and not by the Small Penalties Act.
2. That the power to award specific performance be transferred to the County Courts.

On the 1st of January a letter appeared in the *Times* from Mr. R. Raynsford Jackson, the Chairman of the Executive Committee of the Federation of Employers, dealing with my two letters, and making some very important admissions. He says, "I agree that breach ot contract, pure and simple, ought not to be a crime; that it ought not to be directly punished by imprisonment, or even by fine; that it can be adequately met by the award of damages; and that in cases of simple breach of contract, imprisonment is justifiable as process only, and to enforce payment of damages." Now this admission in effect concedes everything that we demand. It is true that Mr. Jackson still urged the retention of the 9th section, without alteration, but upon an untenable reason, which was altogether unpractical. I wrote an answer to his letter, replying at length to his arguments, which letter with with the rest of the correspondence, I am ready to present to the Congress. I did not send it to the *Times*. I sent a short letter to say that I would not enter into any communication, direct or indirect, with him or the Federation he represented, until a certain calumnious passage contained in the memorial had been formally and publicly withdrawn. You will find it at page 17. No answer appeared until January 10th, when he refused to withdraw it, and therefore I did not print my letter.

Next, as to the 14th section. The interpretation put upon it by the Federation of Employers cannot be supported. They say in this memorial: "The 14th section of this statute is, however, the one which is chiefly complained of, and upon this we submit that there is not a word in the section which can be construed as having reference to a breach of contract." Let the statute itself answer. Here is the preamble which governs the whole Act. "Whereas it is expedient to alter in some respects the existing enactments relative to the determination of questions arising between employers and employed under contracts of service." The whole Act refers to contracts and breach of contracts, and the 14th section itself begins thus: "Where on the hearing of an information or complaint under this act." I think I need say no more on this point. It is, of course, perfectly true that nobody can be punished for breach of contract under the 14th section alone. But it is quite clear that on the hearing of a summons under the 9th section, punishment by virtue of the 14th section can be awarded. My condemnation of the section is that it does not define a single crime, and that it gives a loose and extraordinary power of punishment to a civil tribunal; that on the hearing of a civil suit, without any allegation of crime, without any of the proper formalities and proceedings of criminal courts, the magistrate may proceed to pass sentence of hard labour, a procedure which is analogous to giving a judge at *Nisi Prius*, or a Vice Chancellor power to say: "I give judgment for the plaintiff, and I sentence the defendant to 10 years penal servitude." Without entering into the question of what actions should be criminal and what not, we may take our stand upon the broader ground that whenever actions are to be made criminal, the law which constitutes them crimes must be clearly defined and administered by the proper criminal procedure. And surely a more reasonable proposition could not be stated. No answer of any sort has been given to this. The Master and Servants Act does not define the crimes it punishes; it violates the rules and safeguards by which we rightly surround our criminal justice, and, in fact sins in practice against the very distinction between civil and criminal responsibility which the Federation of Employers, through Mr. Jackson, have now admitted in principle. And, therefore, if the 14th section had nothing to do with contract, it would be a still greater violation of the protection which all other laws afford to individual liberty.

The words of this 14th section are so vague and obscure that no man can say what they include or exclude. Let us dissect the section and examine the offences it creates one by one. First.—*Injury inflicted to person or property*. There is no injury that can be inflicted either to person or property which is not stringently punished by the statutes referring to offences against the person and property. Let the employers specify a single instance. Next—*Misconduct*. Let us suppose a bad case of misconduct by a servant, of disobedience, of the most insolent behaviour, and the setting a whole establishment in confusion. That I distinctly say ought not to be a criminal offence. But misconduct might mean anything. If there is any kind of misconduct which ought to be criminal it must be defined, and not made penal in such a way that any misconduct is a crime. At present misconduct by workmen is a crime punished by three months' hard labour. *Ill-treatment*. I do not know what ill-treatment by a servant of a master can mean. Ill-treatment by a master of a servant ought not to be a crime. The word is as vague and as absurd as misconduct. Unkindness might just as well be made criminal. *Misdemeanour* is the next word, but that is the name of a class of offences. Libel, conspiracy, riot, and perjury are misdemeanours. Do they wish that there should be summary jurisdiction over such crimes under the Master and Servants Act, or is such a word in such a section simply nonsense? Such is this precious section, which a large association of employers think it right to support and to insist on its being a part of the criminal justice of the country.

As to Conspiracy, the Memorial presents us with one paragraph and one observation of the most extraordinary kind. They say that "The Law of Conspiracy is undoubtedly intricate, and because it is impossible to predict or define the possible methods in which the men *may intend to agree* to injure others, the tribunals which deal with cases as they arise must not only retain the power of determining whether a crime has been committed, but also what is the proper punishment." To this I answer, there is no such Law of Conspiracy. Happily, the intention to agree is not yet a crime. So miserable a blunder would not need to be noticed, were it not coupled with the statement that the employers express the deliberate wish that the law should remain vague and undefined, and that the judges should retain their present or even a greater legislative power than they now have. This needs no comment. We have simply to urge the exact opposite, and declare that judicial legislation in penal law has proved injurious, and is contrary to the enlightenment and spirit of the time.

Such is the Memorial of the Federation of Employers. It can

have no prejudicial effect upon the cause we have at heart. It is too feeble, too one-sided, too ignorant of the principles and practice of the Criminal Law. The Memorial is a total failure of that which the employers pretended to accomplish. Already they have had to modify their views; and they will find, as they examine these questions and our conclusions, that the justice of the matter is wholly on our side. From the very first we have known how the land lay, and they have not understood it. From the beginning of the agitation we have never put forward any immoderate demand, or any that we were not prepared to substantiate by fact and argument. We have fought a fight against misrepresentation and abuse. And now we have reached that pleasant stage when it is not unbecoming in me to congratulate the Congress on the victory we see before us, the triumph of as just a demand as was ever made by a people to its rulers.

12

THE CORRESPONDENCE.

TO THE EDITOR OF THE TIMES.

SIR,—The letter of your correspondent "F." in the *Times* of to-day, in reference to a case decided under the Master and Servants Act, misrepresents the views and claims of the Trades Unions to such an extent, that I beg to be allowed to reply to his statement and explain his error. The issues raised are so important, the aspect now assumed by the conflict between labour and capital is so serious, that writers upon the one side and upon the other ought to be very careful not to exaggerate the views of their opponents, and more careful to avoid statements which admit of flat denial. Nowhere and at no time have the Trades Unions put forward any such demand as that for which your correspondent says they are now agitating—namely, for "civil penalties without the alternative of imprisonment." The case, which he says they disapprove, is precisely one which they approve. It serves as an excellent example to show the force of their argument and the strength of their position. It was a case in which damages were awarded against two men for breach of contract, and the amount of the damages was equally divided between the two. Imprisonment followed in default of payment, and as process to enforce payment. The Trades Unions have not objected to this; on the contrary, they have openly accepted it. The deputation to Mr. Lowe declared they would be perfectly satisfied with the entire repeal of the Master and Servants Act, which would have the effect of bringing all these contracts within the jurisdiction of the County Courts, and of making them subject to that extremely stringent process. The formulated demand of the Leeds Trade Union Congress, "No imprisonment for breach of contract as enacted by the Master and Servants Act," was drawn in that form on purpose to negative any assertion that the Congress did not wish that contracts should be enforced. The objection is not to imprisonment as process, but to the direct sentence of fine or imprisonment, instead of or in addition to damages awarded. The case cited, unquestionably, was one of misconduct on the part of the men. For such a breach of contract judgment awarding damages was right, but a penal sentence would have been wrong. The penal law ought not to be used for the purpose of making men work well or behave properly. In this very case the employer chose to contract with men who had behaved badly before, and to enter into such contract without taking any precautions like the retaining a portion of wages or instant dismissal. Workmen ought not to be punished penally for bad work so long as their misconduct does not take the form of maliciously injuring property, which is an offence punishable summarily with imprisonment under another Act, and to which no sensible person could possibly object.

But it may not unfairly be asked, Has there not been a proposal made to alter the Small Penalties Act and the method of enforcing payment of penalties by imprisonment? How is that to be reconciled with what you say? My answer is, that the very fact that the workmen have insisted upon a reconstruction of the Small Penalties Act, as pressing with harsh and undue severity upon the poor, is evidence that they accept the principle of enforcing judicial awards of pecuniary penalties even by imprisonment. And the view that at present imprisonment is too hastily inflicted—that it follows too closely upon the infliction of the fine, without proper consideration of the defendant's circumstances or means of paying—is not inconsistent with accepting imprisonment as a last resort. There is, in truth, a great deal to be urged against the too frequent use of imprisonment. Imprisonment, when used as process, should always be the last resort. There are many ways in which payment of fines might well be enforced by delaying an impending imprisonment, by allowing payments in small instalments, by giving facilities for finding sureties, and so forth. We insist that such improvements in the procedure would be wise and politic, by rendering imprisonment less common, and therefore more degrading and effective, when used as punishment.

Besides all this, there are other serious practical evils connected with the enforcement of penalties, which constitute a great grievance, which cause irritation, which injure the administration of justice, and which have not as yet received proper consideration from Parliament. Let me state one of these, which appears to me to be very irritating and little known. When a fine is inflicted by a magistrate, the person fined has to pay costs besides. The costs are not a fixed quantity, but vary with different circumstances, and especially with the distance; so that a man whose offence is committed a long way from the place where the court is held is more severely punished than one who lives close by. Sometimes the costs are more than the fine, and an enlightened magistrate does often apportion his fine so as to make it less when the costs are great. But even then this is a real evil, because not only is punishment made to appear very uncertain, but, if a man is fined 2s. 6d. and £1 costs—however just it may be that he should be fined £1 2s. 6d. altogether—yet he inevitably treats the 2s. 6d. as the right punishment for his offence, and the £1 costs as downright robbery; and the injustice appears greater when he goes to prison for a month because he cannot pay the costs.

All the evils of this kind are inseparably blended with such laws as that of the Master and Servants Act; and this tends to aggravate and increase the evils which the great imperfection of our summary administration occasions. The time has, in fact, arrived for a thorough examination into this branch, as well as into other branches of our magisterial system.

In conclusion, let me make an assertion that among the leaders of the Trades Unions there is a real desire for justice and a disposition to accept any solution that is practically just. I have thought ever since 1868 that there ought never to have been any agitation on these working-class questions; and I still think that a little more generosity and a little more trust on the part of employers and of Parliament would easily remove the dangerous and mischievous feelings which have been aroused against our laws and our system of justice. It is lamentable to think that the attitude of the employers at the present moment, when all might yet be healed, must necessarily tend to aggravate the difficulty, and to stimulate that which has already gone too far.

I am, Sir, your obedient servant,

HENRY CROMPTON.

15, Phillimore-gardens, Dec. 19.

14

TO THE EDITOR OF THE TIMES.

SIR,—It is rather a hopeless and disheartening task to maintain a controversy with one like your correspondent "F.," who represents me to-day as saying exactly the opposite of what I did say in my letter to you on Saturday. He begins his answer to me thus:—"Mr. Crompton asserts that the Trades Unionists have never agitated for the abolition of all criminal penalties for breach of contract." Whereas I asserted precisely the contrary. That is just the purpose of the agitation. Really "F." is too bad. I must implore him to be more careful. The distinction we stand upon is a very plain one, and not of our inventing,—between civil and criminal; between civil process and punishment for crime; between imprisonment as the indirect result of process, and imprisonment or fine as the direct punishment for crime. Nor is there on our part any shrinking from or shirking of the question of the impunity of workmen who by breach of contract injure their employer. As far as the Criminal Law is concerned, we say that unless there are other circumstances which, like fraud or certain kinds of culpable negligence, may well make a difference, breach of contract, pure and simple, ought not to be a crime. The workman, as every citizen, should be civilly responsible, and all proper means that are just and not cruel should be employed to render the civil remedy effective.

The proposition about the County Courts was not volunteered by the Trades Unionists, but was an answer to a question put by Mr. Lowe to the deputation as to whether such a plan would meet with their approval. Unquestionably there would be difficulties connected with the County Court jurisdiction in such matters, though I cannot think, with "F.," that it would be a "mockery of justice." The Master and Servant Act, as administered in many parts of England by the magistracy, has a very beneficial side. That is the result of inquiries which I have made from persons in different places, whose judgment I trust, and while in some respects the County Court would be by far the most satisfactory tribunal, especially wherever a decree for specific performance is required, it cannot be doubted that the decision and settlement of disputes on the spot is a great advantage to many of the employers and employed.

But, Sir, I would earnestly ask you and the public generally whether compromise on this question is not perfectly feasible. I do not profess to represent Trades Unions, but I believe that the views stated in my former letter are the views of the great majority of the Unionists; at all events several leading men among them have approved that letter. Therefore, feeling very anxious that all these questions should be settled, and believing that my recommendation will be fairly examined and weighed by the men, I propose the following solution and invite discussion and criticism:—

Repeal of the 14th section.

Retention of the rest of the Act in its present form, with the following exceptions:—

1. That the power to fine be struck out of the 9th section, which would leave the magistrates power to award damages up to the amount of £20, such damages to be enforced by the provisions of the 11th section, as intended (in my judgment) by the Act, and not by the Small Penalties Act.

2. That the power to award specific performance be transferred to the County Courts.

No doubt there are further questions relating to this matter. I have this morning a letter from an enlightened lawyer, saying that imprisonment ought only to be used as process where there has been a wilful disobedience to an order of Court; that the mere inability to pay is not sufficient. I myself feel great difficulty in coming to that unqualified conclusion. I do see that any distinction in process between a middle-class man who cannot pay his judgment debt and a workman who cannot do the like is quite untenable. There is, however, no reason why the decision of that point need interfere with the settlement of the present question. Certainly not, if, as I believe is the case, the workmen do not raise that objection. I was, I confess, not a little astonished to find that the memorial presented to Mr. Lowe by the Masters' Association defended the 9th section upon the ground that imprisonment could not be given under that section except for wilful disobedience to an order of Court. The memorial misses our point altogether—namely, that a workman can be directly fined, irrespective of actual damage, and in default be sent to gaol, and that seemingly without the protection afforded by the 11th section. If the only difference between the employers and my proposal were that concerning the 9th section, the difference would be comparatively small; but they object altogether to the repeal of the 14th section. The reason they urge for its retention is a very curious and startling one—that the 14th section does not refer to contract at all; and if I understand the memorial, they assert that it was introduced on purpose to deal with offences irrespective of contract. I cannot for a moment believe that such an interpretation could possibly be put upon the Act. If, however, I am wrong, I can only say that it is the strongest argument that has yet been adduced for the repeal of the section, and the abolition of the unconstitutional and anomalous procedure, which they desire to retain.

I am, Sir, your obedient servant,

HENRY CROMPTON.

15, Phillimore-gardens, Dec. 22.

TO THE EDITOR OF THE TIMES.

SIR,—Perhaps it is not too much to assume that the correspondence which has appeared in the columns of the *Times* is, to a certain extent, an illustration of one of the ways in which the action of the Federation of Associated Employers is calculated to throw light upon, and to facilitate the decision of, important questions in the relations of employer and employed. Among those letters two signed "Henry Crompton," remarkable for their moderation of tone, and, from his point of view, for their fairness of argument, are conspicuous. These are the reasons why, being Chairman of the Executive Committee of the Federation of Employers, I address you to-day.

Mr. Crompton, in his letter of December 19, speaking of the Master and Servant Act, says, "The objection is not to imprisonment as process, but to the direct sentence of fine or imprisonment, instead of or in addition to damages awarded;" and in that of December 22 he says, "The distinction we stand upon is a very plain one, and not of our inventing, between civil process and punishment for crime—between imprisonment as the indirect result of process and imprisonment or fine as the direct punishment of crime." "Nor is there on our part any shrinking from or shirking of the question of the impunity of workmen who, by breach of contract, injure their employer." "As far as the

16

criminal law is concerned, we say that unless there are other circumstances which, like fraud or other kinds of culpable negligence, may well make a difference, breach of contract pure and simple ought not to be a crime."

Now, Sir, there is a great deal in these arguments which I can accept. I agree that breach of contract pure and simple ought not to be a crime; that it ought not to be directly punished by imprisonment or even by fine; that it can be adequately met by the award of damages; and that in cases of simple breach of contract imprisonment is justifiable, as process only, and to enforce payment of damages.

On the other hand "culpable negligence" should surely be punished by fine, and even more so when accompanied by breach of contract. The danger is that, when breach of contract and culpable negligence are punished together, the punishment allotted to the latter is regarded only as given in consequence of the former.

It may be asked, How is culpable negligence, which Mr. Crompton agrees should make a difference, to be defined? The aggravation of breach of contract which amounts to culpable negligence, and would justify a fine, can only be left to the judgment of the tribunal which has all the circumstances of the case in view.

But the 9th clause of the Master and Servant Act harmonises with all the above conditions, excepting when fulfilment of the contract is ordered. It gives to the tribunal the power to adopt one of several alternatives equivalent to assessment of damages, and to order payment, "or else, when no amount of compensation or damage can be assessed, or where pecuniary compensation will not, in the opinion of the Court, meet the circumstances of the case," as where culpable negligence is proved (for fraud should be dealt with by other Acts), "to impose a fine." Should the award of the Court, however, not be carried out, then imprisonment may follow as process under the 11th section.

So far I cannot see what cause of quarrel Mr. Crompton can have with the 9th section. But it also authorises the Court to award fulfilment of contract, and to require the defendant to find good and sufficient security, in default of which he may be imprisoned until he complies with the order, but so that the imprisonment shall not exceed three months. It is true that as regards this portion of the 9th section the defendant has not the protection afforded by the 11th section, and I can also conceive that he might be required to find sureties beyond the reach of a working man, as the Act supplies no limitation to their amount.

I think, Sir, that to prevent the possibility of injustice, a limit might be placed upon the amount of security demandable under the section. I should have liked to add that the process should in this case also be taken under the 11th section, which requires distraint to precede imprisonment; but this might, in the instance of lodgers and many other cases, be equivalent to complete impunity. Then, as to the 14th clause, Mr. Crompton cannot believe that our interpretation of the clause—that it does not apply to breach of contract, and that it is intended to deal only with the other offences defined in it—is correct. His opinion is evidently that also of the deputation which waited upon the Home Secretary on the 5th of November, their attack upon the Master and Servant Act being principally concentrated upon the 14th clause, on account of its supposed applicability to breach of contract.

We adhere to our opinion that breach of contract cannot properly be punished under the 14th clause, and that no man can primarily be sent to prison for breach of contract under the Master and Servant

Act; but we say, further, that we are prepared to co-operate in limiting, if need be, the operation of this clause, so as to harmonise with the view we take of its real scope.

I cannot understand Mr. Crompton's opinion that our interpretation of this section of the Act supplies the strongest argument for its repeal. Surely he does not think that injury inflicted on private property, or misconduct, misdemeanour, or ill-treatment of an aggravated character—the offences punished by this clause—should go scot-free? Surely none of those in whose interest he writes desire to be able to commit these offences? Mr. Crompton is too clear a thinker not to be able to see that excessive eagerness to remove the penalties attached to certain offences can only be regarded as proceeding from a belief that a purpose may be served, occasionally, by the liberty to indulge in them with impunity.

I shall look with interest for letters in the *Times* advocating the repeal of the Criminal Law Amendment Act, and defending picketing. Hitherto the strongest argument I have seen is that the Trades Unions will never rest until it is repealed, followed by the threat that candidates for Parliament, and even Ministers, shall be taught submission.

Against whom do the Unions desire to use picketing? It is fellow-workmen who tremble before it. But is it in defence of the rights of working men, or to supply an engine of Union power, that the unchecked use of this species of intimidation is so fiercely demanded? With our present light we support the Criminal Law Amendment Act, and oppose picketing without compromise, because we cannot see any other than an oppressive motive in the demand for its repeal.

Before I close I must allude to the comments of the press upon the National Federation of Associated Employers. Its aim and machinery have in some cases been misunderstood, as was inevitable upon its first appearance, and, with one or two exceptions, the comments made upon it have been entirely fair. They have been read with deep and studious interest by the members of the Federation. Public opinion is our jury. Its support will be our principal power. We desire—by our advocacy of moderate and just laws only; by seeking rather to diminish than to increase the pressure of punishments; by aiding to render the laws, so far as may be possible, deterrent rather than punitive; and by consulting, to the utmost extent consistent with the rights and freedom of independent working men and with the industrial interests of the Empire, the wishes of the work-people, whether independently advanced by themselves or brought before the public by Trades Unions—to promote cordiality between the employers and the employed; and this we believe our organisation will tend to produce, and not, as has been prophesied by some of our critics, to embitter the relations of capital and labour.

I am, Sir, your obedient servant,

R. RAYNSFORD JACKSON.

Blackburn, Dec. 30.

TO THE EDITOR OF THE TIMES.

SIR,—The letter of Mr. R. Raynsford Jackson, the chairman of the Executive Committee of the Federation of Employers, which you published to-day in answer to my two letters, is so conciliatory and encouraging, both in its tone and matter, that I am very anxious to join issue with him and meet him on the points which he raises for argument and explanation. Upon the Master and Servant Act I have plenty to say in answer to him. With regard to the Criminal Law Amendment Act and Law of Conspiracy, the memorial of the employers presented to Mr. Lowe makes various statements which I am ready to answer—some by argument, others by contradiction.

But before I do so—before I treat Mr. Jackson's letter as the courteous challenge which it appears to be—I want to know whether he and the Federation of Employers maintain or withdraw a certain passage contained in the memorial, and which I regard as inconsistent with his letter in the *Times* to-day. If it is withdrawn, I shall gladly accept the withdrawal, and devote myself to the discussion of these questions with all the energy I possess. If not, I must decline to have any communication with the Federation direct or indirect. I cannot agree with anybody who imputes vile motives to me, or who says that I am guilty of things which my soul abhors. And even if I were content to put my own self-respect aside, I ought not to let the matter pass while the same imputation is made upon those with whom I am associated, and whom, to a certain extent, I represent in writing.

The passage of which I complain is as follows :—

"It appears to us that the main object of the promoters of the repeal of the existing statute is to legalise this practice of picketing, to facilitate the employment of spies, and to authorise Trades Unions to institute a machinery of intimidation to be used against industrious working men, women, and children. We inquire what honest object can such a proposal have ?"

If the employers had confined themselves to saying that our actions were unwise, ignorant, or foolish, or that they were calculated to produce injurious results, that would have been a legitimate expression of opinion, of which I should not have complained ; but the imputation of improper motives stands on a different footing. I have no hesitation in declaring the above statement to be a calumny without foundation. I have been associated with the leaders of Trades Unions on these matters since 1867, and I have found them most anxious for a settlement, and ready to accept any solution practically just. Therefore, I protest against this utterance, and refuse to meet the Association of Employers or their representatives until it be withdrawn.

Nor must I be thought wanting in respect when I say that such words are entirely unworthy of men like Sir T. Bazley, Mr. Edmund Potter, Mr. Robinson, and Mr. Jackson, of their position, their characters, and their lives.

I am, Sir, your obedient servant,

HENRY CROMPTON.

15, Phillimore-gardens, Jan. 1.

TO THE EDITOR OF THE TIMES.

SIR,—It seems somewhat late to answer Mr. Crompton's letter, which appeared in the *Times* of the 2nd of January, but the delay is owing to postal arrangements. Although I have written once before, I venture to trouble you with this communication, being urged by several influential members of the Federation to give a public reply to Mr. Crompton.

I wish to remind that gentleman that in his letter of the 24th of December he asked whether compromise on the question of Master and

Servant Act was not perfectly feasible, and, having proposed a solution, he invited discussion and criticism. Regarding his challenge as a fair one, and desiring that all such subjects should be discussed in an amicable spirit, I addressed you on the 30th of last month, giving my views upon the Master and Servant Act. They certainly did not, in all points, coincide with his own, for had they done so there would have been a surrender, and no compromise, as his proposal embodied the most extreme demands.

Thinking, however, that the Criminal Law Amendment Act, even more than the Master and Servant Act, required public discussion, and that the grounds upon which its repeal is demanded necessitate a more thorough explanation than has been vouchsafed as yet, I invited a public discussion upon this subject also. I cannot think that Mr. Crompton consulted the interests of those in whose behalf he writes when he declined to discuss these questions with the representatives of the Federation of Employers, excepting upon conditions which he must have known, when he made them, could only render discussion impossible.

I regret his choice, because I acknowledge the ability with which he writes; but, even to secure for the public advantage the further light which he might help to throw upon these questions, I cannot accept his *dictum* that the words used in the Federation's memorial to Mr. Lowe, to which he objects, are "a calumny." This is a strong word, and one which a man seldom uses excepting when he desires to render all courteous intercourse impossible.

The Federation cannot withdraw the paragraph with which Mr. Crompton quarrels; it can see nothing but a desire to be free to oppress, without dread of the law, in the bitterness with which the sections in the Criminal Law Amendment Act relating to picketing are attacked; and it regards the statement that picketing is a means of giving information as one which will not bear examination.

I regret to feel constrained to refuse the concession which Mr. Crompton demands, and trust that he, having launched his protest, will lose sight of this episode, and resume his interesting and able letters upon the subject at issue.

I am, Sir, your obedient servant,

R. RAYNSFORD JACKSON,
Chairman of the Executive Committee.

Blackburn, Jan. 8.

TO THE EDITOR OF THE TIMES.

SIR,—Mr. Jackson forces me to reply to his letter in the *Times* of to-day, when I would fain have remained silent. Before I wrote, I had been asked to prepare a report upon the memorial presented to Mr. Lowe by the Federation of Employers for the Trades Union Congress which meets next week at Sheffield. I had hoped that the employers would have withdrawn their accusations, and that I might have been able to pass it over without notice. Therefore, I say that the course I have adopted was both right and conciliatory. All that I hear of Mr. Jackson is so favourable to him that I am very sorry to find him in the position of having made an imputation against many honourable men, that in what they have done they have been actuated by dishonourable motives, and then, when pressed to withdraw it, taking refuge in the assertion that the imputation is not a calumny.

20

Mr. Jackson further urges me to set all this aside and go on with the discussion. To which I reply, Does he assert that when I advocate the repeal of the Criminal Law Amendment Act, my motive is "to facilitate the employment of spies and to authorise the Trades Unions to institute a machinery of intimidation to be used against industrious men, women, and children?" And that my proposal is not honest?

If he answers in the negative, then I say that he is bound, as an honourable man, to withdraw the passage in the memorial which asserts the contrary. If, on the other hand, he still maintains that I am actuated by such vile motives, then, Sir, I can only deny it, and resent it in the strongest terms consistent with the courtesy I owe to the Editor and readers of the *Times*, and then decline all further intercourse with Mr. Jackson or the Federation of Employers. But I would, in conclusion, strongly urge him to adopt the one manly and straightforward course—which is, without further delay, to withdraw that which, being an allegation of motive against many men, cannot be maintained.

I am, Sir, your obedient servant,

HENRY CROMPTON.

15, Phillimore-gardens, Jan. 11.

P.S.—My answer to Mr. Jackson's first letter has been written long ago, and would have been published if he had withdrawn the statement complained of.

TO THE EDITOR OF THE TIMES.

SIR,—The letter published by Mr. R. Raynsford Jackson, on the part of the Federation of Employers, in the *Times* of Jan. 1, is a great advance towards the final settlement of the Master and Servant difficulty. Our agreement as stated by him is very considerable; but, if I am right, his admission goes much further, and involves a still larger agreement than he supposes. He has drawn out the issue, on which he relies, with precision; he has appealed to his tribunal, and stated his case. Let me now state mine. From my point of view, all that I wanted was the complete recognition of the practical and political distinction between civil and criminal responsibility. My present argument is founded on the valuable acknowledgment which Mr. Jackson makes when he says that he agrees with me "that breach of contract pure and simple ought not to be a crime, that it ought not to be directly punished by imprisonment, or even by fine; that it can be adequately met by the award of damages; and that in cases of simple breach of contract imprisonment is justifiable as process only, and to enforce payment of damages." And from several passages in Mr. Jackson's letter, indicating a humane feeling on his part towards one so amerced in damages, I think I may infer that he adopts my phrase, that imprisonment is only justifiable when used for the purpose of process, as a last resort. So that, if the matter ended here, there would be complete agreement, and he would be compelled to accept my proposal as to the modification of the 9th section or something analogous. And if I succeed in showing that the reasons which make him dissent from my conclusion are untenable, then, Sir, I claim the verdict.

Mr. Jackson's objection arises out of a further agreement between us. He adopts my statement that breach of contract coupled with other circumstances, like fraud or certain kinds of culpable negligence, may well constitute a crime. He takes up the latter of these, and says, "culpable negligence should surely be punished by fine, and

21

even more so when accompanied by breach of contract," and "the aggravation of breach of contract, which amounts to culpable negligence and would justify a fine, can only be left to the judgment of the tribunal which has all the circumstances of the case in view." To which I reply, that even if I accepted this proposition (which I do not, because I maintain that such penal laws should be very clearly defined and not left to be constructed by the tribunal) that would not prove either the propriety of the law, or of the process, or of the tribunal, which we directly challenge. And Mr. Jackson himself feels the pinch of this in respect of breach of contract coupled with fraud. For he says, in another part of his letter, "fraud should be dealt with by other Acts." Why? What other Acts? Why is fraud to be dealt with by a different procedure to culpable negligence? He must point out the grounds for such a distinction, which I say cannot be maintained. All breaches of contract coupled with fraud are not criminal offences; for example, fraudulent warranties that do not come within the terms of our laws against fraud. When they are criminal offences they are properly defined: they are not punished summarily except by the consent of the accused, and not always then. Some, as those relating to fraudulent trustees, can only be tried before the superior judges, and no prosecution can be instituted without the Attorney-General's fiat. Then, I ask, does Mr. Jackson mean to assert that questions of fraud are more difficult to deal with than those of culpable negligence? If so, I assert the contrary, and confidently appeal to those who have practical experience in our criminal courts, whether the exact opposite is not the truth? The law as to culpable negligence is extremely vague and most difficult to reduce to practice: in cases of culpable negligence there is almost always conflict of evidence and legal difficulties as to evidence; in truth, they are just those very cases of mixed law and fact which, of all others, should not come within the summary jurisdiction of magistrates. And if so, much more are they unsuitable for the anomalous procedure of the Master and Servant Act. Mr. Jackson admits this where fraud is concerned; I say that it is still more so in cases of culpable negligence. That is my answer to his argument on the 9th section.

* * * * * * *

The rest of this letter, which was not published in the *Times*, has been embodied by Mr. Crompton in that part of his paper which deals with the 14th section of the Master and Servant Act, and which it is not necessary to reprint.

TO THE EDITOR OF THE TIMES.

SIR,—To Mr. Crompton's question—viz., whether I assert that in his advocacy of the repeal of the Criminal Law Amendment Act his motive is "to facilitate the employment of spies and to authorise the Trades Unions to institute a machinery of intimidation to be used against industrious men, women, and children," and that his "proposal is not honest,"—I answer, without hesitation, that I make no such accusation, while, on the other hand, I assume that what he advances upon these subjects is the result of his honest conviction.

It is, however, too much to say that I shelter myself under the assertion that the imputation in the memorial of the Federation, of which he complains, is not a calumny. He may recollect that he so styled it, and that I simply repelled the accusation in terms almost excessive in courtesy.

If I had imitated his example, I might have demanded the withdrawal of this offensive word, and I might have argued that to stigmatise a body of honourable men, like those composing the Federation of Employers, as calumniators because they formed an opinion different from his of the motives impelling certain proposals, exceeded the bounds of courteous correspondence.

I thought, however, that time, attention, and your space were being wasted upon a side issue, and, as I am a business man, with pressing occupations of my own, and have no taste for this species of conflict, I preferred not to engage in anything approaching to an exchange of personal recriminations.

I think Mr. Crompton is scarcely entitled to be so exacting on behalf of those whom he represents, nor to expect such excessive reserve in the expression of opinion with reference to their *animus* and motives. I think he has no right to require for them a measure of consideration which they do not mete out to employers or to the Federation. I refer him to two letters written by habitual contributors to the Trades Union organ, the *Beehive*, of the 27th December, in one of which the employers composing the Federation are described as held together "merely by the lowest and most material of motives—naked selfishness;" and in another of which (alluding to a statement I made to the Home Secretary, and which I reaffirm), it was said, that I "might invent a fable."

I beg also, in support of this view, further to call attention to the tyrannical Trades Union action mentioned by Messrs. Holland and Hannen in the *Times* of to-day, and to ask if nothing is to be said of the motives of the Trades Union leaders should they not exercise their authority to check this cruel act of oppression against a non-Unionist workman?

The merits of the Master and Servant Act, of the Criminal Law Amendment Act, and of the Law of Conspiracy, are subjects of public interest likely to engage a prominent share of Parliamentary attention, and with reference to which the endeavour to assist the formation of a correct and healthy public opinion is a legitimate aspiration.

If the representatives of the Federation can contribute towards this great *desideratum* they desire to do so; but neither they nor I will engage in further controversy upon a subject of so little interest to the public, to whom we both appeal, as any purely personal issue.

I am, Sir, your obedient servant,

Blackburn, Jan. 14. T. RAYNSFORD JACKSON

LONDON:
PRINTED BY H. W. FOSTER, BEAR ALLEY, FARRINGDON STREET.

THE AGRICULTURAL LOCK-OUT OF 1874

WITH NOTES UPON

FARMING AND FARM-LABOUR IN THE EASTERN COUNTIES

BY

FREDERICK CLIFFORD

OF THE MIDDLE TEMPLE, BARRISTER-AT-LAW

"Nec semper feriet quodcunque minabitur Arcus." HOR.

WILLIAM BLACKWOOD AND SONS
EDINBURGH AND LONDON
MDCCCLXXV

THE AGRICULTURAL LOCK-OUT OF 1874.

INTRODUCTION.

ORIGIN OF TRADE-UNIONISM IN EAST ANGLIA—DEMAND OF HIGHER WAGES, SEPTEMBER 1872—NEWMARKET FARMERS' ASSOCIATION FORMED, OCTOBER 1872—OBJECTS AND RULES—UNION AGITATION IN ESSEX AND SUFFOLK—FARMERS' DEFENCE ASSOCIATION AT SUDBURY—SPRING OF 1873—FIRST AGRICULTURAL LOCK-OUT IN EASTERN COUNTIES—SUCCESS OF FARMERS—RATE OF WAGES—EXNING—ANOTHER CIRCULAR—FURTHER DEMAND, FEBRUARY 1874—STRIKE—LOCK-OUT BY NEWMARKET FARMERS—BISHOP OF MANCHESTER'S LETTER—REPLY BY MR R. TEMPLE—SIR E. KERRISON—THE LINCOLNSHIRE LEAGUE (FEDERAL UNION)—STRIKES AND LOCK-OUT IN LINCOLNSHIRE.

SOME preliminary sketch of the Union agitation in the Eastern counties is necessary to explain the narrative which follows. The National Agricultural Labourers' Union was established at Leamington in May 1872. The movement, begun by Mr Arch, soon extended to East Anglia; and in some districts the men joined the Union in considerable numbers. After harvest, in the autumn of 1872, notices were served upon many employers; and the following written notice, signed by seventeen labourers, was received by farmers in Exning, a Suffolk parish on the borders of Cambridgeshire :—

"EXNING, *September* 26, 1872.

"SIR,—We, the undersigned, do hereby jointly and severally agree to call your attention to the following requirements for our labour—namely, fourteen shillings for a week's work, and no longer to conform with the system of breakfasting before going to work during the winter quarter.

"Hoping you will give this your consideration, and meet our moderate requirements amicably.—Your humble servants, * * * *"

This demand would not perhaps, of itself, have alarmed the farmers. The frequent visits and speeches of the Union delegates, however, and the growth of Union sentiments and influence in the villages, gave the Exning notice a look of coming trouble. A few Cambridgeshire and Suffolk farmers around Newmarket therefore convened a meeting for the purpose of conferring upon the best mode of dealing with this communication. On the 15th of October 1872 the farmers assembled at Newmarket, and resolved to form an Association, with the following objects :—

"1. To raise a fund to enable members of the Association to meet any emergency which may arise between employers and employed, or otherwise. 2. To agree upon the maximum rate of wages to be paid. 3. To provide facility of communication between employers of agricultural labour and those in their employ, and for other purposes that may arise."

The committee appointed at this meeting came to the conclusion that it was "most undesirable, and in fact impracticable, to fix a maximum rate of wages," and recommended instead that the Association should merely try "to obtain a uniform rate so far as is practicable." The rules drawn up by the committee were agreed to by the Newmarket Association with this amendment, November 12, 1872. The preamble recited that "it is desirable, in consideration of the altered circumstances and conditions of agricultural labour, and especially in consideration of the formation of Unions among agricultural labourers, that an Association, to be called the Agricultural Association, be formed, having for its objects:—

"To take joint and common instead of individual action in all matters relating to labour, and having a mutual interest. 2. To raise a fund to enable members of the Association to meet any emergency that may arise between employers and employed, or otherwise. 3. To agree upon a uniform rate of wages to be paid as far as practicable. 4. To provide facility of communication between employers of agricultural labour and those in their employ, and for other purposes that may arise, with a view to preserving good feeling, by discouraging agitation, and generally ameliorating the conditions of labour."

An entrance fee of 10s., and a subscription equal to 1½d. in the pound on the rateable value of the assessment to poor-rate, were the terms of membership, the Association being of course limited to farmers and other employers of agricultural labour; and one of the rules enacted "that no member shall make any general alteration in the rate of wages he is at any time paying to his labourers, nor any other general alteration in the terms upon which he engages his labourers, without previously giving the committee due notice thereof, and acting in concert with them."

On the whole, the rules adopted and the objects avowed by the farmers in their new organisation were reasonable and conciliatory. Moreover, it was an organisation obviously called forth by the combined movement among the labourers, and the probable need for a common stand against aggression. At this time the Newmarket farmers seem to have had no thought of locking out, nor did the men then strike in the district. Elsewhere, however, peace was not preserved. At about the same period, after harvest in 1872, notices of substantially the same character were received by farmers in Essex and on the borders of Suffolk; and strikes occurred, though not simultaneously, in the parishes of Weatherfield, Sible Hedingham, Newton, Boxted, Finchingfield, Cavendish, Glemsford, and other places. The result was the formation, in October 1872, of "the Essex and Suffolk Farmers' Defence Association," of which Mr James S. Gardiner, of Borley Lodge, Sudbury, was chairman; Mr G. P. Weybrew, secretary; and Mr W. Mills, treasurer. Their rules were more aggressive than those of the Newmarket Association. For example, by Rule 4 "the Association pledges itself to render every possible help and assistance to its members in cases of strikes taking place amongst the labourers employed by them, supposing they are paying the customary wages of the parish in which such strike takes place." Rule 5 also says "that the members of the Association shall not in any way acknowledge the Labourers' Union by entering into any contract with such Union, or employ a unionist on strike without the consent of the acting committee."

Another Union Notice. 5

One of the notices now in my possession, and served upon an employer at Sible and Castle Hedingham in 1872, seems to have justified the Essex farmers at that time in supposing that they had to do, not so much with their own labourers, or even with the local branch, as with the central organisation of the Union at Leamington. It is a printed circular with a heading which sets forth the title of the "National Agricultural Union;" the address of the central offices at Leamington; the names of the central committee, the officers, and trustees: and it has all the appearance of a common form of notice-paper issued under the authority of the central committee. The dates are left in blank. So also is the amount of increase in wages to be demanded, with the name of the branch committee:—

"*November* 6, 1872.

"DEAR SIR,—The agricultural labourers of this branch of the National Agricultural Union in your employ beg respectfully to inform you that on and after Friday they will require a rise in their wages from 20d. to 26d. per day, and a general conformity to their rules, a copy of which we enclose.

"Being desirous of retaining good relations between employer and employed, and to assure you that no unbecoming feelings prompt us to such a course, we invite you (if our terms are not in accordance with your views) to appoint an early time to meet us, so that we may fairly consider the matter, and arrange our affairs amicably.—Your obedient servants,

"THE COMMITTEE,
North Essex Branch."

Many similar notices were served upon other employers in the district. An address was given, apparently with a view to any communication which the farmers might be disposed to open up, and the name filled in with ink here was "Charles Jay, Honorary President, Cadham Hall, near Braintree." The printed form contemplated a demand for an increase of weekly wages, but by a manuscript correction the request was made for an increase in the daily rate of pay, which would have raised the weekly wages from 10s. to 13s. A copy of the rules was at the same time enclosed for the farmer's information. These rules are prefaced by an address to the members, which is interesting, for reasons obvious on the face of it, and also as being the first address of its kind made after the Leamington Union was established:—

"In submitting to their brethren the rules of the 'National Agricultural Labourers' Union,' the members of the 'National Executive Committee' have added certain supplementary rules for the use of districts and branches. These rules are not regarded by the National Executive as exhaustive, but simply as fundamental. It is felt that districts and branches should have perfect liberty to frame such laws for their own guidance as their own special circumstances may suggest. That liberty is freely accorded, and the National Executive hope it will be exercised on the basis of the rules for districts and branches, and in harmony with the general rules of the National. The National Executive hopes soon to see a branch Union in every parish, and a district Union—that is, a combination of branches—in every county or division, all communicating with a common centre, all observing the same principles, and all working for the same end. In the early stages of our movement let the branch and district meetings be frequent, that enthusiasm may be kept alive, information be dispersed, and the Union be perfected. We must have no local jealousies, no self-seeking, no isolation. Unity of action is above all things necessary; and this can be secured only as all the branches and districts work through a common representative and executive committee. We must have money, and we must have it in one central fund, to which all shall contribute, and from which, in time of need, all shall in turn be aided. The strength of the great trade societies is in their central funds. If we have a balance here and another there, it will be simply impossible to support a number of men in any emergency that may arise. We must have a common treasury large enough, through the payments of all, to support the demands that may be made in the interest of all.

"The funds of a branch or district would soon be exhausted if a number of men were thrown upon them; but the national fund—the fund of all—would be rich enough to meet any demands which the National Executive might entertain, and to support our members through any crisis. Let it be clearly understood, then, that the branch remits its funds to the district; that the districts remit three-fourths of their receipts to the National; and that any branch or district failing to do this has no claim whatever on the general resources of the Union. The fourth allowed to be retained by the districts can be disbursed at the discretion of the district committee, in meeting current expenditure and in promoting the general objects of the Union. For the working expenses of branches an incidental fund is recommended, which may easily be realised by a small payment from each member. Our movement has begun well. Success is, under God, in our own hands. Let us cleave to, and work for, the Union. Let peace and moderation mark all our meetings. Let courtesy, fairness, and firmness characterise all our demands. Act cautiously and advisedly, that no act may have to be repented or repudiated. Do not strike unless all other means fail you. Try all other means. Try them with firmness and patience —try them in the enforcement of only just claims; and if they all fail, then strike, and, having observed Rule 10, strike with a will. Fraternise, centralise! With brotherly feeling, with a united front, with every district welded into a great whole, with a common fund to which all shall pay, and on which all shall have the right to draw, the time will not be distant when every agricultural labourer shall have, what few as yet have enjoyed, a fair day's pay for a fair day's work. Nine and a half hours, exclusive of meal-times, as a day's work, and 16s. as a week's pay, are not extravagant demands. Society supports you in making them, and they will be met soon. Brothers, be united, and you will be strong; be temperate, and you will be respected; realise a central capital, and you will be able to act with firmness and independence. Many eyes are upon you; many tongues are ready to reproach you; your opponents say that your extra leisure will be passed in the public-house, and your extra pay spent in beer. Show that their slander is untrue! Be united, be sober, and you will soon be free!

(Signed) "JOSEPH ARCH,
Chairman of the National Executive Committee."

There is much in this document with which every one must sympathise. From the farmers' point of view, however, supposing that they were giving the market rate of wages, Mr Arch's manifesto showed that the Union meant to try to force up the existing rate, by means of combination, to 16s., at the same time diminishing hours of labour. The 10s. which, according to the foregoing notice, was paid by an individual employer in Essex, was a low rate of wages in itself, and low also contrasted with the minimum wages prevailing elsewhere in the Eastern counties. It will be seen presently that the Essex and Suffolk farmers in this district did not recognise 10s. as the normal rate of wages.

Active warfare began in the spring of 1873. On the 17th of April 1873 the farmers met at Sudbury, the headquarters of the Essex and Suffolk Association, and passed the following resolutions: "That the members of the Association pledge themselves not to pay more than 2s. per day of twelve hours, including breakfast and dinner, for day-work. That in the opinion of this meeting the members of this Association should resist the interference of the National Labourers' Union by discharging the men in their employ belonging to the said Union, after giving them a week's notice to withdraw." The leading farmers in the district who belonged to the Association acted upon this appeal; the men as a rule clung to the Union;* and

* "Dissenters," writes a farmer, "were almost universally the chief propagandists of the Union. Wherever in my district there was the most Dissent, there also were the most Unionists and the greatest tenacity. The propagandists were itinerant preachers, mostly labourers, who on Sundays travel from village to village, and in fine weather address the people on village greens or other open spaces, and in bad weather preach in the Primitive Methodist chapel, or some hired room."

the result was the first agricultural lock-out in the Eastern counties. The farmers appealed to the landowners to help them in resisting the Labourers' Union, urging that the landowner's and the occupier's interest was one upon this question; that if labour could not be obtained at a fair and reasonable rate, land must go out of cultivation or be badly farmed, while it would decrease in value, and by the non-cultivation of roots the price of meat would be raised; and lastly, that this was a communistic movement, and one which, if not "stopped in its infancy," would "lead to confiscation of property, tearing down all rights except the might of the masses." The answer to this appeal was not encouraging. The farmers here, as afterwards around Newmarket, had to fight their own battle. It was indeed a battle in which landowners, however disposed to help their tenants, could do little more than look on: the result depended chiefly upon the organisation and determination of the farmers themselves.* In this instance they were completely successful. The lock-out extended over an area of about 18 miles by 14; and after the resolution of April 1873, about a thousand men were locked out, while many others gave up their tickets† and continued in their em-

* "The great landowners," says a farmer, "did not answer our appeal. The backbone of our Association consisted of men who farm their own land, along with land belonging to others—owning, say, from 100 to 1000 acres. As owners and occupiers, accustomed to look at things from both points of view, we had to measure the strength of our opponents; and we did so without fearing the action of labourers, or the lukewarmness of landlords. We found many tenant-farmers afraid to join us for fear of giving offence to their landlords, though they were glad enough to see the cudgels taken up by us."

† The Union ticket is inscribed, "National Agricultural Labourers' Union. Member's Contribution Card." Four lines below are left for the member's name and number, the name of the branch, the dis-

ployment. The funds of the Union, then in its infancy, were severely taxed.* The lock-out was not recognised or supported by the farmers in other parts of East Anglia, though the Essex and Suffolk Association urged kindred organisations elsewhere to adopt it. Some of the members suffered from a want of labour, but only until after hay-time in 1873. The resolve to dispense with Union hands was generally kept; and the men, on their side, seem to have suffered more than their employers.

"Many labourers who went northward in search of work," says an employer in this district, "returned with illusions dispelled, telling others of the hardships they had undergone, and convinced that higher pay often means greater poverty." In the end the labourers who did not emigrate or migrate left the Union and came back to work, or their places were filled up. The farmers had tided over their difficulty.†

trict in which it is situated, and the signature of the secretary. On the reverse side are columns for "contributions," "fines," and "levies," arranged so as to include a year's fortnightly payments under each of these heads.

* "So many were thrown on the National Union," says a farmer, "that the relief pay was not continued for more than four or five weeks."

† "Without detracting from the deserts of the Newmarket farmers," writes a member of the Essex and Suffolk Association, "and fully recognising their determination and success in fighting the Union, it is due to us to remember that at the time of their lock-out our pioneer Association had already shown the way to victory. In other districts the farmers were for a long time lotus-eating. They temporised in the early stages of the movement, while we were in fierce conflict; and when the storm afterwards burst upon them, we were scudding along under easy sail." Here, as elsewhere, it will be understood that I give the language and opinions of correspondents without necessarily endorsing either.

It is so far creditable to the Essex and Suffolk farmers to find that, after their victory was won, they resolved (March 19, 1874) to rescind the resolution passed the year before, pledging the members of the Association not to exceed 12s. a-week for day-work; and it was understood that each member should be "at liberty to pay such wages as were general in the parish in which he occupied any land." In East Suffolk and in parts of Norfolk this Union agitation had by this time led to partial strikes and frequent demands, and often reasonable demands, for increased wages. In the spring of 1873 the Exning labourers came again to the front, claiming a rise of 3s. a-week and a readjustment of hours of labour, though the change they proposed in the latter respect was not specified. The following (printed) circular was sent to the Exning employers:—

"*March* 1, 1873.

"DEAR SIR,—The agricultural labourers of this branch of the National Agricultural Union in your employ beg respectfully to inform you that, on and after March 7, they will require a rise in their wages of 3s. a-week—a week's work to consist of   hours. Being desirous of retaining good relations between employers and employed, and to assure you that no unbecoming feelings prompt us to such a course, we invite you (if our terms are not in accordance with your views) to appoint an early time to meet us, so that we may fairly consider the matter and arrange our affairs amicably.—Your obedient servants,

"THE COMMITTEE,
Exning Branch."

It will be observed, that this notice is almost identical

with the one given in 1872 in Essex. At this time, as also when the first "notice" was given by the Exning men, their wages were 12s. a-week. The circular of March 1, 1873, was laid before a committee of the Newmarket Agricultural Association by the employers who had received it; and as it bore no signature, a decision was come to that no notice should be taken of it. A resolution, however, was passed at a full meeting of the Association to raise wages to 13s., from March 15, 1873. *Post hoc propter hoc.* The men accepted this increase, naturally attributing it to the influence of the Union, in spite of the farmers' repudiation. Work was therefore resumed as usual. Nothing further was done on either side until February 28, 1874, when the men demanded, in the same way, a rise of 1s. in their weekly wages, and "struck," after the usual week's notice, when their demand was again ignored or rejected, their minimum weekly pay being then 13s., with extras. On March 10 the Newmarket farmers held a general meeting, at which resolutions were come to that no other alteration should be made, either in the rate of wages or the hours of labour. Further, they resolved, on the ground that they might otherwise be beaten in detail, "that all Union men be locked out, after giving one week's notice, such notice to begin on the next payday of each of the members respectively, and that such lock-out continue so long as the men continue on strike." In conformity with this resolution, all Union labourers employed by members of the Association were dismissed; and at a subsequent meeting, held March 24, the farmers passed the following further resolution: "That, taking into consideration the inflammatory and abusive language used by the delegates of the National Agricultural Labourers' Union at their meetings in the neighbourhood, and the persecutions to which labourers who are not members of the Union are daily subjected, the members of this Association shall not in future employ any men to work for them who are members of the Union."

The men were locked out on the 21st of March, three weeks after the strike. Little attention had been aroused outside the district by the lock-out in North Essex and Suffolk. But soon after the lock-out around Newmarket became known, great personages mingled in the fray. The Bishop of Manchester asked, "Are the farmers of England going mad?" "Can they," he continued, "suppose that this suicidal lock-out, which has already thrown 4000 labourers on the fund of the Agricultural Union, will stave off for any appreciable time the solution of the inevitable question, What is the equitable wage to pay the men? The most frightful thing that could happen for English society would be a peasants' war. Yet that is what we are driving to, if insane counsels of mutual exasperation prevail."* This letter was as the beginning of strife, at least in the newspapers. A host of correspondents entered the lists against the Bishop. Mr R. Temple replied † to Dr Fraser's points one by one:—

"1. 'Are the farmers of England going mad?' Not necessarily madder in their lock-out than the peasants in their strike. The strike is a legitimate weapon of attack; the lock-out of defence. Which is the stronger, time, not prophecy, will show. At any rate, the labourers began.

"2. 'The inevitable question, What is the equitable wage?' What labour will fetch in the market. Acts of Parliament in days gone by failed to create a legal wage, and impassioned rhetoric at present will as surely fail to fix

* The Times, April 2, 1874. † Ibid., April 8.

an equitable wage. No man is bound to give more for a commodity than it is worth, be that commodity ploughs or ploughmen. Moreover, if there is an equitable *minimum* there should be an equitable *maximum*, and the workman would most reasonably object to that.

"3. 'A disturbance of the equilibrium of the labour market.' Has the Bishop forgotten Æsop? Does he remember the wolf and the lamb? The labourers have been striking for months; and now, because at last the farmers resort to necessary defence, they 'disturb the equilibrium of the labour market.' It was the peasant who destroyed the balance by throwing into his scale the sword of agitation.

"4. 'Every one must have noticed that the language of the leaders of the movement, at first studiously moderate, has become more violent, and in some cases insurrectionary and menacing.' That is, as long as a man is reasonable let him alone; but when he turns ruffian, traitor, and incendiary, give him what he wants. The farmers are madder than the Bishop thinks them if they are frightened into paying more for labour than it is worth.

"5. 'A peasants' war.' I confess I think no more of this threat than of Mr Chowler's black flag of 1846; but the way to keep the peace is certainly not to make the peasant think himself a tragic victim when he is only acting the prosaic part of an ordinary mortal who, like most men, wants higher pay and will take it if he can get it.

"6. 'Can a man, at the present prices of the necessaries of life, maintain himself and his family?' Coals and meat are dear, chiefly through the prosperity of the working class; but bread, tea, sugar, and clothing are much cheaper than they were forty years ago: so that I think a labourer's expenses are rather less, on the whole, than they were then, while his wages have risen from 8s. per week to 13s., 14s., or 15s., and he is the only man in the country who gets better dwellings, at least on all large estates, with no increase of rent.

"7. 'If farmers can prove the truth of the statement that they cannot afford to pay this rate of wages with their present rentals, rents must come down—an unpleasant thing to contemplate for those who will spend the rent of a 300-acre farm on a single ball or a pair of high-stepping horses.' Rents will fall if the demand for land diminishes; if it increases they will rise. There is no more an 'equitable rental' than there is 'an equitable wage.' As to the rest of this passage, I suppose the Bishop is hardly prepared to say that a man who can afford to give a ball or buy a horse, and does what he can afford, is a grievous sinner; or to deny that no class was ever more munificent in its charities than the higher class in England is now. Be this so or not, he will, I hope, forgive me for saying that this talk about balls and horses is *nihil ad rem*.

"8. 'I am no lover of the principles of trades unionism; but they have been forced upon the working classes by the inequitable use of the power of capital.' When, where, and how? Capitalists have wisely sought to buy labour cheap; workmen have wisely sought to sell it dear. Trades-unions are not objects of love or hate; they are combinations of men who want more money than they have got. They are perfectly legitimate and natural; they are neither praiseworthy, wicked, nor heroic. On a like footing stands the combination of employers; and I cannot for the life of me see why the farmer is the only man in England who is to be called almost a felon, and quite a madman, because to the legitimate attack of a strike he opposes the legitimate defence of a lock-out."

Mr John Algernon Clarke pointed out that, while Dr Fraser had volunteered his opinion that the minimum wage per man anywhere should be 15s. or 16s. per week, in North Lincolnshire men who were receiving 18s. per week had struck for 21s. Lady Stradbroke also wrote that Suffolk agricultural labourers had been offered 17s. and 18s. a-week all the year round, and had declined the offer; * reminded the Bishop of the low-rented cottages, and other advantages which supplement wages in the Eastern counties; and complained that his letter tended to promote strife and dissension.

The Bishop of Manchester, in a longer letter,† reaffirmed his belief that the policy pursued by the East Anglian farmers was "ill-advised and suicidal," though he admitted that the demand of the Lincolnshire labourers for an advance of 3s. a-week (from 18s. to 21s.) was "immoderate and unreasonable." He still, however, held, with Mr Brassey, to the possibility of "an equitable scheme of prices" for labour. "The apostle knew what he was talking of when he said, 'Masters, give unto your servants that which is *just and equal;*' while another principle of his, 'The husbandman that laboureth must be first partaker of the fruits,' expresses a law of natural equity which we have too often inverted. If it be not only true, as Mr Temple says, and as I do not wish to deny, that 'the strike is a legitimate weapon of attack, the lock-out of defence,' but we must wait patiently and calmly to see 'which time will show to be the stronger,' I can foresee nothing but disaster, and indeed ruin, to thousands of homes, both of farmers and labourers, as the ultimate issue of the struggle."

* The Times, April 16. † Ibid., April 14.

18 *Suggestions by Sir E. Kerrison.*

The war of words went on at great length, but the only other letter which need be reproduced is one by Sir Edward Kerrison,* of whom the Bishop of Manchester afterwards wrote, from his experience as an Assistant Commissioner in the Agricultural Inquiry of 1867—"There is no one who, as a resident landowner on a great estate, which he watches over with the most judicious care, has a greater right to be heard in this matter." The following was Sir Edward's contribution to the controversy:—

" In a desert of strikes and locks-out, my property forms a sort of oasis. In Suffolk, for many miles surrounding it, general locks-out have either taken place or are impending; but in Norfolk, two miles from my house, an unasked-for increase of wages has within the last week been given. I venture, as one of the few landlords in the district, the views of whose tenants as a body are in unison with his own as to the general inexpediency of locks-out where no advance of wages has been demanded, to make some suggestions upon the present state of things.

" It is evident to us that unless some steps can be devised to put an end to the internecine war now raging in the Eastern counties between employers and employed, some of the best bread-producing land in the kingdom will be left without labour to cultivate it, and the question now agitating the district thus assumes a national aspect. The farmers wage war against the Agricultural Unions, but, so far as I can judge, without making any serious attempt to obtain a modification of the rules by which those associations are governed. Farming cannot be carried on successfully without a continuous supply of labour at all seasons of the year, and cannot be left liable to the arbitrary action

* The Times, April 18.

which those rules have set in motion against farmers. One rule, which vests all power of ordering strikes at a week's notice in the executive, should be altered before arbitration can be successful. At least one month's notice in writing should be given by either party to the other in case of an advance or lowering of wages being thought necessary. To obviate the unjust system of striking at farmers in detail, a district (which in a county consisting of small parishes like this should not be less than five parishes) ought alone to have the power of action. In case no agreement can be come to, reference might be had to a system of arbitration, to be established upon an agreed basis in each county, or possibly district. The grounds for any change should be set forth in writing by those desiring it.

" Men should be paid by the hour or by piece-work, with the exception of special harvest arrangements, when higher wages are always paid.

" It seems to me that if these—which I believe to be the principal points at issue—can be adjusted, there is no reason why immediate arbitration should not take place, under which smaller matters of detail might be arranged. The Eastern counties locks-out have been directed principally against those who have advised action upon the objectionable rules, and I believe that if these rules are modified, agricultural Unions may proceed in amicable relations with employers, and without the necessity of resorting to extraneous agency. The whole labour question, as now existing, must be divested of all those benevolent or charitable adjuncts which, with the most praiseworthy but most mistaken views, are imported into it; they only divert attention from the real question at issue, that of wages, which of necessity must henceforth be based upon *commercial* principles.

"I have the authority of the farmers on my property to say that the working men employed by them, whether members of Unions or not, have throughout these trying times conducted themselves in the most satisfactory manner. I write this in the confident hope that enlightened public opinion will assist those who are endeavouring to deal fairly with the working-man, and who, while desirous of avoiding locks-out and strikes, object to some of the present rules of the agricultural Unions, which, in their opinion, are calculated to act alike injuriously to the interests of Unionists as to those of farmers."

In East Suffolk and parts of Lincolnshire a lock-out was resorted to by the farmers at about the same time as that begun around Newmarket. In Lincolnshire, and Norfolk and Suffolk, many of the labourers in union belonged to the Lincolnshire League, a branch of the Federal Union; and on April 9th, the general secretary of this League reported that about 2000 members would be locked out at the end of the current week, while many more soon expected to be in the same position. Appeals were issued to the public, and subscriptions from private persons and from workmen's Unions in the towns flowed in freely. The Amalgamated Society of Engineers alone voted £1000, and the public subscriptions were apportioned equally between both Unions in proportion to the number of locked-out members. This apportionment was effected by means of a committee of subscribers: the National and Federal Executives did not act in unison, and indeed from the first appeared to be at variance.

The total cost of the lock-out to the National Labourers' Union may be gathered from the following authentic list of grants made to various districts by the

Central Executive at Leamington, between the months of March and August 1874:—

Newmarket (Exning) District,	£14,984	10	7
Wisbech,	1,550	0	0
Bedford,	980	0	0
Halstead, Essex,	1,460	0	0
Sawston, Cambs,	1,931	0	0
Market Rasen, Lincoln,	858	0	0
Luton,	162	0	0
Aylesbury,	205	0	0
Old Buckenham, Norfolk,	585	0	0
Norwich,	135	0	0
E. Dereham,	205	0	0
Wolverton,	256	0	0
Banbury,	283	0	0
Spalding,	59	0	0
Dorset,	400	0	0
Market Harborough,	164	0	0
Andover,	90	0	0
Farringdon,	85	0	0
Alton,	40	0	0
	£24,432	10	7

But these figures do not represent the whole expenditure, as some of the districts used the receipts of the branches, in addition to the grants from the National Executive, though such an application of money is contrary to the rules.

According to the 'Labourers Union Chronicle' (October 24, 1874), the total number of men locked out belonging to the National Union was 3116, of whom 694 migrated, 429 emigrated, 415 were then still unemployed, 402 left the Union, "and 1176 went back to work retaining their Union cards." It is difficult to reconcile the last statement with the facts. The men may possibly have continued their allegiance to the Union in secret, but it is idle to suppose that employers would have dispensed with Union labour during 18 weeks, including harvest-time, and then have knowingly surrendered the whole principle for which they had been contending by taking back Union men when the Union funds were exhausted. The farmers must at any rate have supposed that the men had ceased their connection with the Union. As to the number of men belonging to the Federal Union who were locked out, Mr Banks, the general secretary of the Lincolnshire and Suffolk Federal districts, at a meeting of trade delegates in London, held April 21, said that about 6000 labourers were locked out in the Eastern counties, of whom about 2500 were members of the Federal Union. The totals here given seem to be exaggerated; and indeed the leaders of each Union charged the others with exaggerating the number of their locked-out members. On May 23, Mr Banks told the Executive Council of the Federal Union that the whole number of Federal labourers locked out in Lincolnshire did not exceed 130; while in Suffolk and the neighbouring districts the number of Federal labourers then locked out, after deducting those who had migrated, emigrated, and found other work, was 1318. Mr Shipton, the general secretary of the Federal Union, afterwards visited the lock-out districts in Cambridgeshire and Suffolk, and reported to the council on June 3 that the total number of Federal labourers locked out was 829; children (?), 810: and the total amount paid in relief was £387, 10s. per week. It seems that the withdrawal of the lock-out in Lincolnshire only relieved the Federal funds to the extent of £60 a-week.

4

EXTRAORDINARY MEETING OF COUNCIL

HELD IN THE MECHANICS' INSTITUTION, DAVID STREET, MANCHESTER, 24TH FEBRUARY, 1875.

MR. ALEXANDER M'DONALD, M.P., IN THE CHAIR.
MR. NORMANSELL, IN THE VICE-CHAIR.

BUSINESS OF THE MEETING.

THE LOCK-OUT IN SOUTH WALES.

The CHAIRMAN said he regretted that the Council had been called so hastily together, but the fault was not his. He received a communication from Mr. Crawford on Saturday evening, as late as six o'clock, announcing that it was the request of his district, which was a very large and important one, consisting of some 50,000 men, that a meeting should be called for Tuesday, or Wednesday at the latest, and he had determined on complying with that request. He would now call the attention of his hearers to what had brought about the present state of things in South Wales. As he understood, some time towards the close of last year the men submitted to a reduction of 10 per cent. on their wages. When they did so there was a common understanding that before any further reduction took place they would be called together and told the reason why, and that it would be a matter for consideration. Instead of a resolution of that description being carried out—for there was undoubtedly such an understanding, as it appeared in one of the public journals and was not denied—the men were called together and were tcld they were to submit to a reduction without consultation, without any conversation, or anything of that kind. Simply, they were to submit, and there was nothing else for them but submission. Now, while that state of things might be right in a bygone age, he ventured to say that it was a state of feeling that ought not to be exhibited by employers at the present day. He held the employers to be buyers of the workmen's labour, and that being so, such a proceeding as that of the South Wales employers was an arrogance and an assumption that the working classes, not alone of this country but of the world at large, could not be put too soon under foot. What would those employers say if their customers were to come and tell them that they would give them so much and nothing more, and if they did not accept the terms they offered they would cease to deal with them further, and, as far as they could, drive them out of house and home? No body of men would resent that more than the coal owners and mine owners of South Wales. They might be, and were, told that those men were capitalists, and

5

being capitalists they had a right to do with their own what they thought fit. He for one did not dispute that they were capitalists, but he seriously disputed their right of doing what they thought fit, and, furthermore, he disputed their right of doing what they might be perfectly legally entitled to do. He further said that the capital which they possessed—when it was considered how it had been collected—was not to be used, and should not be used, to crush the men that had been the means of collecting it. (Hear, hear.) They were told that there was an old aristocracy with their broad acres that did what they liked. It was true that they might, but there was no analogy, he ventured to say, between the old aristocracy of the feudal laws and the men of yesterday. The one class received their vast possessions from a bygone age. Those possessions were given at a time when no one contributed to them anything whatsoever, and whether the feudal rights were to be regarded as right or wrong, the lands came from the hand of God, and no one contributed to them; but if they looked for a moment at how the capital of South Wales was acquired, they saw a very different state of things. They found four or five families notably in the front—the Crawshays, the Fothergills, the Vivians, the Davieses, and a few others, ruling in this despotic manner. What was the position of those men? Not a century ago the dingles and valleys of Wales possessed primæval peace and quietude. One or two men entered those valleys, and began to work there, not with an immense capital, but with small means. By means of the accumulated labour of the workmen they were able to develop their operations still further, but had it been left to the arm of a Crawshay, or a Fothergill, or a Vivian—for those men were not more inclined for work than their successors of the present day—no wealth would have been accumulated in South Wales to-day. It was the working ironminers, the working coalminers, and the navvies that gave those men the wealth they now possessed. It had come from the accumulated capital which the working men had foolishly allowed to be drawn from them, instead of dividing it among themselves by means of co-operative production, which they should have done years ago. Let them take Cardiff itself at that moment. They had a port there filled with flags from every nation in the world. Who, he would ask, made that port? He granted that Lord Bute, or those connected with him, had directed it, but it was the money drawn from the miners, and developed by those hardy sons of toil, that enabled them to do it, and but for them the Welsh mountains would have lain as quiet and would have been as far from producing wealth to-day as the mountains of Colorado or India. The wealth of a few great families in Wales had been collected from workmen and produced by workmen; and he questioned the right of men to turn round with that wealth and try to starve the workmen, their families, and children. (Hear, hear.) He could scarcely find language adequate to describe the conduct of those men. Within the last quarter of a century there had been names rendered infamous in history, and he trusted as time rolled on that those names would become still more infamous. There were the names of Haynau, the Austrian general; and during the American war a name which was rendered infamous in every country in the world was the name of Butler, who flogged women and children, or at least committed indignities on women in New Orleans. These parties who were ill-treated were aliens, persons that they were at war with, and yet society and the world at large had rendered the names of those to whom he had referred infamous; and he said again he hoped they would become doubly infamous. But what had they to-day? The mineowners of South Wales were making war, not against the men, who were fair game to fight, but they were infamously attempting to carry their designs by making war and bringing destitution upon the women, and carrying misery and starvation to the children. If, therefore, the Haynaus and Butlers had rendered their names infamous, the conduct of these men was doubly infamous when the position from which they came was duly considered. He (Mr M'Donald) thought that he could not too strongly express his feelings. Here were men who had drawn their wealth and erected their castles that they had everywhere throughout the country from the blood and toil of those workmen, and of their fathers and grandfathers, and as a reward for tamely permitting them to do this they turned round and starved children and wives at the very first opportunity. He did hope that when the social history of England was written, as it would, he hoped, be written some day, names like those of Crawshay and Fothergill, Vivian and Davies, and others, would go down to posterity with that infamy which poltroons who would attack women and children would ever receive at the hands of an enlightened age. They were not men who attacked women or children, but poltroons, for he was not a man who would infamously strive to carry out his ends by starvation. Those men could wander through their halls and say, "They are ours," but a voice would follow them which he ventured to affirm would say that if they were theirs they starved those who enabled them to build them, as their reward. His advice was, after those present had heard the report which would be read, that they should give expression to their own feelings as he had done; and, if they took his advice, they would go back to their own districts determined to give every assistance in their power, and to excite their fellow-workmen to do the same. The masters had chosen their own ground—the ground of starvation of women and children. Let the Association, therefore, go to the rescue, and try by every means in their power to

7

scatter the hideous Association of the masters, and treat them as they should be treated, by not recognising them as men at all. Looking over their numbers the previous night he found that they were about 140,000. If only 2s. per week per head could be obtained from these it would reach the handsome sum of £14,000; and, even with 6d per week, they would be able to raise at least £3,500. This latter sum he hoped they would try to raise, and obtain 1s per week where they could. The masters, as he had said, had chosen their own battle ground, and the Miners' Association should endeavour to aid the men until their work was done, and they had scattered the masters' conspiracy to the four winds of heaven. (Hear, hear.)

Mr. P. CASEY then read the report drawn up by himself and Mr Crawford, which was as follows:—

THE STRIKE AND LOCK-OUT OF COLLIERS AND IRONWORKERS IN SOUTH WALES.

To the Members of the Miners' National Association.

GENTLEMEN:—We were appointed by your Council, at their meeting in Manchester, to inquire into the whole circumstances of the ruinous conflict now raging in South Wales. We have made a full and personal inquiry into the causes which have led to the present complications, together with the present position of the workmen, and their future prospects. The result of those inquiries we now have the honour of submitting for your consideration, concealing nothing that has come under our observation, and colouring nothing to suit the views of either party. Such being the case, we at once proceed to give our report, believing that the information contained therein will be of more interest than any lengthened introduction we might think proper to make.

In the month of March, 1871, the Coalowners in the steam coal, house coal, and iron making Collieries, requested a reduction of wages from their workmen, amounting to ten per cent., which was accepted, and this reduction was the precursor of a state of prosperity in the coal and iron districts of England, Scotland, and Wales, that was never imagined even by the most sanguine amongst us. Coal and iron, but particularly coal, advanced in the market until prices became absolutely fabulous; and Colliery Owners, who up to that period had found it in many cases not over profitable, were, in two or three years, in the possession of splendid fortunes. This state of trade naturally led the men to expect and look for a share of the general prosperity. We therefore find that in the month of May in the same year, the men gave notice for the ten per cent. reduction they had submitted to in March to be returned. The employers in the Steam Coal Collieries refused to give it, and the refusal entailed a stoppage of the Collieries, which lasted for three months, and was eventually settled by the workmen resuming work at the rate they struck at, and the employers depositing five per cent. of wages in the Banks, which should be subjected to Arbitration; and Mr. Bidder, on behalf of the Owners, and Mr M'Donald, on behalf of the men, were appointed as Arbitrators to decide the question. The result of this Arbitration was, that the men were awarded two-and-a-half per cent. for five back months, which was duly paid; and we find that by this action the principle of Arbitration has been accepted and worked upon in Wales, and makes it, therefore, very difficult for us to understand why the employers are now rejecting it.

During our inquiries, we on two occasions availed ourselves of the opportunity of calling upon Mr Dalziel, the Secretary of the Coalowners' Association, for the purpose of ascertaining the reasons why the principle of Arbitration was refused by the Owners, and of advancing arguments in favour of getting, if possible, the matter settled by that means. We were very courteously received, and talked the question over very fully, showing that if the case stood as the Owners stated—namely, that they were losing money, and could not possibly afford to go on at the old prices, no Arbitrator would attempt to ask them to continue, and that the better their case stood in that direction the less reason they had to reject an impartial inquiry. We also pointed out that nearly every dispute of this description in the North was now settled by this mode; that an Arbitration case of this character was settled in reference to the Durham Miners in November last, affecting more men than are now out in Wales, and the decision of the Umpire was honourably accepted by both sides; that a dispute was now in existence between the Cleveland Mineowners and their workmen, which had been sent to Arbitration during the last week; that the Coalowners of Northumberland were seeking a reduction of wages, and this had also been referred to Arbitration; and, with all those cases before them, in addition to their own previous acceptance of the principle, we could not but regard their doings with astonishment, and must draw our own conclusions therefrom. After putting our views before Mr Dalziel, he said he would communicate with the Chairman of the Coalowners' Association, Mr Fothergill, M.P., and a few of the members, and if we were remaining a few days he would communicate with us further on the matter. We saw Mr. Dalziel again on Wednesday, the 17th inst., when we were informed that the Chairman, Mr. Fothergill, M.P., and the other employers, could not see their way to alter their mode of proceeding—that it had not been hastily or thoughtlessly entered into, and any arguments we might use would not have the slightest effect; that, in fact, until two-thirds of the men now on strike agreed to resume work the locked-out men would not

9

be allowed to commence, and with this very emphatic but very despotic statement ringing in our ears, we left Mr. Dalziel.

It is, therefore, plain to us that the owners are determined to do their utmost to starve the men into submission, and it is a useless waste of time to endeavour to bring about any mediation. The men themselves have tried it, we have tried it; but the determination among the large employers is fixed and final—"submission or starvation"—they being altogether impervious to reason.

Having, therefore, endeavoured to negotiate for a settlement without success, we proceeded to inquire into the number of men affected—their position and prospects—in order that some accurate knowledge might be circulated upon this subject, and the following information may be relied upon as correct:—

In the month of November, 1874, the employers requested their workmen to accept a reduction of 10 per cent. in their wages, to which the men objected, but offered to refer the case to Arbitration, and if the Umpire decided they should accept it, they would do so. This the employers refused, and gave them a month's notice to enforce the reduction, which notice expired on the 31st of December. At the end of this month the men working on the Rhondda and Aberdare Valleys refused to accept the reduction, while the men at Merthyr, Dowlais, and other places, numbering nearly 20,000, did accept it and resumed work. When they had been working a month they were locked out, as the employers say, to prevent them giving any support to the strike, and starve all into submission.

The number of men on strike and locked out are as follows:

NUMBER OF MEN ON STRIKE.

Rhondda Valley and District, ...	12,000
Aberdare District, ...	11,000
Cwmbran and Pontypool, ...	2,000
Abersychan, ...	2,500
Blenhaven, ...	2,000
Abertillery, ...	2,000
New Tredegar, ...	1,300
Swansea Vale, ...	1,500
Neath, ...	700
Longhor, ...	500
Blackwood and Argoed, ...	500
Pengam, ...	500
Darram and Vockriaw, ...	500
Caerphilly, ...	300
Ogmore, ...	300
Millfran, ...	200
Begelly, ...	70
Maesteg, ...	50
TOTAL, ...	37,920

10

NUMBER OF MEN LOCKED OUT.

Merthyr and Dowlais, ...	6,000
Old Tredegar, ...	3,500
Rhymmey, ...	2,700
Ebbw Vale, ...	2,300
Sirhowy, ...	1,800
Abernant, ...	1,500
Aboravon, ...	1,200
Longhor, ...	500
Gwendrath, ...	200
Hentlys, ...	100
TOTAL, ...	19,800

The number of men working at the old rate of wages are from twenty to twenty-two thousand, and this accounts for the whole number of men employed in connection with the Collieries of Monmouth and Glamorganshire before the dispute commenced.

The men put their case as follows:—

In March, 1871, they submitted to a 10 per cent. reduction; since then they have been advanced—

Steam Coal Colliers, ...	42½ per cent.
Ironwork Colliers, ...	45 ,,
Ironstone Workers, ...	45 ,,

The above are only nominal advances, which means that those advances have only been placed upon some descriptions of work, and not upon all, and they have submitted to above one-half of that going off already.

That in 1871 the selling price of coal in Cardiff was—

Steam Coal, 11s. to 13s. per ton.

In November, 1874, when the notices were given for a reduction of wages, the price in Cardiff was—

Steam Coal, 16s. to 17s. per ton.

That the rate of wages and the selling price of coal bear no proportion to each other, as the following figures, given for the purpose of being made public, will show—

Wages at Merthyr and Dowlais before the reduction was made—

Steam Coal Colliers, ...	4s. 9d. per day.
Ironwork Colliers, ...	4s. 0d. ,,
Ironstone Miners, ...	3s. 4d. ,,

They divide the Merthyr and Dowlais men as follows—

	£	s.	d.
3400 Steam Coal Colliers, at 4s. 9d., ...	£807	10	0
1600 Ironwork Colliers, at 4s., ...	320	0	0
1000 Ironstone Miners, at 3s. 4d., ...	166	13	4
	£1294	3	4
6000 men at 4s. 3¾d. per day, ...	£1294	3	4

11

The above figures were given to us with the knowledge that they would be used in a public manner, and the men declare it is at least two pence per day more than they receive, and with the increased and continually increasing price of every necessary of life they cannot live for less.

They also contend that when the last dispute occurred it was agreed that, if any further complication occurred, the men should be called together and consulted before any alteration was made or proposed; and this has not been carried out, as the owners met and decided upon a reduction before the men were called to meet them, and they only sent for a deputation, to inform them what they had decided to do, and it was not a meeting for consultation at all. That in this particular the old agreement was broken and trampled under foot; they are aggrieved in consequence, and resist for this breach of faith.

The employers state that the Coal and Iron trade is in such a condition that they cannot possibly pay the old rate of wages, that the Ironstone Miners and Colliers could and did earn more than they state; and Mr Hussey Vivian, in a letter published on the 16th instant, gives the wages at from four and elevenpence up to six and ninepence per day, and that if the men sent out as much material as they did in 1870 they would earn considerably more. The employers could have proved their case, if it was a good one, if they had agreed to Arbitration. Men who refused such a course are clearly not worthy of credit.

It will be remembered, however, that a new Mines Regulation Act has been passed since then, and the boys are not allowed to work so long by Act of Parliament as formerly they did. This in some measure has a bearing upon the working hours of the men, which are now less than they were previously, and which makes the latter part of Mr Vivian's argument altogether useless. Again, the experience we have had in lists of wages published from employers' books in times of strikes and lock-outs, convince us that such lists are totally unreliable, and that it is utterly impossible for the employer to give them correctly. For instance, John Jones is credited with a certain amount of money, and William Edwards is credited with a certain amount more. Now this money is divided between the two, and gives a certain amount per day, whereas very likely the two men men named have another man working for them whose name is not in the employer's book, because he is working for the men, who are set down as contractors, and they have him to pay out of the wages they are set down as receiving for themselves. This has been done over and over again, and we cannot therefore place any reliance on the statement, more particularly seeing that we have had the pay-sheets of men working in the Blenavon district produced, showing by unmistakeable figures that the wages paid to the best workmen, when taken to do day work, was three shillings and elevenpence per day.

We find that the wages paid in Monmouth and Glamorganshire, before the proposed reduction, were very low; but whether they are more than the employers can afford to pay, we have no means of judging. Of this we are however satisfied, that in refusing Arbitration, and insisting upon the old and barbarous method of locking-out the workmen, the employers have committed a grievious wrong—a wrong that deserves the severest condemnation, and that every Trade Society in the entire country should support the men of Wales in the conflict that is now forced upon them. We believe the dispute will continue longer than is anticipated by many people, and we hope the various districts connected with the National Association will call Meetings at every village and at every Colliery, to awaken sympathy and procure assistance for the men of Wales, in order to show the employers that the old and cruel system of locking-out their workmen, in case of a dispute, will not now be tamely submitted to; that workmen everywhere will show this by rallying to aid the men, women, and children, and thereby make this the last lock-out of the century.

The position of the miners of Wales may be the state of the miners of any part of England in a brief space of time. We are not without indications that it will come if the Welsh Mineowners succeed in their cowardly work. The terms of the Mineowners of Wales are the most abject, the most unqualified submission—a dependence more degrading than those offered during the worst periods of feudal vassalage. The feudal lord fed his dependants; they were to him a part of his own existence. The Fothergills, the Crawshays, the Menalaws, and others have drained the life's blood of the fathers of the men of Wales, and they now turn round, with that wealth which their labour has created, and starve, as far as they can, their wives and their children. Could savagery go farther than this? Will conduct of this description not rouse the working men of England—nay, of Europe, from its centre to its outer limits?

There exists among the working men of Wales and their families great poverty in many places. This has been occasioned by the struggle they have been engaged in for some time. It is stated that women and children have fallen down in the street from sheer want! To their rescue one, to their rescue all, and by thus becoming able to knock to pieces this giant conspiracy, the cause of labour will be emancipated.

We are, Gentlemen, yours respectfully,

PHILIP CASEY.

WILLIAM CRAWFORD.

13

The CHAIRMAN—I want to put one or two questions—First, as regards the assumption of the employers, and what they are trying to make the public believe, that the giving in of one portion of the men was a concerted action with the other portion, and that that was the cause of their being locked out; did you find in your visit anything of that description?

Mr. CRAWFORD—Quite the contrary.

The CHAIRMAN—I understand from what I read in the papers that there was a strong feeling on the part of the men who did remain out on account of the others giving way, and that so far from there being concerted action it was quite the reverse. Did you find in your visit that the men who had gone in had been giving support to the men who were on strike?

Mr. CRAWFORD—Not a penny.

The CHAIRMAN—And those are the statements that the employers would hoodwink the public with!

Mr. CASEY said he wished to explain that the men who were on strike were principally connected with the export collieries. They had been getting higher wages than the others, and in addition they were the steadiest and most careful of the population. They had, therefore, better homes, and were in a better position for standing out for six months than any other of the South Wales men. Those men refused to accept the reduction, and declared that they would remain on strike for six months, even if they got no pecuniary assistance from anywhere, should the employers persist in their refusal to refer the matter to arbitration. The men that were locked out were mostly connected with the iron-making collieries, who had been receiving less wages, and were in a worse position for standing out than the others. They could not see their way to a strike like the Rhondda Valley and Aberdare men, and they therefore accepted the reduction. They went to work, and requested the other men, for their sakes, to accept the reduction and wait for a future time for the purpose of getting it back again. They did all they could to induce the men on strike to resume work, but their influence was of no avail. They were locked out, and now the Rhondda Valley and Aberdare men said, "If you"—meaning the men locked out—"had stuck to us at first you would have convinced the owners that they would have to give way, and they would have given way before now. The owners have locked you out; they refuse to let you go to work unless we go; but we will not go to work, and we will make you smart for accepting the reduction." That was the position of the men locked out. They were helpless and hopeless; they could exercise no influence whatever in inducing the men on strike to accept the employers' terms, and all that remained to them was downright starvation. It was no question of their accepting the reduc-

14

tion; they had always been and still were willing to accept it, but the employers would not allow them to resume work. The consequence was that thousands were applying to the parish for assistance; they were employed on the roads breaking stones, where they had been earning, many of them, something like 4d or 5d per day; but the Poor-law Board was now, he understood, compelling the Guardians to see that those men should not starve. He had seen many a lock-out and strike, but he had never before seen such tyrannical conduct on the part of employers. To say to 20,000 men, who were willing to go to work and to accept the reduction, that they would not allow them to do so unless somebody else whom those men were utterly unable to influence did so too, was something perfectly monstrous. (Hear, hear.) When Mr Fothergill, M.P., was trying to oust Mr Bruce in Merthyr, he stated at a public meeting that when he stood on the hill where his mansion was situated and saw his men going to work in the morning, he felt as if he could take every man to his breast and hug him like a brother. Now, when he was a member, the men he could have taken to his breast for the purpose of hoodwinking them to send him to Parliament were left to starve, and could not help themselves. The case of the South Wales miners was one of the worst that had ever come under his observation. There was not a collier working in England or Scotland but ought to send something to those people in Wales, and he and Mr Crawford were there that day to ask for assistance for them. As far as the Amalgamated Association of Miners was concerned, it seemed to be altogether broken up in Wales; it was unable to support those men, and there was, therefore, no hope except from the National Association. (Hear, hear.)

The CHAIRMAN said, besides those great employers to whom reference had been made, there was another class of employers, namely, the limited liability companies. A short time ago they paid their confiding shareholders dividends in reports which gave accounts of the smoking of cigars and the drinking of immense quantities of champagne. He was afraid that if they went on at that rate they would have to give dividends in the form of strikes and lock-outs.

Mr. CRAWFORD said every inquiry was made as to whether or not the locked-out men had subscribed to the maintenance of the men on strike, and they were informed that not a single penny had been subscribed, nor was it anticipated that the men would subscribe. He never saw such a state of despotism and terror as was existing in Wales. Starvation and want was upon the faces of hundreds and thousands, and the case was one which deserved earnest support.

Mr NIXON (Northumberland), urged that all classes of working people should be asked to assist the locked-out men, and he moved—" That this meeting of the Council of the Miners'

15

National Association, having heard the report laid before them by the deputation appointed to visit South Wales, desire to express their condemnation of the refusal on the part of the masters to arbitration, and of the despotic lock-out now in operation in the two counties of Monmouth and Glamorgan, and calls upon the working men in every branch of industry to assist the thousands now starving in consequence thereof."

Mr WATTS (Brierley Hill), seconded the resolution, which was passed.

Mr PICKARD (West Yorkshire), moved—"That this meeting also desires to express its deep sympathy with the miners of South Wales and their starving families; and they at the same time request the officials in every district to take every step immediately to collect money to aid them in their distress."

Mr COLLEDGE (Warwickshire) seconded the resolution, which was passed.

Mr BEECH (Oldham), moved a resolution to the effect that the report be printed as early as possible, and circulated amongst the members of the different lodges.

The resolution was seconded by Mr SIMPKINS (Derbyshire), and adopted.

A committee, consisting of Mr Brown (South Staffordshire), Mr S. Stansfield (Ashton), and Mr J. Colledge (Warwickshire), having been appointed to confer with the Secretary and Treasurer as to the receipt and distribution of the money collected, Mr CASEY moved that a sum of £1,000 be voted at once from the funds of the Association to relieve the present distress, to be refunded from the levies.

RESOLUTION.

Mr DIXON moved that in order to allow districts to have the Report of the deputation to Wales as soon as possible, all districts that choose may print a sufficient number of copies to suit themselves.—This was seconded by Mr BEECH and agreed to.

AN ADDRESS.

It was then moved and agreed to that, if necessary, an address be printed and circulated among all the trades of the kingdom on the subject of the Wales' lock-out.

A NATIONAL CONFERENCE.

The PRESIDENT stated that Mr Crawford had stated to him that he was instructed to ask that a National Conference should be called upon an early day to discuss the wages question and also the state of the trade of the country. He thought that Mr Crawford would do well to explain his views.

Mr CRAWFORD briefly stated that there was a strong feeling among the men of Durham that employers would go on reducing wages till they were quite as low as ever they had been.

16

The men were of opinion that this could only be stopped by a great national effort.

Mr NIXON stated there was a similar feeling among the men of the county of Northumberland. They thought that such a meeting should be held soon, it being agreed at the Conference held in Barnsley.

Mr DIXON thought there was no need for reductions if the men would only be firm.

Mr PICKARD of Wigan, who was present, asked leave to be allowed to say a word or two. He said that the resolution just met the views of the men of Wigan. They had withdrawn from the Amalgamated Association, but saw there were great evils ahead if care was not taken.

Mr Beech and others followed.

Mr CRAWFORD then moved—"That a National Conference of Miners' Delegates be convened by this Council from all parts of England, Scotland, and Wales, to take into consideration the state of trade and how to avert reductions; and that the meeting take place in the Court-House, Leeds, on Tuesday the 30th of March next, at 11 o'clock a.m."—The resolution was agreed to.

PROGRAMME OF THE MEETING.

It was moved, and agreed to, that all parties desirous of placing motions for discussion on the programme should send the same to Mr Alex. M'Donald, M.P., the President of the Association, not later than the 6th of March next, to enable Mr M'Donald to make out a programme, which shall be adhered to by the meeting.

It was further agreed that Mr M'Donald draw up an address to the miners of the United Kingdom on the questions to be discussed at the meeting. He was also authorised to see that the meeting was advertised in several influential papers in the country for some days prior to the meeting.

ASPARTRIA.

On the request of the Chairman and some others, Mr BROWN made a statement of the condition of things there. He stated that the employers still refused to allow the subject of dispute to go to arbitration; and also, that they would have to submit to more reductions if they went to work. Mr Brown thanked the Association for what they were doing, but wished they could do a little more.

Mr DIXON, West Yorkshire, stated that they were going now to do something more.

On the motion of Mr CRAWFORD a vote of thanks was awarded to the President for his service in the chair.

THE WAGES AGITATION ON THE CLYDE.

25,000 MEN TO BE LOCKED OUT.

Reprinted from the GLASGOW HERALD, *Saturday, May 12th, 1877.*

The agitation among the workmen employed in the Clyde shipbuilding trade for an advance of wages has resulted in the masters agreeing to close their establishments, and make a general lock-out all over the Clyde. This important resolution was adopted at a large meeting of shipbuilders held in Glasgow yesterday, and is to take effect on the 19th curt. The number of workmen who will then be thrown out of employment will be very large, there being at the very lowest estimate 25,000 men engaged in the shipbuilding trade from Glasgow to Greenock. This step has been taken in the belief that by no other means can satisfactory relations be re-established between masters and workmen.

The movement among the workmen began early in March. The shipwrights requested an advance of 1d. per hour, and the ship joiners a ½d. per hour, while the rivetters, fitters, and caulkers sent in a demand for an increase of 10 per cent. on time and 10 per cent. on piece work. To each demand the masters gave a decided refusal, stating that their contracts having been taken at the present rate of wages they could not see their way to grant the advances sought. Subsequently the joiners and rivetters, fitters and caulkers resolved to continue at work at their present rate of wages, but the shipwrights unanimously agreed to come out on "strike," and continue out until their request was granted. Their "strike" has now lasted five weeks, and there is apparently no disposition on their part to return to work. The progress of the movement, its effect on the Clyde shipbuilding trade, and the position taken up by the masters have been fully reported in these columns from time to time. Its effect on the trade has been most disastrous so far as the booking of new orders is concerned, and much valuable work has passed away from the Clyde. In fact, no new contract of any consequence has been taken in the upper district of the Clyde for the last two months. The position taken up by the masters has already been fully explained, and there is nothing new to add to what has already appeared in our reports of the wages agitation. The last lock-out on the Clyde occurred in 1866, when the masters were successful, and the men submitted to a reduction of wages. In 1872 the masters and workmen came into collision about the number of working hours per week, when the former submitted to the terms proposed by the latter. In this movement the masters were compelled to yield owing to the existence of "penalty" clauses in their contracts, which entailed a heavy loss on shipbuilders if they exceeded the specified time in finishing a vessel. Now this state of matters has been altered, and every contract contains a conditional stipulation or "strike clause," which sets forth that a vessel shall be finished by a certain time, *provided no strike occurs.* This gives the masters a decided advantage in dealing with trade questions, and they enter on the present movement with considerable assurance of success. The following is a report of yesterday's proceedings:—

MEETING OF MASTER SHIPBUILDERS.

A meeting of master shipbuilders was held yesterday afternoon, in the Religious Institution Rooms, 172, Buchanan Street. There was a large attendance of shipbuilders connected with the Clyde and other shipbuilding districts. Mr. Simons, Renfrew, chairman of the Clyde Shipbuilders' and Engineers' Association, presided. In opening the proceedings, the Chairman explained the various steps that had been taken in connection with the present "strike" of the Glasgow shipwrights. At the former meeting of the association it was agreed, "That in order to assist in bringing the present dispute to a termination, it was highly desirable to 'lock-out' the shipwrights of the lower district of the Clyde," and a deputation was appointed to convene a meeting at Greenock to come to some arrangement on the question. The Chairman of the Greenock meeting (Mr. Steele) then reported the result of that gathering. The opinion unanimously expressed was to the effect that were an advance of wages to be conceded in the upper, a similar advance would require to be given in the lower district. The meeting, however, did not think it advisable to "lock-out" the shipwrights at present at work, as it was desirable to have all the branches of the shipbuilding trade working at the same time; but it was agreed to recommend to the association a general lock-out all over the Clyde as the only method of settling the present wages agitation. A gentleman from Dumbarton stated that the opinion there was to the effect that it was high time some decided action was taken in order to put a stop to the demands of the men, and admit of work being carried on in a satisfactory manner. The Secretary (Mr. J. W. Ross) reported that it had been suggested to him that a conference should be held between the men on strike and the employers, and endeavour to come to some understanding regarding the present dispute. The Chairman said the meeting was now in possession of the facts of the case, and it was for it to decide what steps should be taken under the circumstances. A long sederunt followed, the proceedings lasting fully four hours, but complete unanimity prevailed in recommending a general lock-out of all departments of the shipbuilding trade on the Clyde from Glasgow to Greenock. One gentleman stated that no orders had been taken on the Clyde of any consequence for the last two months, as it was impossible to fix a date when contracts could be accomplished while the present harassing policy of the men was pursued. Ultimately, the following statement was drawn up and signed by all the firms on the Clyde, except two :—

"At the largest meeting of Clyde shipbuilders ever held in Glasgow, it was unanimously resolved that, in consequence of the present continuous strike of shipwrights for an advance of about 15 per cent. on their wages, there should be a general lock-out of all workmen employed in shipbuilding on the Clyde, to take effect on the 19th inst. In taking this course they were influenced by the fact that unless the shipwrights returned to their work a general stoppage and great disorganisation of all departments must inevitably occur very shortly, as the ironwork of the vessels in hand will soon be finished, and a general block will be unavoidable, so that the decision arrived at was taken to hasten a settlement of the question. The employers have been reluctantly compelled to adopt this course, with the view of promoting the ultimate interests of all concerned, preventing the trade leaving the district, and avoiding unnecessary distress."

The above resolution was signed by 22 firms.

It was then agreed that notice of the lock-out should be issued on Monday in all the shipbuilding yards whose proprietors had signed the resolution.

The masters are of opinion that the lock-out will last all summer, and there can be little doubt that it will be a determined struggle on both sides.

Palmer & Howe, Printers by Steam Power, 1, 3, and 5, Bond Street, Manchester.

ADDRESS

FROM THE

EXECUTIVE COUNCIL

OF THE

Amalgamated Society of Engineers, &c.,

To their Fellow Workmen throughout the United Kingdom, the British Colonies, and Foreign Countries.

FELLOW WORKMEN,

If there is any time more favourable than another for the cultivation and spread of trade union principles it is such a time as the present. This may appear contrary to the generally accepted opinion, which holds that a time of brisk and prosperous trade is really the best period for making converts to trades unionism. It is true that in a time of good trade workingmen have more money available for paying the entrance fees and contributions of a union; but, at the same time, they feel so well off, so generally comfortable and content with plenty of work and good wages, that they forget the lesson which previous adversity had taught them, and think that because no clouds lower it will be always sunshine. In a period of bad trade, however, wages fall, employment is scarce, and those who are employed look to each succeeding pay-day with fear and trembling lest reduction or discharge should await them. They see also that employers in such times, either from necessity or from a desire to reduce expenses for the sake of profit, deprive them of many of their privileges, and even of things which they had considered rights. Seeing those things, they are impressed as individuals with their powerlessness to act alone, and are therefore struck with a sense of the power of union.

We see this going on every day around us, and we see that in nearly every case the workmen are the losers, entirely because of their want of cohesion, and of their infidelity to each other. Therefore it is that we address you at present, and desire to lay before you some of the leading arguments in favour of union generally; and a few facts relating specially to our Society. We do this with a desire to induce you who are yet non-society men to join us, not only that you may advantage yourselves, but that in so doing you may help us to become more powerful for our great mission. We have toiled and striven for a quarter of a century, and we are not afraid to say that we have during that time materially improved the condition of our trade; neither do we hesitate to declare that as a Benefit Society we have relieved much misery.

We confess to you, however, that we have not done all that might have been achieved had you been with us instead of against us. It is not the mere wealth or numerical strength of a union, which gives it true efficiency as an organisation formed for the protection and improvement of the position of labour. The true element of power or weakness in a union is to be found rather in the relative proportion which its numbers bear to the entire total of men engaged in their trade. In this respect we have been wanting. There are probably 150,000 working engineers in this country, and our whole strength is but 45,000 men. We have thus upwards of a hundred thousand men who, if not actually against us, hinder our efforts, and render us comparatively helpless. In all great trade crises we have this mass of unorganized, unprovided for, and undisciplined workmen to consider and help, and the work of the whole 150,000 is left to be done by the 45,000. A few general remarks will lead up to the facts which show that the lesser number has really been the power which has raised and improved the condition of the whole body, and it will be seen that had the number of unionists been in the reverse proportion to what it is, the progress of our class would have been much greater.

We frequently hear of the "good old times," and no doubt it is good to praise what is old, and what tradition has rendered venerable, but we do not think that many of the most ardent admirers of the days that are gone will be found bold enough to assert, that the condition of the working classes to-day is not immeasurably superior to what it was in byegone periods of our history. During the early part of the present century the bulk of the population of this kingdom lived in a state of comparative wretchedness, and were forbidden by combination laws of the strictest character to organise themselves for any improvement of their condition.

Just, however, as it is the darkest hour which precedes the dawn, so at the time when the working people of this kingdom were in their deepest misery, the means of securing their own release from a situation little better than bondage was at hand. The almost continual wars in which our country had been engaged during the latter half of the eighteenth century, and the first fifteen years of the nineteenth, had much impoverished the nation, and saddled it with a vast national debt. Taxation was consequently heavy, and the resulting poverty of the masses was intense. A poverty-stricken people means too often an abject and a servile nation, but, in our case, it was happily not so. The indomitable pluck and perseverance of our people have been our strong points as a nation, and in no way have these qualities been more grandly displayed than in the slow but sure improvement which our working population has made in its own position.

What are the facts? Previous to 1820 wages were at their lowest, provisions were at their dearest. Combinations for raising wages were prohibited by law; but it was in this forbidden system that the workman looked for relief. The engineering trade was then in its infancy, and the early working engineers were among the pioneers of unionism. The steam engine had been introduced by Watt, and a great impetus had been given to the work of mechanical construction and invention. The locomotive engine had been imagined in, and evolved from the minds of Trevethick and Stephenson, and was promising to become one of the chief agents of civilization and human progress, while at the same time it produced work for thousands, and called into existence what was almost a new industry. The men who built the early steam engines must have been no ordinary men. The difficulties of their work at that time must have been enormous. The magnificent machines of to-day, which render hard labour almost unnecessary, were then unknown, and "hammer and hand" were then the tools most in

3

request. Working men are not slow to draw analogies, and if the engineers of sixty years ago needed something to show them the power of combination they might easily have found illustrations in the work of their hands, which depended for its strength, solidity, and smoothness of motion, upon the perfect union and harmonious working of its parts. Almost naturally, therefore, the engineers were among the first organisers, and even when unions were forbidden they began to form them. What they did in secret prior to 1824 was rendered legal in that year, thanks to the exertions of Mr. Joseph Hume, M.P., the celebrated economist.

From this time progress was easy, and the work of organisation made rapid strides, numerous trade societies sprang up in all directions, and though the whole of the change may not be clearly traceable to their influence, there can be no doubt that from that time there commenced a very perceptible improvement in the condition of the trades that were best organised.

In the engineering trade of to-day we know what wages are, but even at a no more remote period than forty years ago the general rate of wages was very much lower than at present. At that time wages were best in London, and in those provincial towns where trade societies had taken the deepest root; but in the remote parts of the kingdom the old state of things remained in full force. In London, at that period, the wages of engineers had been raised within a brief time to very near their present rate, and in such centres as Bradford, Birmingham, Manchester, and Liverpool, where the work of organisation was far in advance, the wages of our trade ranged from 24s. to 32s. per week. In towns like Leeds, Newcastle, Glasgow, Edinburgh, and Dundee, where the new societies had found little or no foothold, wages were very low. In many places a pound a week was considered good pay, while in other quiet towns from 17s. to 18s. per week was the prevailing rate. At that time, also, the hours of labour were sixty per week; in some cases more than that. Thus it would appear that organisation had reaped for those who practised it many advantages which their undisciplined fellows had been unable to gain, and here was a triumph of a policy of co-operation and union over a simple system of "every man for himself."

It is further to be observed that all movements in favour of larger wages and shorter hours proceeded from the men who started their ideas and formed their plans in the associations in which they met to discuss trade affairs. This must always be the case. The man who stands by himself thinks little of matters that will raise his whole class, and that for the best of reasons. He knows that he is powerless to carry into practice any plan that he may form, and so he forms no schemes which may endanger his situation. On the other hand, organised workmen, when they meet in common, talk over their affairs, and having convinced themselves that their ideas are good they do not hesitate to give utterance to them in the workshop, and if necessary in the presence of their employers, for they know that they have the power and support of their society behind them.

In the largest centres of our trade, progress during the last forty years has been comparatively slow, and wages have not increased as they did during the first ten years after the repeal of the Combination Laws. Closer association had given the workmen power to force wages to their highest natural level, and this fact is clearly proven by the position of the unorganised

4

towns, and unassociated classes of workmen. The position of the regular engineer artisan of forty years ago was one of comfort compared with that of those who worked in the mine, or toiled amid the din and dust of the national cotton mills.

The artisan had learned to live like a man, because he had learned to mix with, and act in concert with, his fellows for the common good. The miners and the mill hands had, on the contrary, gone on, in most cases, each for himself alone, and their condition was deplorable. Sir J. P. Kay-Shuttleworth, speaking in Manchester in 1832, described the mill hands as rising at five in the morning, and working in the mills from 6 a.m. to 8 p.m. Their breakfast was coffee and bad bread, and their dinner was a sort of general mess of potatoes and lard, or perhaps butter or fat bacon.

This was simply the condition from which the artisan had raised himself by combination with his fellows. In his turn the factory worker learned also to combine, and also reaped the benefits of union.

As the unions extended they reached the low paid towns, and those centres of industry began to feel their effects in the increased wages of the trades which organised themselves. The northern towns especially reaped the benefit, and in Dundee, Newcastle, and those towns where wages had formerly been so low, a marked improvement has taken place. It may, in fact, be said that organisation in small provincial towns, even though imperfectly and partially carried out, has raised the wages of our trade from 30 to 50 per cent. But this is not all, for, to come nearer down to our own day, the shortening of the hours of labour is almost clearly traceable to the action of trade societies or their members. The Saturday half-holiday, the weekly payment system, extra remuneration for overtime, and above all the nine hours' movement, have been almost entirely effected by the organised workmen of the trade, who, in some instances, have had to carry their point in the face of the passive indifference, if not active hostility, of their unorganised brethren. These facts prove the one great truth, that "Union is Strength." Even in cases where union has failed to accomplish its objects, it has failed simply because it was not sufficiently all-embracing, and because all who would have reaped the benefits of its work, if achieved, did not join in its efforts.

The old fable of the bundle of sticks is as true now as in the days of Æsop, and it would almost seem as though men were as slow to learn the lesson. We have seen that the engineers were among the first to organise. They, however, organised in sections only, and though those sections or branches of trade were really fair embodiments of the principle of union, they were weak, because they kept only their own special objects in view; and sometimes they lost sight of the interest of the whole trade, in looking after a local or special benefit. This state of things was terminated by an amalgamation of the bulk of the sectional societies in 1851, when our Society was formed under the designation of the Amalgamated Society of Engineers, Machinists, Millwrights, Smiths, and Pattern Makers, by which title it has since been so honourably known. It has now existed as such upwards of a quarter of a century, and may be looked upon with pride by its members as being the grand development and outcome of the principles enunciated by the early engineers, who formed one of the first forbidden trade societies, in the days before the Combination Laws were repealed. Its first pioneers did their work in darkness and in fear, for the wealth and privilege of capital had made the laws which kept the workman down; to-day its deeds are done in the light of freedom before the law, and its members are even proud and boastful that they are unionists. The change which has, in this respect, taken place, is not the least of the things which have been accomplished by union.

So far we have merely glanced retrospectively at some of the good which we believe to have accrued from combination in our trade, preferring in this way to base our claims to consideration rather on work fulfilled than on mere abstract arguments as to the advantage of union. The facts we have referred to may be thus summarised:—

The condition of the working classes can never be materially improved except by the action of that portion of the community itself, acting in a united and disciplined manner.

The condition of the workmen of this country never was improved until after the era of trades' unions; and all their advancement, whether in wages or in better conditions of working, has gone on hand in hand with the extension of trades' union principles.

During the early part of the present century, which may be taken as an example of what the state of the working classes would be without organisation, the wealth of this country enormously increased, but its possession was shared by a very small section of the community, while the poverty of the masses was intense.

Employers are seldom known to raise wages or shorten hours of their own accord, and combination among the workmen was found to be the only means whereby pressure would be brought to bear upon them to effect either object.

Since the era of trades' unionism set in we have seen—

A more equal distribution of the wealth of the country, and a greater participation by the working classes in the fruit of their labour.

Such an alteration in the laws of the country, and in the social position of the workman, as to place labour on an equality with capital in the eye of the law. Also the enactment of laws for the protection of the health and life of the workman in the workshop.

And last, but not least, the shortening of the hours of labour, which has given to the working man increased opportunities for mental development and recreation; the fruits of which, though even now apparent, can only be fairly seen after the lapse of a generation or two.

As to the special claims of our Society upon your support they will be best explained by a simple reference to the benefits which it pays for the contributions of its members. The regular contribution is one shilling per week, and the benefits offered are as follows:—

Donation Benefit.—Should any free member be thrown out of employment, under circumstances satisfactory to the branch to which he belongs, and not disentitled to benefits through arrears, or a non-free member be withdrawn from his situation by a branch district committee, or Executive Council, and continue out for three consecutive days, he shall be entitled to the sum of 10s. per week for fourteen weeks, 7s. per week for thirty weeks, and a further sum of 6s. per week so long as he remains out of employment, making a total of £19 18s. in the year.

Sick Benefit.—Any free member, when visited by sickness (not occasioned by drunkenness or disorderly conduct, or any disease improperly contracted),

6

upon giving due notice to the secretary of his branch, as prescribed by rule, shall be entitled to 10s. per week for twenty-six weeks, and 5s. per week afterwards, so long as he continues ill.

Superannuation Benefit.—Any member fifty years of age, who has been eighteen years successively in the Society, and who, through old age or infirmity, is unable to obtain the ordinary rate of wages, shall be entitled to a superannuation allowance (if he applies for it) of 7s. per week; and if twenty-five years in the Society at the time he first commences to receive superannuation benefit, he shall be entitled to 8s. per week; and if thirty years in the Society, he shall be entitled to 9s. per week; and if forty years in the Society, 10s. per week until his death.

Accident Benefit.—Any free member not more than 16s. in arrears, who may, by losing a limb, or having one disabled by accident or otherwise, or through blindness, imperfect vision, apoplexy, or paralysis, be rendered permanently unable to follow any of the branches of trade mentioned in the preamble of these rules, provided such was not the result of intemperance, or other improper conduct, shall receive the sum of £100 on the production to the Executive Council of satisfactory medical and other testimony of such permanent disablement. Any free member being permanently disabled from working at the trade, through any of the before-mentioned causes, and not having been in the Society eighteen years, or attained the age of fifty, at the time of such disablement, must accept this benefit in preference to any other.

Funeral Benefit.—On the death of a free member £12 shall be paid to his widow or nearest of kin; or should the wife of a member die first he can draw (if he chooses) £5 at such death, leaving £7 to be paid upon his own death.

We now invite the attention of non-members to the vast sums that our Society has expended during the past twenty years, in disbursing the claims that have been made upon our Donation, Sick, and Superannuation, also for those of Accident and Funeral benefits, and then some idea may be formed of the magnitude of the Society's financial transactions.

We would here remark that in addition to these guaranteed benefits, members who are in distressed circumstances are assisted from a Benevolent Fund, from which source, from the date of its origin till the close of 1878, has been expended £37,024 in giving grants proportioned to the necessities of each case.

Amount expended in Benefits during Twenty-eight Years.

Donation	benefit	£789,538
Sick	"	365,503
Superannuation	"	153,497
Accidents, &c.	"	30,700
Funeral	"	118,332
		1,457,570
Benevolent Grants		37,024
Assistance to other Trades		22,892
		£1,517,486

Despite this enormous expenditure for the relief of suffering and poverty, the Society has saved a large fund to meet future contingencies, and at the

7

close of the December quarter of 1878, had a balance in hand of £251,675 8s. 6½d.

Surely all prudent men, anxious for their own welfare, and for that of those who are dependent upon them, will appreciate the advantages of a society of this character, which gives them a hold upon their position in the world, and renders them comparatively independent of the various chances of suffering to which all are liable. Even if the Society had nothing but these individual advantages to offer, the Council believes that thinking men would scarcely feel themselves justified in holding aloof from it; and it is sincerely hoped that those to whom this address is specially directed, after due consideration, will arrive at the same conclusion.

But the trade advantages which may be gained are of even greater importance than those which are secured to individuals. Each man should understand that in the well-being of the trade his own prosperity is inevitably bound up. Enough, and more than enough, evil has been experienced from the selfishness which prompts each to regard his own interest as apart from, and independent of, every other man's. The amelioration of the condition of each individual and the elevation of his social position, is not to be obtained singly, but only by the growing prosperity of the mass. Efforts made without concert are powerless, but power springs from the combined action of thousands. Union wisely directed is the only method of permanently bettering the condition of the labourer. Out of union grows a public opinion affecting the whole trade, which acts with greater force than a law in preventing men from betraying or sacrificing others by the side of whom they work. Through union, customs arise which, made for the many by themselves, tend to regulate the trade and to keep it in a healthy state. By means of union, and in consequence of the necessity for intercourse which it creates, a medium of communication is opened between those who have common wants and common interests. This renders information and counsel accessible to all, gives them the knowledge of where labourers are in greater plenty than the work to be done, where men are wanted, and therefore likely to be better remunerated, while the funds of the Society defray the expense of transit from one place to another in cases of emergency. Thus, members are enabled to avoid the localities in which the labour market is over-stocked, and to place themselves in the most advantageous attainable position.

The time is rapidly approaching, if not already arrived, when Trades' Unions such as ours will be emancipated from mistrust, jealousy, and misrepresentation, and take their stand as recognised institutions by the community at large, and one class of society will be prevented from treading another in the dust. In spite of the efforts of certain economists, in teaching working men that true patriotism consists of meekly submitting to every reduction of remuneration and every device for lengthening the hours of labour or intensifying exertions, they have failed to carry conviction as to the truth of their teaching, and men are disposed, and justly so, to think for themselves on these matters, and from a sense of right and duty have acted on different conclusions. Employers, if true to the duties which belong to their position, will not regard a society like ours with disfavour, and will understand that it is not intended, nor adapted, to damage their interests, but rather to advance them, by elevating the character of their workmen, and proportionately lessening their own responsibilities. The

8

man who is aware that when work fails, or sickness takes from him the power to labour, he will not be altogether deprived of the means of living—who is certain of some provision for the declining days of age—becomes a more contented, as well as a more independent being. With a greater freedom from anxiety he is able to devote more care and attention to his work, and to perform it more skilfully. Besides this, those who are subject to a public opinion such as we are, and who feel that the character of a whole society, as well as their own, depends upon their conduct, will be more anxious than others to fulfil the duties which are confided to them.

In conclusion, the Council of the Amalgamated Society, besides addressing those who have not yet joined them, desires to impress upon all their members that the present likelihood of increasing their numbers as well as the future prosperity of the Society, depends pre-eminently upon their conduct. It is for them to enhance the character of the Society by their demeanour in their workshop as well as in their home lives. It is for them to take care that the nature of their Association and the benefits it confers shall be known as widely as possible, and correctly appreciated. It is for them to cultivate confidence by a readiness to extend to others that help upon which they rely themselves. And it is for them to watch over the young minds among them, and to give them such direction as will tend to make the men of the future honourable, useful, independent, and united members of a powerful and thriving industrial community.

Scale of Entrances.

			£	s.	d.
If not exceeding	25	years of age	0	15	0
„	26	„	0	16	6
„	27	„	0	18	0
„	28	„	1	0	0
„	29	„	1	2	0
„	30	„	1	4	0
„	31	„	1	6	0
„	32	„	1	10	0
„	33	„	1	13	0
„	34	„	1	16	0
„	35	„	2	0	0
„	36	„	2	4	0
„	37	„	2	8	0
„	38	„	2	12	0
„	39	„	2	16	0
„	40	„	3	0	0

By Order,

JOHN BURNETT, *Gen. Sec.*

General Office, 90, Blackfriars Road,
London, *May*, 1879.

Any Branch or District Committee requiring a supply of these Addresses must apply for the same to the General Secretary, and at the same time state the number required.

TRADE-UNIONS:

THEIR NATURE, CHARACTER, AND WORK.

By George Howell.

THE subject for consideration in the following pages is an oft-debated one, and, if we may judge from the never-ending controversies to which it gives rise, it is one upon which there would seem to be an utter impossibility of thorough accord, inasmuch as the conclusions arrived at vary according to the standpoint from which it is viewed. An exhaustive treatment of the whole question within the compass of a short paper is not practicable; but by confining ourselves to the exact points indicated in the heading, by avoiding unnecessary detail, and above all side issues, a general outline may be given which, it is hoped, will be sufficiently clear to be easily comprehended. For the sake of brevity its historical bearings will be omitted, excepting by a passing reference here and there; abundant materials for forming a judgment on this part of the subject are now readily accessible, and may be consulted by those who desire further information on particular points.[1]

It is now generally received as an authenticated fact, that trade-unions are the legitimate successors of the old English guilds; in many ways traces of their early origin may still be found in their forms, ceremonies, observances, methods, and policy. In some cases the unions have discarded that which was bad in the old system, and have retained the good; in others, it is to be feared that they have acted on the contrary principle of clinging to the bad and forsaking the good; such is the perversity of human nature. With the latter phase I have nothing whatever to do. My object will be to present the healthiest aspects of trade-unionism, believing, as I do, that this is the only trustworthy mode of dealing with any matter really worth the trouble of debating. The drift of these remarks will not, I trust, be misunderstood; there is no wish on my part to shrink from the contemplation of the unpleasant side of the picture; by no means; but it would be a pure waste of time to dwell upon merely incidental features and accidental circumstances, on which presumably there would be little difference of opinion, instead of raising those points which constitute the very marrow of the question itself.

In their essence, trade-unions are voluntary associations of workmen for mutual assistance in securing generally the most favourable conditions of labour. Many persons deny the voluntary character of these associations, alleging that they are kept up by coercion. This is altogether a mistake. Coercion is alike opposed to individual liberty and to the constitution of a *bonâ fide* trade-union; no body of men could be held together on such a principle, and least of all a trade society. That there are instances of coercive action cannot be denied, but they are few and far between; when they do occur, however, we hear a good deal about them, for they are reproduced in all the newspapers of the country, and are spoken of as though they represented the normal condition of

[1] For an exhaustive treatment of the question, historically, see *Conflicts of Capital and Labour*, Chapters I. and II., and the essay on *English Gilds*, by Dr. Brentano, Early English Text Society.

things in a trade-union. Pressure is sometimes brought to bear, to an unfair degree, upon those who refuse to join the union; but usually this is exercised over those who have in some way transgressed its rules. This kind of pressure is not confined to trade-unionists; we can find plenty of it among barristers, doctors, ministers, traders and commercial men of all classes. The Reform Club recently gave us an example of its working in political circles; in what is termed 'society' we find it developed to a most unhealthy extent, often it operates without rhyme or reason, to the exclusion of a luckless offender from its ranks. It is not intended by these remarks to defend the practice of the unions; the object is rather to narrow down the issues.

Trade-unions are sometimes described as being the 'hot-beds of tyranny,' the 'instruments of oppression,' and as 'secret organisations for the purpose of assassination.' Secret societies they are not, and never have been, in the sense attributed to them. Whatever of secrecy they may ever have had was forced upon them by unjust laws; and even then they were less secret than some others that were tolerated, and in a sense protected by law. In their darkest days, when penal laws were enacted for their suppression, and cruel sentences inflicted on all offenders, they were generally as open and above board in their acts and deeds as were the political organisations of the same date. Since the partial repeal of the combination laws in 1824, trade-unions have no more deserved the name of secret societies than those established for purely benevolent purposes; in this respect they stand on precisely the same footing as friendly societies, provident societies, building societies, scientific associations, and political clubs. Before that date their members were hunted like wild beasts, and there are men still living who can remember having to bury their books on the town moor for fear of a vindictive prosecution.

The charges of tyranny and oppression which are brought against the unions, must be examined a little more closely. These accusations did not originate with the investigations which took place at Sheffield and Manchester in 1867; they belong to a period much farther back in the annals of their history; but their repetition in our day is almost entirely due to the facts then and there brought to light. The disclosures made before the Royal Commission had no more to do with trade-unionism than rioting and plunder have to do with political meetings; the whole of the evidence given has been sifted, and my conclusions thereupon stated elsewhere.[2] Briefly they are these: that the trade-unions of England should be held guiltless of any participation in those crimes, just as much as you would hold the Government of to-day guiltless of the crimes committed at Peterloo. No good can ever come from the impeachment of a whole class because of the wrong-doing of a few belonging to that class; to brand trade-unionists as criminals for the reason that some union men have dishonoured themselves, and brought disgrace on their fellows, is an outrage on common decency, and a violation of all justice.

It is said that the unionists are a tyrannical majority who rule the minority with an iron hand; this, however, is hardly true in any case; in most instances it is the very reverse of being true. Let the statement be tested by a few facts. The members of the Amal-

[2] See *Conflicts of Capital and Labour*, Chapter VII., for examination of these charges.

gamated Society of Engineers, according to their own reports, do not number *one-third* of those working at the trade. The estimated number of working engineers in the country is stated to be 150,000: of these, 45,000 only are members of the union; so that in this, one of the best organised in the world, the society men are but a small minority. The boiler-makers and iron ship-builders estimate their numerical strength as being fully three-fourths of the whole; the iron-founders number, probably, about two-thirds; the steam-engine makers, one-half; the carpenters and joiners about one-fourth; the masons have, possibly, nearly two-thirds; the bricklayers most likely number three-fifths; the plasterers, one-third; the painters and decorators probably one-sixth; the tailors and shoemakers about one-fifth. In most other trades the proportions are far less than those above given. The idea, therefore, of a domineering majority, crushing the minority under their heel, is a little far-fetched, seeing that, in the examples selected, which are among the strongest and wealthiest societies in the United Kingdom, the proportion stands at twenty-two non-society men to sixteen unionists.

How, then, it will be asked, is it that society men are usually represented as being oppressive towards non-union men? The explanation is not difficult, when all the facts are fairly taken into account. Non-society men, be it remembered, are eager enough to enjoy all the advantages gained for them by means of the union, and as a rule they are equally anxious to avoid contributing to its funds, submitting to its discipline, and taking their share of its risks. At the same time they are often quite as reckless as the men in union in entering into a strike, and not unfrequently even more so, for in the one case an organised body has to be consulted, the majority of whom are not in any way mixed up with the dispute, whereas in the other they can act on the spur of the moment without being called to account. When a strike has actually taken place these men expect to be paid out of the union funds just in the same way as though they were members. When they find that the society men are getting 10*s*., 12*s*., or 15*s*. per week, and that they have no claim to a like allowance, they first grumble, then waver in their resolves, and at last submit to the terms offered by the employers. But there is another class quite different from those before indicated, namely, the unskilled, or only partially skilled, men who would not at any other time be accepted as members by the union, nor would they be employed by the masters if it were not for the strike. It is astonishing how much inconvenience the employers will put up with, at these times, if they can only manage thereby to beat the union, and frustrate the objects of those on strike. Then comes retaliation; those who during the dispute took the places of the 'turn-outs,' as they are called, are gradually weeded out, by two processes; first, by the masters, who rid themselves of the more incompetent workmen as soon as possible; after having served his turn, as makeshifts, and being no longer needed, they have now to make way for better men; secondly by the society men refusing to work alongside of those who helped to destroy their chances of success. Some of them are thus forced, as it were, into the union, while others seek employment elsewhere. The feeling here described, however much we may condemn it, is but natural, and evidences abound of its existence in other classes than those who constitute the members of trade-unions. Still it must not be forgotten that the policy is a shortsighted one; unionists forgive and forget the injuries inflicted on them by the masters much sooner than they do the individual desertions from their own ranks. It is much the same in a conquered country; the victorious invaders are more easily forgiven than the time-servers who change sides and go over to the ranks of the enemy. As a matter of common prudence, unionists ought never to strike against their fellow-workmen, however great the provocation might be; the full force of their energies ought to be reserved for conflicts in which principles are at stake rather than the actions of individual men.

Trade-unions in the first place originated out of the repressive Acts of the Legislature which began as early as 1349 by the Statute of Labourers, 23 Edward III., under which men were consigned to prison for asking for a higher rate of wages. To this succeeded the 25th of Edward III., 1351, followed by the Acts of Richard II. 1388, and again in 1390; and the Act of Henry IV. 1406, and subsequently by the Statute of Apprentices, 5th of Elizabeth, cap. 4, by means of which the former Acts were extended so as to embrace and apply to all the handicrafts of the time. These went on accumulating up to 1720, a period of nearly 400 years.

Combinations, in the modern sense of the term, were the inevitable and necessary outcome of those changes in industrial life which led to the growth of a capitalist class, men who were not craftsmen, as formerly, but manufacturers; they were fostered by the introduction of machinery, the consequent division of labour, the aggregation of large numbers of workpeople in certain localities, and the inauguration of factory life. These were the more immediate causes of the development of trade-unions, and the feeling in favour of them became intensified from the fact that the whole statutory law with reference to labour was in the interests of the masters and adverse to the men. Still there were some provisions which the craftsmen regarded as favourable to them, and they combined to resist the encroachments on their legal rights—rights invaded, be it understood, by the masters: this was the object of all the earlier associations. The masters then sought for, and obtained, further enactments, the intention of which was to suppress the right of association altogether, and to punish rigorously those who dared to join in them. Thus we find, in 1727, the 12th of George I. passed, and again in 1799 the 39th of George III. c. 81, by which all combinations then existing were suppressed; in the year following, 1800, the 40th of George III. was passed, by which all who infringed the statute of the preceding year were to be severely punished. Atrocious as these laws were, they could not repress the tendency to unite, nor did they have the effect of stamping out the unions. Men braved the most cruel penalties rather than submit to the tender mercies of the trading class, and hence we find that coincident with the passing of the latest edict, for intimidating workmen by legal coercion, there grew up the more distinct form of trade-unionism which in a modified shape we see now.

From 1800 to 1824 the unions developed, and were extended enormously; the laws failed by reason of their stringency, for many employers declined to avail themselves of the terrible weapons placed within their reach, and those who did take advantage of them were seldom proud of their victories. That excesses were sometimes committed by the men, must be admitted; and who can wonder?

The marvel is that such a state of things did not produce a craftsmen's war, more disastrous in its destructiveness than the peasants' war of 1381, or the revolt of the weavers in 1756. The repeal of the combination laws in 1824, although slightly modified in 1825, was due entirely to the concentrated force of the unions as they had organised themselves under the very statutes which were intended to suppress them. The result of the removal of these repressive measures was, that the unions grew and expanded under the eye of the law, and they began to take that form which in later years has shown such vast power and resources. From 1825 to 1871 they were still outside the pale of the law, their funds and property being totally unprotected, so that any dishonest officer could rob the society with impunity, and snap his fingers at those whose hard earnings he had embezzled. The Act of 1871 removed the most crying evils, but it remained for the Acts of 1875, known as the Labour Laws, together with the Act of 1876, to complete their emancipation. It is necessary that so much of their history should be perfectly understood, in order to estimate their character.

The leading features of trade-unions are best seen in their acts. Their constitution is essentially democratic; as a member no man is greater or less than another. Each has an equal voice in all matters pertaining to the union. In all essential particulars they are constituted and governed on a basis exactly similar to that of a social club. Candidates for membership have to be proposed by one member and seconded by another; these have to speak from personal knowledge as to their fitness, abilities as workmen, their capabilities of earning the current wages of the district, their general character, and their desire to be initiated as members of the union. On the night of election, all these matters have to be vouched for in open lodge; if any member has anything to allege against the candidate he is bound to state it openly. If no objections are raised he is admitted by the vote of the lodge, and is welcomed as a brother by the president in the name of the members. If, on the contrary, something is said to his prejudice he is asked by the president to give an explanation. A private member is at liberty to ask any further question; if the replies are considered satisfactory he leaves the lodge, as a matter of form, while the votes are being taken; after which he is readmitted, and led to the chairman, whose duty it is to read to him certain rules, having reference for the most part to his voluntary submission to the regulations of the society, his conduct as a member, and his behaviour in the lodge during business hours.

Should the replies of the candidates be regarded as evasive or dubious, his admission is sometimes delayed for a week, so that inquiries may be made; if on the second occasion there is still some opposition manifested towards him, the question is put to the vote, and he is admitted or rejected, as the case may be. No man, however, is blackballed in the dark; if he is refused as a member, he knows the reason, and has ample opportunity afforded to him of explaining anything that may be alleged against his candidature.

The question of apprenticeship is sometimes raised, but not often; in olden times legal apprenticeship was invariably insisted on as an indispensable condition of membership, but it is so no longer; in one or two societies an attempt is made to enforce it, but the rule is inoperative—practically dead, except on paper. The only absolute conditions of membership now are, that a man shall know his trade, that he gets his living by it, and that he is able to earn the current wages of the town or district. If a man can manage to pick up his trade, no matter how, he will find no difficulty in gaining admission into the union,[3] and no obstacles are placed in the way of his earning a living by the trade which he has chosen.

Every member on being elected has a copy of the rules, by which he is bound, given him; the absurd notion as to bye-laws, passed by a committee, is altogether a fiction. Whatever laws, or bye-laws, there may be, they are printed, and each member has a copy of them, and he is at liberty to suggest or propose amendments as soon as he is a full member. Bye-laws, where they exist, are simply local regulations for facilitating the transaction of the business of particular lodges or branches; in societies having branches all over the country these are often necessary; inasmuch as each local branch conducts its own affairs in its own way, it reserves to itself the right of making its own bye-laws; and these are never interfered with by the executive council, *so long as they are not at variance with the general laws of the society*. The other kinds of bye-laws, of which we hear so much, exist only in the fertile imaginations of those ingenious writers whose chief amusement seems to be to endeavour to discredit trade-unions in the estimation of the public.

Again, every member is entitled to know everything appertaining to the management of the society, its funds, and its action. Nothing can be done without his knowledge; he has to vote for all officials, for or against every strike, for the amendment of all laws, for grants of money for any purpose, and can exercise his influence and express his opinion upon every point brought forward for discussion. Moreover, all reports, circulars, balance-sheets, and printed matter of every kind are open for his inspection, and he is entitled to a copy on payment of the trifling fee charged for them Fourpence is the highest charge for a copy of the annual report in any society, and some of them are big volumes consisting of from 280 to over 300 pages, large octavo.

The chief objects of trade-unions are the protection of their members in all matters pertaining to wages, hours of labour, and conditions of employment, such as overtime, piece-work, and the like. These may be said to be their primary and fundamental aims; all others are accessories. To say that the unions use their funds for trade purposes is equivalent to saying that they are used for the specific purposes for which they are contributed; to assert that the unions use the funds contributed for friendly society objects for other objects not named is incorrect. The friendly society purposes of the union are in all cases subordinate to the main object—namely, the protection of trade privileges. The man who ventures to employ an argument founded on such a misconception as the one above given, is totally unacquainted with the constitution and aims of trade-unions, and therefore disqualified for the task of criticism.

The trade objects being paramount, it will be well to direct attention to a few of the more important ones. (1) Wages. It is not true that the unions insist on a uniform rate; it is not true of any society. What they endeavour to do is to fix a minimum rate; the masters, on the other hand, strive to fix a maximum rate, so that between the two there is a tendency to create a mean rate always ap-

[3] [This would somewhat surprise any member of an old English Guild.—Ed.]

proaching to uniformity. Beyond this the unions do not go. It would be quite as true to say that the masters insist upon a uniform rate, as to say that the unions attempt to do so. In neither case would it be correct. All that we can safely say is, that wages gravitate towards a uniform standard by reason of the operation of two opposite forces; beyond this we are in the region of speculation. There is absolutely no uniform rate imposed in any trade; in the engineers, iron-founders, smiths, boiler-makers and iron ship-builders there are marked variations not only in the same towns, but even in particular shops. Still there is a tendency to uniformity of price for the same kind of work in all trades and professions, quite apart from the action of trade-unions; we see its operations and effects on all hands, whether there are or are not combinations among the workers; it might almost be said to be the result of a natural law. In Government offices copying is paid for at the rate of tenpence the hour; newspaper work is three-halfpence the line; professional fees of all kinds are similarly fixed, not so much by law as by custom. It is in the very nature of things that it should be so, for it applies to prices just as much as it does to wages and salaries.

(2) It is precisely the same with regard to working hours; in certain businesses the office hours are from ten to four, in others from nine to five, in others again from eight to six; in the mechanical and artisan trades they are from six A.M. to five P.M. And this regularity is most useful and convenient to all concerned. When the men sometimes ask for a reduction of working hours, without a corresponding curtailment of their wages, it is said they are demanding 'ten hours' pay for nine hours' work;' but the masters very often try to enforce a reduction, which means 'ten hours' work for nine hours' pay.' Where is the difference? If the men are wrong, are not the masters equally so? This way of putting the question, however, is only an evasion of the real point at issue. From beginning to end it is purely a question of value; and here it is pertinent to ask, 'Who should fix the value, masters or men?' Labour, we are told, is a commodity, and as such is governed by the same laws as all other commodities. But in the sale of other commodities it is the seller, not the buyer, who fixes the price. If the owner of certain articles wishes to dispose of them, and an intending purchaser desires to buy but will not give the price asked, bargaining takes place; this is what Adam Smith terms the higgling of the market. But in all such cases the two parties occupy equal ground; there is no assumption of authority on either side, or if there is it is the seller who occupies the vantage ground, and not the buyer. If workmen seek to take the same position with regard to their labour, it is denounced as presumption on their part, and yet, be it remembered, they are the sellers of labour, which is their commodity. Still, the assumption on either side of the sole right of determining what shall be paid leads to dictation; this is the point to be avoided. And the day is dawning when masters and men will see that their disputes can best be settled by bargaining, that is by some method of conciliation, or if terms cannot be arranged by this means, then by resorting to an equitable system of arbitration.

The question as to overtime needs no lengthened remarks; systematic overtime is a curse to the men, and of doubtful benefit to the masters. Its evils were recognised by the Legislature when they prohibited the practice in factories and workshops, and they are not less baneful to men than they are to women and children; in the one case they are prohibited by law, in the other the men are a law unto themselves. Occasional overtime is quite another matter, and few men will object to work extra hours in cases of emergency. Nor do the laws of the union forbid this; the only thing they do is to fix a higher rate for the time so worked than that paid for the ordinary working hours; the intention of such rules is to restrict overtime to occasions of actual and real emergency.

With reference to piecework, there is a great deal of misapprehension in the public mind; it is thought that the unions, as a rule, are opposed to it; this is a mistake. Something like seventy-five per cent. of all the work of the country is done on the principle of piece-work, and that, too, under the rules and regulations of trade societies. In some trades, it is true, the men combine to resist its introduction; but, rightly or wrongly, their contention is that the practice is unsuited to the particular industry. This might be a matter of dispute, but in all instances each case must be taken on its own merits, and reasoned out by the full light of all the facts connected therewith.

What the unions do not do may be summarised thus: they do not interfere with individual work, curtail individual labour, pare all workmen down to a dead level of mediocrity, or demand that the same wages shall be paid to all alike, whether bungler or craftsman.

In all matters relating to labour trade-unions are protective of the workman's interests; but they have other and not less important functions to perform—namely, they provide the necessary funds for succouring their members in sickness, maintaining them when out of work, compensating them in cases of injury, keeping them in old age, and burying them and their wives when dead. In these respects they are friendly societies and mutual assurance societies combined; and indeed they are much more, for they have one purpose above and beyond any of those, such, for example, as the weekly allowance to members when out of work through depression of trade or other causes.

The whole of those benefits are secured by a payment of one shilling per week, and we venture to affirm that in no other kind of association can anything be found equal to it or like it. By way of example, let us take four of the largest and best organised societies to see how it works. In these the contributions, benefits, constitution, and rules, are in all essential features similar throughout:

Society	Payments	Sick Benefit	Funeral Benefit	Do. Wife	Accident Benefit	Superannuation	Out of Work	Emigration	Strike Pay
	s.	s.	£	£	£	s. d.	s.	£	s.
Engineers .	1	10	12	5	100	7 0	10	6	10
Iron-founders	1	9	10	5	150	5 6	9	5	11
Boiler-makers	1	11	12	6	100	6 0	8	0	12
Carpenters .	1	12	7	5	100	8 0	8	6	15

In some societies extras are given for children under a certain age, and benevolent grants to distressed members, caused by long illness, heavy domestic affliction, or being out of work for some time, and who in consequence thereof are unable to make both ends meet from the smallness of the weekly allowance guaranteed under the rules.

No adequate idea can be formed of the value of those institutions by a mere description on paper: it is indeed difficult to realise all the advantages which they offer to the struggling artisan and mechanic. Some notion, however, may be conveyed by a brief summary of the amounts disbursed under the several heads above given during the past year in those four societies alone. The one phase of unionism which is altogether unique, and to which therefore attention should be specially called, is the out-of-work fund. If judged merely by what it saves parochially in the shape of poor's rates, this feature is worthy of notice and of the highest commendation. The moral and social advantages of this one benefit cannot be over-estimated; it preserves the workman's self-respect, and saves his family from the degradation of pauperism. The following table gives the payments for the year 1877:

Society	Out of Work	Sick Benefit	Superannuation	Accident Benefit	Funeral Benefit
	£ s. d.	£ s. d.	£ s. d.	£ s. d.	£ s. d.
Engineers . .	54,470 17 11	23,257 16 7	13,858 15 8	2,200 0 0	7,659 2 3
Iron-founders	26,027 2 6	6,377 13 2	3,452 1 6	943 10 5	2,010 0 0
Boiler-makers	28,623 9 5	12,727 1 4	1,219 15 8	1,500 0 0	1,929 10 5
Carpenters .	17,455 4 4	8,409 13 8	280 1 10	1,560 0 0	1,702 4 3
Totals . .	126,576 14 2	50,772 4 9	18,810 14 8	6,203 10 5	13,300 16 11

The grand total paid for the before-mentioned benefits, by those four societies alone, in one year, exclusive of strike pay, was, 215,664*l.* 0*s.* 11*d.* During the last eleven years they have expended considerably over a million and a half pounds sterling, under the same heads; and, in addition thereto, they have given in benevolent grants to their distressed members nearly fifty thousand pounds. Whatever difference of opinion may exist with regard to the economical action of trade-unions, all will agree in commending so useful an expenditure as that above tabulated.

Some idea can be formed as to the extent, numerical strength, and wealth of the unions by the following table:

Name of Society	Branches	Members	Funds in hand
			£ s. d.
Amalgamated Society of Engineers .	396	45,071	275,270 0 2
Friendly Society of Iron-founders. .	112	12,612	59,525 2 5
Boiler-makers and Iron Ship-builders.	172	19,660	37,280 8 2
Amalgamated Carpenters and Joiners	330	16,829	74,248 7 8
Totals: four societies . . .	1,010	94,172	£446,323 18 5

The influence exercised by associations possessing such vast power and resources must be great indeed, and it is satisfactory to know that disputes between employers and employed are less frequent in those trades, and when they do occur are not, as a rule, so bitter, as they are in other branches of industry where the organisations are not so complete and wealthy as they are in those enumerated. If a sound and equitable system of arbitration could once be agreed upon, we believe that it would be possible to avoid those industrial conflicts which involve so much misery on the one side, and inflict such sad losses on the other.

It must be self-evident that the

most elaborate administrative machinery is required to conduct associations of such magnitude as the four selected as examples, with branches not only in most of the larger towns of the United Kingdom, but in our colonies, America, and various other countries. And yet the branches are, to a very great extent, self-governing communities, subject, of course, at all times to the executive council, chosen annually, or half-yearly, by the whole of the members, through the several lodges to which they belong. Each officer is elected by popular suffrage, by ballot or voting papers, and every member is entitled to vote not only on the question of officers, but also on all matters affecting the welfare of the general body, and the progress and well-being of the society. There are, indeed, but few, if any, institutions in this country in which the members have and exercise a larger *direct* share of control in the management thereof, than in those known as trade-unions. The wide-spread notion that they are governed by a set of lazy, self-elected demagogues, is simply a popular delusion; a casual examination of the carefully compiled reports issued monthly and yearly by the managing committees, will show that the officials are, and must be, the very reverse.[4]

In proportion as the unions increase in number, and extend the sphere of their operations, so do they gather experience, and effect improvements in their methods of conducting their business. The consequence of this is, that the older societies are able to avoid some of the errors into which the newer associations fall. This is strikingly manifest in the matter of trade disputes, which are fewer in the large, long-established, and consolidated unions, than they are in those of more recent growth. That mistakes are often committed, we admit and deplore; no human agency is entirely free from them; if once we acknowledge the good to be found in those organisations under discussion, we shall be all the better qualified to point out their faults, and to suggest the necessary remedies.

[4] In order that there shall be no possibility of misconception with regard to my advocacy of trade-unions as a means of self-help, I may perhaps be allowed to state that I am not, and never have been, a paid officer of any union. My only official connection with them was solely with reference to Parliamentary work, and this ceased more than two years ago. When, therefore, I speak in terms of respect of their officials, I do so as an outsider, but one who has had unusual opportunities of coming to a right conclusion.—G. H.

EXPLANATORY NOTES

George Howell, 'To the Trades' Unionists of the United Kingdom', *Operative Bricklayers' Society's Trade Circular*

p. 2, ll. 6–7: *lent soldiers to a private contractor at Chelsea*: In 1861, during a strike of London building workers, the War Department had sent in sappers of the Royal Engineers to replace striking workmen at Chelsea barracks. It convinced Howell and others that political representation was necessary: 'Whilst Capital is heard on all sides of the House, Labour has no voice there; how then can we expect impartial laws from one-sided men?' (*Operative Bricklayers' Trade Circular*, 4 (December 1861)).

p. 2, l. 42: The Reform League: The Reform League was formed in February 1865 to campaign for manhood suffrage and the ballot. It used some trade-union resources to support Liberal candidates who were sympathetic to trade unionism.

p. 2, l. 46: *GEORGE HOWELL*: George Howell (1833–1910). Howell went on to re-organize the London Bricklayers' Society and became secretary of the London Trades Council. From 1865 until 1869 he was secretary of the Reform League; from 1871 to 1875 he was general secretary of the Trades Union Congress; from 1885 to 1895 he was MP for N.E. Bethnal Green. See F. M. Leventhal, *Respectable Radical: George Howell and Victorian Working Class Politics* (London: Weidenfeld and Nicolson, 1971).

Mr. Potter and the London Trades Council

p. 3, l. 1: *MR. POTTER*: George Potter (1832–93), secretary of the Progressive Society of Carpenters and Joiners; secretary of the Building Trades' Conference which co-ordinated the campaign for a nine-hours' day. In October 1861 he launched the *Bee-Hive* newspaper.

p. 3, col. 1, ll. 13–14: *the Masters' and Workmen's Act*: A campaign against the Master and Servant Acts had been launched by the Glasgow Trades Council in 1860 and a Master and Workmen's Act Reform Committee was set up. The London Trades Council called a four-day conference from 30 May until 2 June 1864 to co-ordinate the campaign. For master and servant legislation see Douglas Hay and Paul Craven (eds), *Masters, Servants and Magistrates in Britain and the Empire, 1562–1955* (Chapel Hill: University of North Carolina Press, 2004).

p. 3, col. 1, l. 21: *calling trades' delegate meetings*: Potter had called a meeting in March 1864 to protest against the proposed Annuities Bill that Gladstone had introduced. He regarded it as a threat to friendly societies (*Beehive* (12 March 1864)). Gladstone reacted with an attack on trade unions to which the London Trades Council responded with a deputation to him, 'To disabuse the mind of the Chancellor of the Exchequer as to there being any organized opposition to the principles of the bill by the trade societies, as societies'. From then on relations between Potter and the leaders of the Trades Council were embittered. The relationship is discussed in S. Coltham, 'George Potter, the Junta and the Bee-hive' in *International Review of Social History*, 9:3 (1964) and 10:1 (1965).

p. 3, col. 1, l. 27: *the* Beehive *newspaper*: Launched by Potter in October 1861, this was the main vehicle for reports on trade-union activities and it gave Potter great influence.

p. 3, col. 1, l. 28: *a meeting of trades' delegates*: Up until this time Potter had been acting within the London Trades Council, but in calling a delegate meeting in support of the North Staffordshire Ironpuddlers he was clearly challenging the Council's position.

p. 3, col.1, l. 46: *E. COULSON*: Edwin Coulson (d. 1893) was secretary of the London Order of the Operative Bricklayers' Society from 1860 until 1890. He was active in the London Trades Council in the 1860s.

p. 3, col. 2, l. 16: *Mr. Allan*: William Allan (1813–74) was an Ulster-born engineer who worked in Liverpool and Crewe. He was secretary of the Journeymen Steam Enginemakers' Society before becoming general secretary of the Amalgamated Society of Engineers in 1851, a position that he held until his death. He was a highly efficient administrator.

p. 3, col. 2, l. 43: *Mr. Cope*: Cope was a boot-closer by trade.

p. 4, col. 2, l. 65: *G. Odger, Secretary*: George Odgers (sometimes Odger) (1813–77), ladies' shoemaker, secretary of the London Trades Council, from 1862 to 1872; member of the General Council of the International from 1864 to1871.

p. 5, col. 1, l. 8: *Ld. Lichfield's proposal*: The second Earl of Lichfield, a substantial coalowner, was one of a small group who were advocating arbitration, by which was largely meant collective bargaining between employers and unions, to settle disputes.

p. 5, col.1, l. 34: *G. Newton*: George Newton (1831–67), a potter, was secretary of the Glasgow Trades Council and secretary of the Master and Workmen's Act Reform Committee from 1864 until 1867.

p. 5, col. 2, l. 18: *Mr. Dronfield*: William Dronfield (1826–1914), a printer, was secretary of the Sheffield branch of the Typographical Association from 1855 until 1890, and secretary of the Sheffield Association of Organized Trades from 1859 until 1871.

p. 6, col. 2, l. 22: *Mr. Cremer*: William Randal Cremer (1838–1908) member of the Amalgamated Society of Carpenters and Joiners, member of the General Council of the International Working Men's Association from 1864 to 1866 and a Liberal MP from 1885 to 1908. A Nobel Peace Prize winner in 1902, he was knighted in 1906.

p. 8, col. 1, l. 1: *Troup, the Printer and Editor*: George Troup (1810–79) was a professional journalist who edited the *Beehive* from its launch in 1861 until he was forced out in 1863 because of his pro-South sympathies in the American Civil War.

p. 8, col. 2, l. 3: *judgement went by default*: The charges against Potter were investigated by a group of middle-class allies of trade unions and most were found 'not proven'. He was, however, condemned for his conduct over the North Staffordshire dispute and found guilty of packing delegate meetings. See *Beehive*, 24 June 1865.

p. 9, col. 1, l. 50: *Mr. Guile*: Daniel Guile (1814–83) was general secretary of the Friendly Society of Iron Founders from 1863 until 1881. Together with William Allan, Robert Applegarth, Edwin Coulson and George Odger he was part of the group that Sidney and Beatrice Webb described as the 'Junta'. Guile, however, often showed some sympathy with Potter's position.

Trades' Societies and Lock-Outs. Report of the Conference of Trades' Delegates of the United Kingdom, Held in the Temperance Hall, Townhead Street, Sheffield, on July 17th, 1866, and Four Following Days

p. 13, l. 26: *Mr. M. Lawrence*: Matthew Lawrence had been one of the founders of the Glasgow Trades Council before moving to London to become president of the Operative Tailors' Protective Association. He was a member of the International Working Men's Association.

p. 18, l. 2: *the Iron trade, in Staffordshire*: There was a long-drawn out dispute of Staffordshire iron puddlers in 1865 during which they refused arbitration. Reactions to the dispute divided George Potter and the Junta.

p. 18, l. 3: *the Lock-out on the Clyde:* A strike of 3,000 shipwrights on the Clyde in May 1866 for an advance in wages and a reduction in hours from sixty to fifty-seven led to a general lock-out of 20,000 workers.

p. 18, l. 4: *the Staffordshire Potteries*: The campaign among the potters was mainly for an end to the annual hiring system.

p. 23, ll. 9–10: *the great Preston Lock-out*: In 1853 there was a thirty-week strike and lock-out of weavers after a demand for a wage increase. See H. I Dutton and J. E. King, *Ten Per Cent and No Surrender: The Preston Strike, 1853–54* (Cambridge: Cambridge University Press, 1981).

p. 23, ll. 10–11: *the Iron Trade of the north*: There were a series of disputes at this time in the shipyards of Tees-side and Tyneside for a reduction in working hours to a nine-hour day.

p. 30, ll. 39–40: *the London Working Men's Association*: George Potter had formed the London Working Men's Association in February 1866 in rivalry to the Reform League and to the London Trades Council. It ostensible aim was to obtain the insertion of a clause in the Franchise Reform bill giving the vote to lodgers.

p. 40, l. 8: *Lord St. Leonards*: Edward Sugden, first Baron St Leonards, a former Conservative Lord Chancellor, was a leading advocate of arbitration in industrial disputes.

p. 42, l. 3: *the Masters' and Workmen's Executive*: Encouraged by Alexander Campbell of the *Glasgow Sentinel*, the Glasgow Trades Council launched a campaign for reform of the Master and Servant legislation. A Master and Workman's Act Reform Committee, with Newton as secretary, was set up. It persuaded the London Trades Council to call a national conference on the issue in June 1864.

p. 43, l. 14: *Lord Elcho*: Richard Cobbett had introduced a private member's bill for reform of the Master and Servant Act, but when he failed to get re-elected in the election of 1865, the Master and Workman's Act Reform Committee turned to Lord Elcho. Elcho had been a useful ally to the miners' leader Alexander McDonald in the campaign for the Coal Mines' Regulation Act of 1860. However, at this time, he was a noted opponent of the proposals for an extension of the franchise.

p. 51, ll. 6–7: *a national association formed some years ago*: The National Association of United Trades for the Protection of Labour was formed in 1854 to being about a 'general federation or union of trades'. Its success was limited and it largely fades from view in the 1850s.

p. 51, ll. 13–14: *the late Mr. Thomas Duncombe*: Thomas Slingsby Duncombe (1796–1861), radical MP for Finsbury from 1834 until 1861, was a supporter of moderate Chartism and was one of the key figures behind the National Association of United Trades for the Protection of Labour.

p. 51, l. 15: *the late Mr. Wakley*: Thomas Wakley (1795–1862) was editor of the *Lancet* and Duncombe's fellow MP for Finsbury from 1835 until 1852.

p. 53, l. 24: *the Protection Society of Liverpool*: A Liverpool Trades Guardians' Association, formed in 1848, has some claim to be regarded as the oldest trades council. It survived until 1861 when it merged with the Trades' Protective Association. Charles O. Williams was secretary of the Liverpool United Trades' Protective Association from 1861 until 1867 and general secretary of the National Association of Operative Plasterers from 1862 until 1885.

p. 64, l. 51: *meeting held in St. Martin's Hall*: The London Working Men's Association was planning a national conference that eventually met in March 1867.

p. 65, l. 17: *140 Mrs. Partingtons*: According to Sydney Smith, during a flood in 1824, Mrs Partington had tried to stop the sea coming into her cottage with a mop and bucket.

'The Tailors' Strike and the System of Picketing', *Operative Bricklayers' Society Monthly Report*

p. 87, l. 9: *the offence of conspiracy*: Under the Molestation of Workmen Act of 1859 workmen were allowed 'peaceably' and 'in a reasonable manner' to persuade others to strike work. In July 1867 there was a great strike among Edinburgh and London Tailors, during which employers tried to break the strike by recruiting workers from Germany. Tailors' workshops were picketed. The officers of the London Operative Tailors' Society were charged with conspiracy and Lord Bramwell, in the case of Regina *v.* Druitt, held that even watching non-strikers and giving them 'black looks' was intimidation, as was the very presence of large numbers.

p. 87, l. 25: *The men of Sheffield*: In October 1866 a can of gunpowder was put down the chimney of a non-unionist in Sheffield. It again brought into the public domain a history of violent action against non-unionists in Sheffield. See Volume 4, pp. 307–10. See also K. D. Brown, 'Nonconformity and Trade Unionism: The Sheffield outrages of 1866', in E. F. Biagini and A. J. Reid (eds), *Currents of Radicalism. Popular Radicalism, Organised Labour and Party Politics in Britain, 1850–1914* (Cambridge: Cambridge University Press, 1991), pp. 86–105.

Edward Spencer Beesly, *The Amalgamated Society of Carpenters and Joiners*

p. 89, l. 6: *EDWARD SPENCER BEESLY*: E. S. Beesly (1831–1915), academic and editor of the *Positivist Review*. He had supported the campaign of the London building workers in 1859–60 for a shorter working day. For an examination of the relations between the Positivists and the London trade unionists see Royden Harrison, *Before the Socialists: Studies in Labour and Politics, 1861–1881* (London: Routledge & Kegan Paul, 1965).

p. 91, ll. 16–17: *Mr. Harrison's*: Frederic Harrison (1831–1923), barrister and president of the English Positivist Committee from 1880 to 1905.

p. 92, l. 21: *Mr. Applegarth*: Robert Applegarth (1834–1924), a joiner, born in Hull, became general secretary of the Amalgamated Society of Carpenters and Joiners from 1861 until

1871. See A. W. Humphrey, *Robert Applegarth: Trade Unionists, Educationist, Reformer* (Manchester: National Labour Press, 1913).

p. 92, ll. 33–4: *the masons alone of the building trades*: The Operative Stonemasons' Union dated from 1833.

p. 96, ll. 28–9: *the drunken venal being known to Mr. Lowe*: The Liberal Robert Lowe had opposed the extension of the franchise during the debates on the Reform Bill in 1866 by saying that the evidence was that the poorest were the most corruptible. His words roused many trade unionists to political action.

p. 105, l. 2: *The "General Union"*: Formed in 1827, the General Union of Carpenters and Joiners, under the secretaryship of Robert Last from 1862, remained a rival to the Amalgamated Society.

p. 106, l. 19: *marched with Mr. Beales*: Edmond Beales (1803–81) was a radical lawyer who began campaigning for the extension of the franchise in 1864 and who convened the meeting in February 1865 that formed the Reform League.

Report of the Various Proceedings Taken by the London Trades' Council and the Conference of Amalgamated Trades, in Reference to the Royal Commission on Trades' Unions, and Other Subjects in Connection Therewith

p. 115, l. 6: *Mr. Broadhead*: William Broadhead (1815–79) was secretary of the Saw Grinders' Union from 1848 until 1869. He was also a publican. His role in organizing assaults on non-unionists eventually emerged after he and his associates were granted immunity from prosecution.

p. 116, ll. 13–14: *Council of the 'Organized Trades' of Sheffield*: The Association of Organized Trades of Sheffield and Neighbourhood was formed in 1859 after a dispute in the office of the *Sheffield Times*. William Broadhead was its first treasurer.

p. 119, l. 1: *Mr. Thomas Hughes, M.P.*: Thomas Hughes (1822–96), a barrister and author of *Tom Brown's Schooldays*, was a Christian Socialist. He was secretary of the Social Science Association investigation of trade unions in 1860, MP for Lambeth from 1865 until 1874 and a member of the Royal Commission on Trade Unions.

p. 119, l. 44: *MR. ROEBUCK*: John Arthur Roebuck (1802–79) had been MP for Sheffield since 1849. He had been a utilitarian radical in the 1830s and 1840s but by the 1860s had lost much of his radicalism and had been very critical of Sheffield trade unionism.

p. 121, l. 9: *Mr. Hadfield*: George Hadfield (1787–1879), Liberal radical MP for Sheffield from 1852 until 1874.

p. 132, l. 35: *the Friendly Societies' Act*: This act of 1855 gave protection to the funds of registered societies and a number of trade unions, including the Boilermakers' Society, had registered.

p. 135, l. 41: *Mr. Tidd Pratt*: John Tidd Pratt (1797–1870) was registrar of friendly societies from 1855 to 1870.

p. 138, l. 26: *Mr. Neate*: Charles Neate (1806–79) was Drummond Professor of Political Economy at Oxford from 1857 until 1862. He was elected MP for Oxford in 1863. He published *Two Lectures on Trade Unions* in 1862, a text that was sympathetic to unionism.

p. 139, l. 28: *Mr. J. Stuart Mill, M.P.* John Stuart Mill (1806–73) was MP for Westminster from 1865 until 1868. The author of the influential *Principles of Political Economy*, first published in 1848, was, by the 1860s, challenging the wage-fund theory that denied that it was possible for trade combinations to raise wages and accepting that unions had a role to play. See W. Hamish Fraser, *Trades Unions and Society: The Struggle for Acceptance* (London: George Allen & Unwin, Ltd, 1974), Chapter 7.

p. 144, l. 19: *that in 1864*: Soon after this meeting, when a deputation of trade-union leaders had pressed him as Chancellor of the Exchequer to allow unions to deposit their funds in the new Post Office savings bank, Gladstone claimed to have been so impressed that he came out in favour of an extension of the franchise to working men.

p. 147, l. 36: *Colonel Dickson*: Colonel Dickson was an associate of Edmund Beales in the Reform League.

p. 147, l. 37; *Mr. Ludlow*: John Malcolm Ludlow (1821–1911), a barrister, who was one of a group of Christian Socialists who took up the cause of trade unionism and co-operation. He became Chief Registrar of Friendly Societies 1875–91. See Fraser, *Trade Unions and Society*, pp. 76–83.

p. 150, l. 11: *Mr. Dunning*: see Volume 4, pp. 185–236.

p. 151, ll. 38–9: *by wealthy men in Jamaica*: An uprising in October 1865 over conditions in the Jamaican sugar plantations was suppressed with great brutality by Governor Eyre. Hundreds were summarily executed and flogged. He was dismissed as governor but had support from the plantation owners and many defenders in Britain, including Thomas Carlyle.

Report of St. Martin's Hall United Kingdom Trades' Conference Committee on the Trades' Union Inquiry Commission till 7th August, 1867

p. 155, l. 4: *Trades' Conference Committee*: This had emerged from the conference called by George Potter in March 1867. The conference stemmed from resentment at the Conference of Amalgamated Trades that had made clear that its concern was getting protection for the funds of the large national societies. It attracted support from provincial societies and from the smaller London unions.

p. 155, l. 21: *Mr. CONNOLLY presiding*: Thomas Connolly was an associate of George Potter.

p. 155, ll. 29–30: *the Glasgow Trades' Council*: The Glasgow Trades Council had been formed in 1858.

p. 157, ll. 23–4: *Mr. KANE of the Iron-Workers*: John Kane (1819–76) was general secretary of the National Association of Ironworkers from 1868 until 1876. Although the leader of a national union and occasionally associated with the Conference of Amalgamated Trades, he is more usually found in sympathy with Potter.

p. 157, ll. 24–5: *Mr. PROUDFOOT of Glasgow*: John C. Proudfoot, a joiner, was secretary of the Glasgow Trades Council from 1867 until 1871.

International Working Men's Association, *To the Trades' Unionists' of Great Britain and Ireland*

p. 179, ll. 8–9: *the International Working-men's Association*: The International Working Men's Association was formed in September 1864. See Volume 4, pp. 403–4.

p. 182, ll. 5–6: *the London Conference of 1865*: Convened on the insistence of Karl Marx, this conference was attended by nine delegates from France, Switzerland, Belgium and Britain.

p. 182, l. 8: *the Congress of 1867*: held in Lausanne.

p. 182, l. 31: *H. IUNG:* Hermann Jung (1830–1901), a Swiss watchmaker living in London.

p. 182, l. 32: *R. SHAW*: Robert Shaw (d. 1869), a painter by trade.
p. 182, l. 33: *J. GEORGE ECCARIUS*: Johann Georg Eccarius (1818–89), a German tailor, living in London.

J. M. Ludlow, 'Sir W. Erle on Trade Unions', *Operative Bricklayers' Monthly Report*

p. 183, l. 1: *Sir W. ERLE*: Sir William Erle (1793–1880), judge, responsible for the controversial judgement on the Wolverhampton tinplate workers in 1851, when leaders of the National Association of United Trades were sentenced for conspiracy to molest and intimidate an employer. He was Lord Chief Justice of Common Pleas from 1859 to 1866. After his retirement from the bench he chaired the Royal Commission on Trade Unions and Employers' Associations from 1867 until 1869.
p. 184, l. 2: *"Hilton* v. *Eckersley"*: a case in 1855 before the chief justice, Lord Campbell.
p. 185, ll. 20–1: *restraint of trade*: This common-law concept had been used in the case of Hornby *v.* Close in 1866 when the Bradford justices threw out a case in which a branch of the Boilermakers' Society had sought protection of their funds under the Friendly Societies' Act. The justices argued that provisions in the rule-book, regulating piece-work, supporting strikes and fining members, were illegal because they were in restraint of trade. The Court of Queen's Bench upheld that ruling in January 1867.

Trades' Unions and the Cost of Labour. Speech Delivered by Thomas Brassey, Jun M.P. in the House of Commons 7th July 1869

p. 187, l. 7: *THOMAS BRASSEY, JUN., M.P.*: Thomas Brassey (1836–1918) was the son of the great civil engineering contractor of the same name. He was Liberal MP for Hastings from 1868 until 1886.
p. 195, l. 13: *Mr. Connolly of the Masons' Society*: Thomas Connolly of the Operative Stonemasons' Society was associated with George Potter.
p. 198, l. 12: *adopted by Messrs Briggs*: Henry Briggs and Son Ltd were coalowners at the Whitewood Colliery near Wakefield. They had a profit-sharing scheme. As with most such schemes, it was dependent upon the men not joining a trade union.
p. 203, l. 3: *Sir Sydney Waterlow* was a printer and philanthropist. He was an alderman of the City of London and later Lord Mayor.
p. 211, l. 8: *Mr Lothian Bell*: Sir Lowthian Bell (1816–1914), a leading iron, steel, coal and chemical manufacturer in North-East England.
p. 215, l. 17: *Messrs* Hornby: the leading Blackburn cotton spinners.
p. 218, ll. 30–1: *the Canada Works at Birkenhead*: Thomas Brassey, sen., built the works at Birkenhead to supply all the material for the Canadian Grand Trunk Railway of which he was the main contractor.
p. 226, l. 4: *'Handicraftsmen and Capitalists'*: H. Herries Creed and W. V. V. Williams, *Handicraftsmen and Capitalists: Their Organisation at Home and Abroad* (Birmingham: Henry Wright, 1867).
p. 227, ll. 8–9: *Admiral King Hall:* supervisor of the Devonport Dockyard.
p. 248, l. 7: *the practice of ticketing*: Presumably the reference is to picketing. The term seems to have come into widespread use in an industrial context in the 1860s.

Robert Jobson, *Trades Unions: An Inquiry into Their Rules and Working, Based on the Evidence before the Royal Commission; Showing the Folly of All Attempts to Raise Wages by Violence and Strikes, and the Beauty and Excellence of the Divine Law Governing Workmen and Employers*

p. 254, l. 28: *the discovery of gold in California and Australia*: This was in 1849 and 1851 and it contributed to the economic growth of the early 1850s.

p. 257, l. 46: *Mr. R. Harnott*: Richard Harnott (1807–72) was general secretary of the Operative Stonemasons' Society from 1847 until 1872. See R. W. Postgate, *The Builders' History* (London: The Labour Publishing Company Ltd, n.d), chapter 7.

p. 262, l. 42: *Mr. Trollope*: G. F. Trollope of the London building firm George Trollope and Sons.

p. 267, l. 29: *Mr. Ernest Jones*: Ernest Jones (1819–69) had kept alive the aims of the Chartist Movement in the 1850s and 1860s.

p. 280, l. 31: *Mr. Rupert Kettle*: Rupert Kettle (1817–94), barrister and county-court judge, along with A. J. Mundella, was one of the main campaigners for a system of conciliation and arbitration in industrial relations.

W. P. Roberts, *Trade Union Bill, 1871. A Letter by Mr. W. P. Roberts, (of London and Manchester, Solicitor) to Mr. George Potter, Mr. William Allan, Mr. Alexander M'Donald and Others Interested in Trades' Societies*

p. 283, l. 5: *Mr W. P. ROBERTS*: William Prowting Roberts (1806–71), an active Chartist, he was legal adviser to the Northumberland and Durham Miners' Union , earning the sobriquet of 'the miners' attorney general'. See R. Challinor, *A Radical Lawyer in Victorian England: W. P. Roberts and the Struggle for Workers' Rights* (London: Tauris, 1990).

p. 286, l. 24: Lord Elcho's Act: Elcho, a leading opponent of franchise extension in 1866–67, was sympathetic to the plight of many miners and had carried an amendment to the earlier Master and Servant legislation with the Master and Servant Act of 1867. It still made breach of contract by a worker a criminal offence.

Report of the Conference of Trades Delegates, Held in the Town Hall, Leeds on December 2nd 1871. Called to Consider the Statement made by Wm. Newmarch, Esq., in His Address to the Social Science Congress Lately Held in Leeds

p. 301, l. 1: *LEEDS AND DISTRICT TRADES COUNCIL*: This had been formed in 1862.

p. 301, l. 33: *the Social Science Meetings:* Founded in 1857, the National Association for the Promotion of Social Sciences held annual meetings where a range of topics was discussed. In 1860 it published *Trades' Societies and Strikes: Report of the Committee on Trades' Societies.*

p. 301, l. 34: *William Newmarch, Esq.*: William Newmarch (1820–82), a statistician and banker, with a taste for controversy. He was a leading member of the Statistical Society.

p. 302, l. 20: *Sir John S. Pakington*: John S. Pakington (1799–1880), Conservative MP for Droitwich from 1837 until 1874. He had a particular interest in educational reform. In 1871 he was attracted to the 'new social movement' that sought to link trade unions with Conservative aristocrats (see Fraser, *Trade Unions and Society*, pp. 161–4) As chair of the Leeds meeting he was accused of 'giving encouragement to wild theories' of a socialist tendency.

W.H. Wood, *The Advantages of Trades Unions*

p. 313, l. 6: *W. H. Wood*: William Henry Wood, a printer and secretary of Manchester Typographical Society. He was secretary of the Manchester and Salford Trades Council from 1866 to 1870. With his associate S. C. Nicholson he was responsible for calling the founding meeting of the Trades Union Congress in 1868.

p. 313, l. 8: *MANCHESTER & SALFORD TRADES COUNCIL*: This had been formed by Nicholson and Wood in 1866.

p. 314, l. 10: *a "National Federation of Employers"*: The National Federation of Associated Employers of Labour was formed in 1873 largely as a response to the trade-union campaign against the Criminal Law Amendment Act.

p. 316, l. 2: *Professor LEONE LEVI*: Levi's *History of British Commerce and of the Economic Progress of the British Nation, 1763–1870* had been published in 1872.

p. 316, l. 43: *Many failures in the Shipbuilding trade*: There had been a rapid collapse of the mainly wooden shipbuilding trade on the Thames after 1867 as production moved to Clydeside and Tyneside.

p. 325, l. 18: *Sir A. ALISON*: Archibald Alison, the sheriff of Lanarkshire, had been the scourge of trade unions in the 1830s and 1840s and responsible for the arrest of the Glasgow cotton spinners' leaders in 1837.

p. 325, l. 25: *Mr J. RUSKIN*. John Ruskin had been publishing monthly *Fors Clavigera: Letters to the Workmen and Labourers of Great Britain* since 1871, calling for a more moral political economy.

p. 325, l. 43: *Mr W. R. CALLENDER*: William Romaine Callender (1825–76) was a large cotton spinner and merchant in Manchester, but unlike most of his fellow manufactures, was a Conservative in politics and a supporter of Tom Hughes' Christian Socialist projects.

Proceedings of the Miners' National Association Conference, Held in Leeds, on 18, 19, 20, 21 and 22 November 1873

p. 328, l. 12: *Mr M'Donald*: Alexander McDonald (1821–81), founder of the Coal and Iron Miners' Association of Scotland. He remained president of the National Association until 1881 and was MP for Stafford from 1874 to1881. See Gordon Wilson, *Alexander McDonald Leader of the Miners* (Aberdeen: Aberdeen University Press, 1982).

p. 328, ll. 25–6: *the Judes, the Grays, the Wallows and the Tetlows*: These were among the founders of the Miners' Association of Great Britain and Ireland formed at Wakefield in 1841. Martin Jude was the leading figure.

p. 329, l. 6: *HARTLEY CATASTROPHE*: In 1862, 199 miners were trapped in the Hester Pit of the Northumberland village of New Hartley. None survived.

p. 329, l. 12: *Mr Towers*: John Towers, editor of the *British Miner*, and someone who became increasingly critical of McDonald's tactics.

p. 329, l. 38: *Messrs Crawford, Sheldon and Normansell*: William Crawford (1833–90) was general secretary of the Durham Miners' Association and later also secretary of the Miners' National Association from 1877 until 1890. He was elected MP for Mid-Durham in 1885. John Normansell (1830–75) was secretary of the South Yorkshire Miners' Association.

p. 330, l. 28: *the* Advocate: Towers' paper, the *Miner* became, in 1863, the *Miner and Workman's Advocate* and then, from 1865, the *Workman's Advocate.*

p. 331, ll. 2–3: *Messrs. Pickard and Halliday*: William Pickard (1821–87) was miners' agent in Wigan and vice-president of the National Association of Miners in 1863. He helped form the Amalgamated Association of Miners; Thomas Halliday (b. 1835) was secretary of the Lancashire Miners' Association and secretary of the Amalgamated Association of Miners 1869–75 that was a rival to McDonald's Miners' National Association. The two Associations merged in 1875 to form the Miners' National Union.

p. 332, l. 11: *Mr. Walpole and others*: Spencer Walpole was the Conservative Home Secretary in 1867–8.

p. 332, l. 34: *a Royal Commission was appointed*: the Royal Commission into the Truck System, 1870–1.

p. 333, l. 12: *true and just weighing*: The Coal Mines' Regulation Act of 1860, among the other things listed, allowed miners to elect representatives to check the weight of coal on which their payments were based.

p. 335, l. 14: *under Mr Lowe*: Robert Lowe, Chancellor of the Exchequer from 1868 until 1873, became Home Secretary in 1873.

p. 335, l. 19: *THE CRIMINAL LAW AMENDMENT ACT*: This act which accompanied the Trade Union Act of 1871 still made it an offence to use 'threats' or 'molestation and obstruction' to coerce an employer or a workman. There was a concerted trade-union campaign to get the Act repealed.

p. 335, l. 29: *Mr Bruce*: H. A. Bruce was Liberal Home Secretary from 1868 until 1873.

p. 335, l. 36: *the discharge note*: Requiring all workers to get a note from a previous employer was regularly used as a means of identifying trade-union activists.

p. 336, l. 36: *the case of the gas stokers of London*: Mr Justice Brett, in December 1872, delivering judgement against five leaders of the London gas stokers who had been on strike, applied the common law of conspiracy rather than the Criminal Law Amendment Act and defined 'molestation' very broadly as something 'which would be likely to have a deterring effect upon masters of ordinary nerve'. The gas stokers received twelve months with hard labour.

The Criminal Law Amendment Act. A Memorial to the Right Honourable Henry Austin Bruce, MP., with an Appendix of Cases under the Criminal Law Amendment Act

p. 339, l. 7: *this year*: 1873.

p. 357, col. 1, l. 63: *Mr. Mundella, M.P.*: Anthony John Mundella (1825–97), Nottingham stocking manufacture; Liberal MP for Sheffield from 1868 to 1895; sympathetic to trade unions and an advocate of conciliation and arbitration.

Henry Crompton, *Tracts for Trade Unionists No IV. The National Federation of Associated Employers of Labour. Report upon*

the Memorial Presented to the Home Secretary by the National Federation of Associated Employers of Labour, December 13th, 1873

p. 359, l. 14: *HENRY CROMPTON, ESQ.*: Henry Crompton (1836–1923), Clerk of Assize of Chester and North Wales circuit, was a member of the Positivist Society and an associate of Frederic Harrison and E. S. Beesly. He was very active in campaigning against the Criminal Law Amendment Act and the Master and Servant Act.

p. 363, ll. 1–3: *THE NATIONAL FEDERATION OF ASSOCIATED EMPLOYERS OF LABOUR*: This was formed in April 1873. The secretary was Henry Whitworth of Liverpool who had previously been secretary of the National Association of Factory Occupiers. The president was John Robinson from the Manchester engineering firm of Sharp, Stewart and Co. It contained many of the leading names in British industry, such as E. J. Harland of Harland and Wolff, Belfast, John Laird of Laird Brothers, Birkenhead, the textile manufacturers, L. J. Crossley of Halifax, Stephen Marshall of Leeds, the Ashworths of Bolton, engineering employers, like David Greig of Fowler and Son, Leeds, Thomas Vickers of Sheffield, and builders such as Trollope. According to its publication *Capital and Labour*, it was to act as 'a rampart to keep back an advancing army of unionists'. It survived until 1881 with a fairly relentless attack on trade unionism. See Fraser, *Trade Unions and Society*, pp. 116–19.

p. 363, l. 5: *the Home Secretary*: Robert Lowe was now the Home Secretary.

p. 364, l. 1: *Turk at Hammersmith*: Turk had been charged under the Criminal Law Amendment Act for picketing. The Nine Hours' League appealed his case to the Court of Queen's Bench.

p. 370, l. 5: *a case decided under the Master and Servants Act*: This referred to two leather japanners in Greenwich who were charged with neglect and ordered to pay £4.10*s.* damages to their employer. They claimed that they were unable to pay the damages.

p. 375, l. 19: *candidates for Parliament*: In a number of places trades councils advised their members not to vote for candidates who would not pledge themselves to repeal of the Criminal Law Amendment Act, the Conspiracy Law and the Master and Servant Act.

p. 376, l. 39: *Sir T. Bazley, Mr. Edmund Potter*: Thomas Bazley (1797–1885) was a cotton spinner and Liberal MP for Manchester from 1858 to 1880. Edmund Potter (1802–83) was a calico printer and Liberal MP for Carlisle from 1861 to 1874. These were the Federation's main spokesmen in Parliament.

p. 376, l. 40: *Mr. Robinson, and Mr. Jackson*: For John Robinson, the president, see above, note to p. 363, ll. 1–3; Robert Raysford Jackson, chairman of the Executive Council, was from a Blackburn cotton firm.

p. 380, l. 25: *tyrannical Trades Union action*: Masons building the new Smithfield market went on strike to get a non-unionist dismissed.

Frederick Clifford, *The Agricultural Lock-Out of 1874 with Notes upon Farming and Farm-Labour in the Eastern Counties*

p. 383, l. 19: *Mr Arch*: Joseph Arch (1826–1919), agricultural labourer from Barford in Warwickshire, founder and chairman of the National Agricultural and Labourers' Union, Liberal MP for West Norfolk from 1885 until 1886, and from 1892 until 1900.

p. 396, l. 10: *Bishop of Manchester*: James Fraser, appointed Bishop of Manchester in 1870, was an active social reformer.

p. 403, l. 18: *'Labourers' Union Chronicle'*: the journal of the National Union of Agricultural Labourers.

p. 404, l. 8: *the Federal Union*: There were a number of other agricultural labourers' unions that appeared at this time and refused to be absorbed in the National Union. In March 1873 the London Trades Council called a conference to try to achieve a measure of unity. Eventually half a dozen of the smaller societies formed a Federal Union.

p. 404, l. 23: *Mr Shipton*: George Shipton (1839–1911), general secretary of the London Amalgamated Housepainters' Society from 1866 until 1889 and secretary of the London Trades Council from 1872 until 1896, acted as secretary of the Federal Union for a time.

National Association of Miners, *Extraordinary Meeting of the Council Held in the Mechanics' Institution, David St Manchester, 24 February 1875*

p. 406, l. 21: *four or five families:* A Crawshay dynasty ran huge coal and iron works at Cyfartha, Merthyr Tydvil; the Fothergills had ironworks in Aberdare; the Vivians had smelters in Ynyscedwin and the Davies Coal Co. controlled most of the upper Rhonda Valley.

p. 408, l. 36: *a state of prosperity*: there was a spectacular economic boom in 1871 thanks largely to American demand.

p. 412, ll. 28–9: *Mines Regulation Act*: This Act of 1872 prohibited the employment in the mines of all girls, women and boys under the age of twelve years, introduced powers to appoint inspectors of mines, and set out rules regarding ventilation, blasting and machinery.

Address from the Executive Council of the Amalgamated Society of Engineers to their Fellow Workmen Throughout the United Kingdom, the British Colonies, and Foreign Countries

p. 424, ll. 8–9: *Sir J. P. Kay-Shuttleworth*: formerly Dr. James Phillips Kay (1804–77), campaigner for improved public health and the extension of educational provision.

p. 428, l. 41: *JOHN BURNETT*: John Burnett (1842–1914), engineer, leader of the Newcastle nine-hour movement 1871. He was general secretary of the ASE from 1874 until 1886, after which he became the first labour correspondent of the Board of Trade.

George Howell, 'Trade-Unions: Their Nature, Character, and Work', *Fraser's Magazine*

p. 429, col. 1, ll. 32–3: *successors of the old English Guilds*: Lujo Brentano's *The History and Development of Gilds and the Origin of Trade Unions* had been published in 1870 and Howell leaned heavily on this in the introduction to his book *Conflicts of Capital and Labour* which had been published in 1877. He seems to have believed that Brentano was making the link between gilds and trade unions, when in fact Brentano had made the point that their roots were in fraternities of journeymen. See Sidney and Beatrice Webb, *The History of Trade Unionism 1666–1920* (London: Longmans, Green and Co., 1920), pp. 12–13.

PERMISSIONS

George Howell, 'To the Trades' Unionists of the United Kingdom', *Operative Bricklayers' Society's Trade Circular*, 51 (1 August 1865), pp. 399–400. By permission of the Modern Records Centre, University of Warwick, shelfmark MSS78/OB/4/1/1.

Mr Potter and the London Trades Council (London: n.p., [1865]). By permission of the Bishopsgate Institute, George Howell Collection, shelfmark 331.

Trades' Societies and Lock-Outs. Report of the Conference of Trades' Delegates of the United Kingdom, Held in the Temperance Hall, Townhead Street, Sheffield, on July 17th, 1866, and Four Following Days (Sheffield, n.p., 1866). By permission of Senate House Library, University of London, shelfmark [G. L.] Burns 4045.

'The Tailors' Strike and the System of Picketing', *Operative Bricklayers' Society Monthly Report*, 76 (1 September 1867), pp. 711–12. By permission of the Modern Records Centre, University of Warwick, shelfmark MSS 78/OB/4/1/3.

Edward Spencer Beesly, *The Amalgamated Society of Carpenters and Joiners* (London: n.p., 1867). By permission of Senate House Library, University of London, shelfmark [G. L.] B.876.

Report of the Various Proceedings Taken by the London Trades' Council and the Conference of Amalgamated Trades, in Reference to the Royal Commission on Trades' Unions, and Other Subjects in Connection Therewith (London: n.p., 1867). By permission of Senate House Library, University of London, shelfmark [G. L.] B.867.

Report of St. Martin's Hall United Kingdom Trades' Conference Committee on the Trades' Union Inquiry Commission till 7th August 1867 [s.l.: n.p., 1867]. By permission of Senate House Library, University of London, shelfmark Burns [G. L.] 4046.

International Working Men's Association, *To the Trades' Unionists of Great Britain and Ireland* [s.l.: n.p., 1868]. By permission of the Bishopsgate Institute, George Howell Collection, IWMA Minutes between pages 145 and 146.

J. M. Ludlow, 'Sir W. Erle on Trade Unions', *Operative Bricklayers' Monthly Report*, 91 (1 March 1869), pp. 854–7. By permission of the Modern Records Centre, University of Warwick, shelfmark MSS 78/OB/4/1/3.

Trades' Unions and the Cost of Labour. Speech delivered by Thomas Brassey, Jun M.P. in the House of Commons 7th July 1869 (London: Longman, Green, and Co., 1870). By permission of the British Library, shelfmark 8282.bbb.68.(7.)

Robert Jobson, *Trades Unions: An Inquiry into their Rules and Working, Based on the Evidence before the Royal Commission; Showing the Folly of All Attempts to Raise Wages by Violence and Strikes, and the Beauty and Excellence of the Divine Law Governing Workmen and Employers* (London: News-agents Publishing Company, [1870?]). By permission of Senate House Library, University of London, shelfmark [G. L.] B.869.

W. P. Roberts, *Trade Union Bill, 1871. A Letter by Mr. W. P. Roberts, (of London and Manchester, Solicitor) to Mr. George Potter, Mr. William Allan, Mr. Alexander M'Donald and Others Interested in Trades' Societies* (London: n. p., 1871). By permission of Senate House Library, University of London, shelfmark B.871.

Report of the Conference of Trades Delegates, Held in the Town Hall, Leeds on December 2nd 1871. Called to Consider the Statement Made by Wm. Newmarch, Esq. in his Address to the Social Science Congress Lately Held in Leeds (Leeds: Leeds and District Trades Council, [1872]). By permission of the British Library, shelfmark 1609/2331.

W. H. Wood, *The Advantages of Trades Unions* (Salford: n.p., [n.d.]). By permission of Senate House Library, University of London, shelfmark [G. L.] Burns 4023.

Proceedings of the Miners' National Association Conference, Held in Leeds, on 18, 19, 20, 21 and 22 November 1873 (Glasgow: n.p., [1873]), pp. 18–29. By permission of Glasgow University Library, Research Annexe, shelfmark W2/2-i.7

The Criminal Law Amendment Act. A Memorial to the Right Honourable Henry Austin Bruce, MP., With an Appendix of Cases under the Criminal Law Amendment Act (London: n.p., [1873]). By permission of Senate House Library, University of London, shelfmark [G. L.] B.871 /5.

Henry Crompton, *Tracts for Trade Unionists No IV. The National Federation of Associated Employers of Labour. Report upon the Memorial Presented to the Home*

Secretary by the National Federation of Associated Employers of Labour, December 13th, 1873 (London: n.p., [1874]). By permission of the Bishopsgate Institute, shelfmark HOWELL/6/19.

Frederick Clifford, *The Agricultural Lock-Out of 1874 with Notes upon Farming and Farm-Labour in the Eastern Counties* (Edinburgh and London: William Blackwood and Sons, 1875). By permission of the British Library, shelfmark 8282. bb. 50.

National Association of Miners, *Extraordinary Meeting of the Council Held in the Mechanics' Institution, David St Manchester, 24 February 1875* . By permission of Glasgow University Library, Research Annexe, shelfmark W2/2- i.7.

Glasgow Herald, *The Wages Agitation on the Clyde. 25,000 Men to be Locked Out* ([Glasgow]: Glasgow Herald, 1877). By permission of the London School of Economics Library, shelfmark BLPES FOLIO FHD4/32.

Address from the Executive Council of the Amalgamated Society of Engineers to their Fellow Workmen Throughout the United Kingdom, the British Colonies, and Foreign Countries ([London]: n.p., [1879]). By permission of the British Library of Political and Economic Science, shelfmark HD6/351.

George Howell, 'Trade Unions: Their Nature, Character, and Work', *Fraser's Magazine*, n.s. 19 (January 1879), pp. 22–31. By permission of the Bodleian Library, University of Oxford, shelfmark Per. 3977 e.200.

Printed and bound by CPI Group (UK) Ltd, Croydon, CR0 4YY
22/10/2024
01777605-0018